CAMBRIDGE TRACTS IN MATHEMATICS

General Editors

B. BOLLOBÁS, W. FULTON, A. KATOK, F. KIRWAN, P. SARNAK, B. SIMON

173   Enumeration of Finite Groups

SIMON R BLACKBURN

Royal Holloway, University of London

PETER M NEUMANN

The Queen's College, Oxford

GEETHA VENKATARAMAN

St Stephen's College, University of Delhi

Enumeration of Finite Groups

CAMBRIDGE
UNIVERSITY PRESS

CAMBRIDGE UNIVERSITY PRESS
Cambridge, New York, Melbourne, Madrid, Cape Town, Singapore, São Paulo

Cambridge University Press
The Edinburgh Building, Cambridge CB2 8RU, UK

Published in the United States of America by Cambridge University Press, New York

www.cambridge.org
Information on this title: www.cambridge.org/9780521882170

First published 2007

Printed in the United Kingdom at the University Press, Cambridge

A catalogue record for this publication is available from the British Library

ISBN 978-0-521-88217-0 hardback

To
Kay and Keith Blackburn,
Sylvia Neumann,
and Uttara Rangajaran

Contents

Preface

This book has grown out of a series of lectures given in the Advanced Algebra Class at Oxford in Michaelmas Term 1991 and Hilary Term 1992, that is to say from October 1991 to March 1992. The focus was—and is—the big question

how many groups of order n are there?

Two of the lectures were given by Professor Graham Higman, FRS, two by Simon R. Blackburn and the rest by Peter M. Neumann. Notes were written up week by week by Simon Blackburn and Geetha Venkataraman and those notes formed the original basis of this work. They have, however, been re-worked and updated to include recent developments.

The lectures were designed for graduate students in algebra and the book has been drafted with a similar readership in mind. It presupposes undergraduate knowledge of group theory—up to and including Sylow's theorems, a little knowledge of how a group may be presented by generators and relations, a very little representation theory from the perspective of module theory and a very little cohomology theory—but most of the basics are expounded here and the book should therefore be found to be more or less self-contained. Although it remains a work principally devoted to connected exposition of an agreeable theory, it does also contain some material that has not hitherto been published, particularly in Part IV.

We owe thanks to a number of friends and colleagues: to Graham Higman for his contribution to the lectures; to members of the original audience for their interest and their comments; to Laci Pyber for comments on an early draft; to Mike Newman for permission to include unpublished work of himself and Craig Seeley; to Eira Scourfield for guidance on the literature of analytic number theory; to Juliette White for comments on the earlier chapters of the book and for help with proofreading our first draft. We would also like to acknowledge the support of the Mathematical Sciences Foundation, St. Stephen's College, Delhi, The Indian Institute of Science, Bangalore and our respective home institutions. Geetha Venkataraman would also like to acknowledge the encouragement and support extended by Uttara, Mahesh and Shantha Rangarajan and her parents WgCdr P. S. Venkataraman and Visalakshi Venkataraman. Professor Dinesh Singh has been a mentor providing much needed support, encouragement and intellectual fellowship. We record our gratitude to an anonymous friendly referee for constructive suggestions and for drawing our attention to some recent references that we had missed. We are also very grateful to the editorial staff of Cambridge University Press for their great courtesy, enthusiasm and helpfulness.

Turning lecture notes into a book involves much hard work. Inevitably that work has fallen unequally on the three authors. The senior author, too happy to have relied on the excellent principle *juniores ad labores* (which he admits to having embraced less enthusiastically when he was younger), is glad to have the opportunity to acknowledge that all the hard work has been done by his two colleagues, whom he thanks very warmly.

SRB, ΠMN, GV: 25.xi.2006

1

Introduction

The focus of this book is the question how many groups of order n are there? This is to be interpreted in the natural way: we define $f(n)$ to be the number of groups of order n up to isomorphism and ask for information about the function f.

The values of $f(n)$ for small values of n are:

n	1	2	3	4	5	6	7	8	9	10	11	12	13	14	15	16	...
$f(n)$	1	1	1	2	1	2	1	5	2	2	1	5	1	2	1	14	...

For $1 \leqslant n \leqslant 16$ the groups of order n were classified well over a hundred years ago, and the value of $f(n)$ clearly follows from this classification. The easiest case is when n is a prime—Lagrange's Theorem shows that a group of order n must be cyclic, and so $f(n) = 1$. When n is in the range of the table above, only $n = 16$ requires a lengthy argument to establish a classification. Note that $f(15) = 1$ even though 15 is not prime.

As n increases, the problem of classifying groups of order n becomes hard. The groups of order 2^{10} have only recently been classified, by Besche, Eick and O'Brien [6]. An appendix to their paper lists $f(n)$ when $1 \leqslant n \leqslant 2000$; in particular when $n = 2^{10}$ they count 49 487 365 422 groups! However, the groups of order 2^{11} have not been classified and it is not known how many groups of order 2^{11} there are. (We will show in Chapter 4 that $f(2^{11}) > 2^{44}$.) So if we are to say anything about $f(n)$ when n is large, we must resort to giving estimates for $f(n)$ rather than calculating $f(n)$ exactly.

Graham Higman [45] showed in 1960 that

$$f(p^m) \geqslant p^{\frac{2}{27}m^3 - O(m^2)},$$

Charles Sims [86] proved in 1965 that

$$f(p^m) \leqslant p^{\frac{2}{27}m^3 + O(m^{8/3})}$$

1

and, as the culmination of a long line of development, Laszlo Pyber [82] proved in 1991 (published in 1993) that

$$f(n) \leqslant n^{\frac{2}{27}\mu(n)^2 + O(\mu(n)^{5/3})},$$

where $\mu(n)$ is the highest power to which any prime divides n. Amplification of these results, and their proofs, forms the main part of the work: the results of Higman and Sims are expounded in Part II (incorporating a modification of Sims' argument due to Mike Newman and Craig Seeley [77], which improves the error term significantly) and Pyber's theorem is the subject of Part III. The proofs use a large amount of very attractive theory that is just beyond the scope of an undergraduate course in algebra. All that theory is expounded here, so that our treatment of the theorems of Higman and Sims and of Pyber's theorem in the soluble case is self-contained. Our treatment of the general case of Pyber's theorem in Chapter 16 is not self-contained, however, because it relies ultimately upon the Classification of the Finite Simple Groups (CFSG).

The asymptotics of the function f tell us much, but far from everything, about the groups of order n. To get a clearer picture we consider related matters. For example, context is given by the questions how many semigroups and how many latin squares of order n are there? These questions are treated briefly in Chapter 2. Detail is given by such questions as: how many abelian groups of order n are there? how many of the groups of order n have abelian Sylow subgroups? how many of the groups of order n satisfy a given identical relation? how many are soluble? how many are nilpotent? Questions of this type are treated in Part IV.

Standing conventions:

- most groups considered are finite—if at any point finiteness is not mentioned but seems desirable, the reader is invited to assume it;
- f has already been introduced as the group enumeration function;
- for a class $\mathfrak{X}$ of groups (or of other structures) $f_{\mathfrak{X}}(n)$ denotes the number of members of $\mathfrak{X}$ of order n, up to isomorphism;
- logarithms are to the base 2;
- maps are on the left;
- p always denotes a prime number;
- if $n = p_1^{\alpha_1} p_2^{\alpha_2} \cdots p_k^{\alpha_k}$, where $p_1, p_2, \ldots, p_k$ are distinct prime numbers, then $\lambda(n) = \alpha_1 + \alpha_2 + \cdots + \alpha_k$ and $\mu(n) = \max\{\alpha_i \mid 1 \leqslant i \leqslant k\}$.

Other notation and conventions are introduced where they are needed.

I

Elementary results

2

Some basic observations

This chapter is devoted to elementary estimates for $f(n)$, the number of groups of order n (up to isomorphism). We begin by looking at some enumeration functions for weaker objects than groups.

Since a binary system is determined by its multiplication table, we find that

$$f(n) \leqslant f_{\text{binary systems}}(n) \leqslant n^{n^2}.$$

At most $n!$ of these multiplication tables are isomorphic to any fixed binary system, since an isomorphism is one of only $n!$ permutations. Hence

$$n^{n^2-n} \leqslant \frac{n^{n^2}}{n!} \leqslant f_{\text{binary systems}}(n) \leqslant n^{n^2}.$$

If we consider binary systems with a unit element, we have

$$n^{n^2-3n+O(1)} \leqslant f_{\text{binary systems with } 1}(n) \leqslant n^{(n-1)^2} = n^{n^2-2n+1}.$$

Recall that a semigroup is a set with an associative multiplication defined on it. For all $\epsilon > 0$,

$$n^{(1-\epsilon)n^2} \leqslant f_{\text{semigroups}}(n) \leqslant n^{n^2}$$

if $n \geqslant n_0(\epsilon)$. To see this, consider the binary systems on $\{0, 1, \ldots, n-1\}$ described by tables of the following form:

5

	0	1	...	$m-2$	$m-1$	m	$m+1$	...	$n-2$	$n-1$
0	0	0	...	0	0	0	0	...	0	0
1	0	0	...	0	0	0	0	...	0	0
$\vdots$	$\vdots$	$\vdots$		$\vdots$	$\vdots$	$\vdots$	$\vdots$		$\vdots$	$\vdots$
$m-1$	0	0	...	0	0	0	0	...	0	0
m	0	0	...	0	0	*	*	...	*	*
$m+1$	0	0	...	0	0	*	*	...	*	*
$\vdots$	$\vdots$	$\vdots$		$\vdots$	$\vdots$	$\vdots$	$\vdots$		$\vdots$	$\vdots$
$n-1$	0	0	...	0	0	*	*	...	*	*

Here the starred entries are arbitrary subject to being at most $m-1$. The associative law holds for this table, since

$$(a_i a_j)a_k = 0 = a_i(a_j a_k).$$

Hence

$$f_{\text{semigroups}}(n) \geqslant m^{(n-m)^2}.$$

(Notice here that we should divide by $n!$, but this again does not make a significant difference.) Setting m to be approximately $n^{1-\frac{1}{2}\epsilon}$ we have

$$f_{\text{semigroups}}(n) \geqslant n^{(1-\frac{1}{2}\epsilon)(n-n^{1-\frac{1}{2}\epsilon})^2}.$$

For sufficiently large n,

$$n^{(1-\frac{1}{2}\epsilon)(n-n^{1-\frac{1}{2}\epsilon})^2} \geqslant n^{(1-\epsilon)n^2}.$$

Thus we get the requisite lower bound.

If we add the condition that all our semigroups contain a unit element, we have similar results to the above.

Daniel Kleitman, Bruce Rothschild and Joel Spencer enumerate semigroups more precisely in [55]. They show that most semigroups can be split into two subsets A and B having the following property: there exists an element $0 \in B$ such that if $x, y \in A$ then $xy \in B$ but if $x \in B$ or $y \in B$ then $xy = 0$. They then use this fact to prove

$$f_{\text{semigroups}}(n) = \left(\sum_{t=1}^{n} g(t) \right)(1 + O(1)), \text{ where}$$

$$g(t) = \binom{n}{t} t^{1+(n-t)^2}.$$

The function $g(t)$ maximises at t_0, where $t_0 \sim n/2\log_e n$. Thus we may improve the lower bound we gave above to

$$f_{\text{semigroups}}(n) \geqslant n^{n^2(1-(\log\log n/\log n)-O(1/\log n))},$$

where (for this inequality only) log should denote the natural logarithm—although, as the astute reader will realise, in fact the base of the logarithms does not matter here.

A multiplication table with inverses is a latin square (i.e., in each row and column of the table, an element appears only once). We have

$$n^{\frac{1}{2}n^2-O(n)} \leqslant f_{\text{latin squares}}(n) \leqslant n^{n^2}.$$

The lower bound was proved by Marshall Hall [40]. Using less elementary methods, the lower bound may be improved: Henryk Minc showed in [69] that

$$(n!)^{2n}/n^{n^2} \leqslant f_{\text{latin squares}}(n).$$

His proof uses the Egoryčev–Falikman theorem [26, 32, 70] establishing the van der Waerden conjecture on permanents. Note that there is a constant c such that $n! > c\,(n/e)^n$, and so $f_{\text{latin squares}}(n) > c^2 n^{n^2(1-1/\log n)}$. Much remains to be discovered about this enumeration function. In 2005, Brendan McKay and Ian Wanless [68] state 'At the time of writing, not even the asymptotic value of $f_{\text{latin squares}}(n)$ is known'.

Returning to the group enumeration function, we see that even very elementary methods are enough to show that there are seriously fewer groups than semigroups or latin squares:

Observation 2.1

$$f(n) \leqslant n^{n\log n}.$$

Proof: For a group G, define

$$d(G) = \min\{k \mid \exists\, g_1,\ldots,g_k \in G \text{ such that } G = \langle g_1,\ldots,g_k\rangle\}.$$

We first show that if $|G| = n$ then $d(G) \leqslant \log n$. Let

$$\{1\} = G_0 < G_1 < G_2 < \cdots < G_r = G$$

be a maximal chain of subgroups. Let $g_i \in G_i \setminus G_{i-1}$ for $1 \leqslant i \leqslant r$. Then $\langle g_1,\ldots,g_i\rangle = G_i$, as one easily sees by induction. In particular, G can be generated by r elements. Now by Lagrange's Theorem

$$|G| = \prod_{i=1}^{r} |G_i : G_{i-1}| \geqslant 2^r. \tag{2.1}$$

Hence $r \leqslant \lfloor \log n \rfloor$. Then by Cayley's theorem $G \leqslant \mathrm{Sym}(n)$ and so

$f(n) \leqslant$ number of subgroups of order n in $\mathrm{Sym}(n)$

$\leqslant$ number of $\lfloor \log n \rfloor$-generator subgroups of $\mathrm{Sym}(n)$

$\leqslant$ number of $\lfloor \log n \rfloor$-element subsets of $\mathrm{Sym}(n)$

$\leqslant (n!)^{\log n}$

$\leqslant n^{n \log n}$

and the result follows.

Recall the notation $\mu(n)$ and $\lambda(n)$ that we introduced at the end of Chapter 1. A factorisation of n has at most $\lambda(n)$ non-trivial factors. Equation (2.1) shows that $r \leqslant \lambda(n)$, and therefore the bound for $d(G)$ can be sharpened to say that $d(G) \leqslant \lambda(n)$. We remark that in fact $d(G) \leqslant \mu(n) + 1$, as we will see in Corollary 16.7, but ignoring this for the moment and feeding the simple bound for $d(G)$ into the above argument we get that

$$f(n) \leqslant n^{n \lambda(n)}.$$

That is about as far as one can go with elementary methods. Nevertheless, it already shows that the associative law and the existence of inverses are separately very much weaker than is their combination.

The aim of the remainder of the book is to prove the better bounds on $f(n)$ given in the Introduction, using more sophisticated methods.

II

Groups of prime power order

3
Preliminaries

This chapter contains a brief account of some of the results we will need in the next two chapters. More specifically, we review some basic commutator identities and results on nilpotent groups, discuss the Frattini subgroup of a group and prove some simple enumeration results concerning vector spaces, general linear groups and symplectic groups. We emphasise that all groups are finite in this section—some of the results (and definitions) differ in the infinite case. We assume that the reader has already met a few commutator identities and the idea of a nilpotent group, and so we have included sketch proofs rather than full detail for some of the results. For more detail, see Gorenstein [36, Sections 2.2 and 2.3].

3.1 Tensor products and exterior squares of abelian groups

As preparation for some of our treatment of commutators we recall (without proofs) the definition of tensor product and exterior square of abelian groups. If A, B are abelian groups (which we write additively here) the *tensor product* $A \otimes B$ is defined to be the abelian group which is generated by all symbols $a \otimes b$ for $a \in A$ and $b \in B$ subject to the relations

$$(a_1 + a_2) \otimes b - a_1 \otimes b - a_2 \otimes b = 0,$$

$$a \otimes (b_1 + b_2) - a \otimes b_1 - a \otimes b_2 = 0,$$

which make the operation $\otimes$ bilinear. We identify $a \otimes b$ with its image modulo the relations and then the map $A \times B \to A \otimes B$ (where here $A \times B$ simply denotes the *set* of pairs), $(a, b) \mapsto a \otimes b$, is bilinear. If A is generated by $a_1, \ldots, a_r$ and B is generated by $b_1, \ldots, b_s$ then

$$\{a_i \otimes b_j \mid 1 \leqslant i \leqslant r, \ 1 \leqslant j \leqslant s\}$$

will be a generating set for $A \otimes B$; moreover, the order of $a \otimes b$ divides the greatest common divisor of the orders of a and b.

The *exterior square* $A^{\wedge 2}$ (sometimes written $\bigwedge^2 A$) of A is defined to be the abelian group generated by all symbols $a \wedge b$ for $a, b \in A$ with the same bilinearity relations as the tensor product and, in addition, the relations

$$a \wedge a = 0 \quad \text{for all } a \in A$$

which make $\wedge$ an alternating function of its arguments. Again, we identify $a \wedge b$ with its image modulo the relations and then the two-variable function $(a, b) \mapsto a \wedge b$ is an alternating bilinear map $A \times A \to A^{\wedge 2}$. Note that the equation $b \wedge a = -(a \wedge b)$ follows easily from the defining relations for the exterior square, and that if A is generated by $a_1, \ldots, a_r$ then

$$\{a_i \wedge a_j \mid 1 \leqslant i < j \leqslant r\}$$

is a generating set for $A^{\wedge 2}$.

One of the main properties of the tensor product is that it is universal for bilinear maps. That is, if C is an abelian group and $f : A \times B \to C$ is a bilinear map then there is a unique homomorphism $f^* : A \otimes B \to C$ such that $f^*(a \otimes b) = f(a, b)$ for all $a \in A$, $b \in B$. Similarly, the exterior square is universal for alternating bilinear maps in the sense that if $f : A \times A \to C$ is bilinear and such that $f(a, a) = 0$ for all a then there is a unique homomorphism $f^* : A^{\wedge 2} \to C$ such that $f^*(a \wedge b) = f(a, b)$ for all $a, b \in A$. Another fundamental property is functoriality: the tensor product and exterior square are functorial in the sense that if A_1, A_2, B_1, B_2 are abelian groups and $f : A_1 \to A_2$, $g : B_1 \to B_2$ are homomorphisms then there are homomorphisms $f \otimes g : A_1 \otimes B_1 \to A_2 \otimes B_2$ and $f^{\wedge 2} : A_1^{\wedge 2} \to A_2^{\wedge 2}$ such that $(f \otimes g)(a \otimes b) = f(a) \otimes g(b)$ for all $a \in A_1$, $b \in B_1$ and $f^{\wedge 2}(a \wedge b) = f(a) \wedge f(b)$ for all $a, b \in A_1$.

3.2 Commutators and nilpotent groups

Let G be a group and let $x, y \in G$. The commutator $[x, y]$ of x and y is defined by $[x, y] = x^{-1} y^{-1} x y$. For $x, y, z \in G$, we define $[x, y, z] = [[x, y], z]$. Throughout this section, we will write x^y to mean $y^{-1} x y$.

Lemma 3.1 *Let G be a group.*

(1) *For all $x, y \in G$,*

$$[x, y] = [y, x]^{-1}. \tag{3.1}$$

(2) *For all* $x, y, z \in G$,

$$[xy, z] = [x, z]^y [y, z] = [x, z][x, z, y][y, z], \tag{3.2}$$

$$[x, yz] = [x, z][x, y]^z = [x, z][x, y][x, y, z]. \tag{3.3}$$

(3) *For all* $x, y, z \in G$,

$$[x, y^{-1}, z]^y [y, z^{-1}, x]^z [z, x^{-1}, y]^x = 1. \tag{3.4}$$

The proof of this lemma is easy: just use the definition of a commutator to express each side of the above equalities as a product of x, y, z and their inverses.

Corollary 3.2 *Let G be a group.*

(1) *For all* $x, y, z \in G$,

$$[x^{-1}, y] = \left([x, y]^{-1}\right)^{x^{-1}} = [x, y, x^{-1}]^{-1}[x, y]^{-1}, \tag{3.5}$$

$$[x, y^{-1}] = \left([x, y]^{-1}\right)^{y^{-1}} = [x, y, y^{-1}]^{-1}[x, y]^{-1}. \tag{3.6}$$

(2) *For all* $x, y, z \in G$,

$$[x, y, z] = \left([z, x^{-1}, y^{-1}]^{-1}\right)^{xy} \left([y^{-1}, z^{-1}, x]^{-1}\right)^{zy}. \tag{3.7}$$

Proof: The corollary follows from Lemma 3.1 by making the appropriate substitutions. To derive (3.5), replace y by x^{-1} and z by y in (3.2). For (3.6), replace z by y^{-1} in (3.3). To derive (3.7), replace y by y^{-1} in (3.4).

Lemma 3.3 *Let G be a group. Let* $x, y \in G$. *Suppose that* $[y, x]$ *commutes with both x and y. Then for all positive integers n*

$$[y, x^n] = [y^n, x] = [y, x]^n, \tag{3.8}$$

$$(xy)^n = x^n y^n [y, x]^{\frac{1}{2} n(n-1)}. \tag{3.9}$$

Proof: The equality (3.8) follows by induction on n, using (3.2) and (3.3) in the inductive step. To establish (3.9), use the fact that $y^i x = x y^i [y^i, x]$.

We will now consider a collection of results related to nilpotency of groups. Let H and K be subgroups of a group G. Then $[H, K]$ is defined to be the subgroup generated by all elements of the form $[h, k]$ where $h \in H$ and $k \in K$. Note that $[H, K] = [K, H]$, by Equation (3.1). The subgroup $[H, K, L]$ is defined by $[H, K, L] = [[H, K], L]$. The following lemma, known as the Three Subgroup Lemma, is often useful.

Lemma 3.4 *Let K, L and M be subgroups of a group G. Then $[K, L, M] \leqslant$ $[M, K, L][L, M, K]$ whenever $[M, K, L]$ and $[L, M, K]$ are normal subgroups of G.*

Proof: Suppose that $[M, K, L]$ and $[L, M, K]$ are normal subgroups of G. The subgroup $[K, L, M]$ is generated by elements of the form $[g, h]$ where $g \in [K, L]$ and $h \in M$. We may express g as a product of commutators of the form $[g', h']$ where $g' \in K$ and $h' \in L$, and then use Equations (3.2) and (3.3) to express $[g, h]$ as a product of conjugates of elements of the form $[x, y, z]$ where $x \in K$, $y \in L$ and $z \in M$. But (3.7) expresses $[x, y, z]$ as a product of a conjugate of an element of $[M, K, L]$ and a conjugate of an element of $[L, M, K]$. Since $[M, K, L]$ and $[L, M, K]$ are normal, we find that each generator of $[K, L, M]$ lies in $[M, K, L][L, M, K]$, so the lemma follows.

The *lower central series* $G_1, G_2, G_3, \ldots$ of a group G is defined by $G_1 = G$ and $G_{i+1} = [G_i, G]$ for every positive integer i. From now on, we will always use G_i to denote the ith term of the lower central series of G. It is not difficult to see, using the definition of the lower central series, that the subgroups G_i are characteristic subgroups of G. Clearly G_i/G_{i+1} is central in G/G_{i+1}. For all normal subgroups N of G, we have that $(G/N)_i = (G_i N)/N$. Moreover, if H is a subgroup of G then H_i is a subgroup of G_i for all positive integers i.

Proposition 3.5 *Let G be a group. Let $A = G/G_2 = G/G'$ and $A_i = G_i/G_{i+1}$. Then A_2 is a homomorphic image of $A^{\wedge 2}$ and A_{i+1} is a homomorphic image of $A_i \otimes A$ for all $i \geqslant 1$.*

Proof: It follows immediately from Lemma 3.1 that the map $A \times A \to A_2$, $(aG', bG') \mapsto [a, b]G_3$ is well-defined and bilinear. It is also alternating since $[a, a] = 1$ for all $a \in G$. Therefore there is a homomorphism $A^{\wedge 2} \to A_2$ such that $aG' \wedge bG' \mapsto [a, b]G_3$ for all $a, b \in G$. This is surjective since G_2 is generated by the commutators $[a, b]$ for $a, b \in G$, and therefore A_2 is a homomorphic image of $A^{\wedge 2}$. The proof that A_{i+1} is a homomorphic image of $A_i \otimes A$ for all $i \geqslant 1$ is similar and we omit it.

Proposition 3.6 *Let G be a group. For all positive integers i and j, we have that $[G_i, G_j] \leqslant G_{i+j}$.*

Proof: We use induction on j. The case when $j = 1$ follows by definition of the lower central series. Assume that $j > 1$ and that $[G_i, G_{j-1}]$ is a subgroup of G_{i+j-1} for any group G and any $i \geqslant 0$. We prove that $[G_i, G_j]$ is a subgroup of G_{i+j} as follows.

By replacing G by the quotient G/G_{i+j} if necessary, we may assume that $G_{i+j} = \{1\}$. Our inductive hypothesis implies that

$$[G_i, G_{j-1}, G] \leqslant [G_{i+j-1}, G] = G_{i+j} = \{1\}, \text{ and}$$
$$[G, G_i, G_{j-1}] = [G_i, G, G_{j-1}] = [G_{i+1}, G_{j-1}] \leqslant G_{i+j} = \{1\}.$$

In particular, $[G_i, G_{j-1}, G]$ and $[G, G_i, G_{j-1}]$ are normal. But now Lemma 3.4 implies that

$$[G_i, G_j] = [G_j, G_i] = [G_{j-1}, G, G_i]$$
$$\leqslant [G_i, G_{j-1}, G][G, G_i, G_{j-1}] = \{1\} = G_{i+j}.$$

Hence the proposition follows by induction on j.

Proposition 3.7 *Let i be a positive integer. Let G be a group generated by a set S. Let T be a subset of G_i whose image in G_i/G_{i+1} generates G_i/G_{i+1}. Then G_{i+1}/G_{i+2} is generated by the set*

$$\{[t, s]G_{i+2} \mid t \in T, s \in S\}.$$

Proof: Write $\bar{S}$ for the image of S in G/G' and $\bar{T}$ for the image of T in G_i/G_{i+1}. Then $\bar{S}$ generates G/G' and, by assumption, $\bar{T}$ generates G_i/G_{i+1}. It follows that the set $\{\bar{t} \otimes \bar{s} \mid \bar{t} \in \bar{T}, \ \bar{s} \in \bar{S}\}$ generates the tensor product $G_i/G_{i+1} \otimes G/G'$ and the result now follows immediately from Proposition 3.5.

The final result of this section will form one of the key steps in Sims' upper bound for the number of isomorphism classes of p-groups of a given order. Recall that a group G is *nilpotent* if $G_r = \{1\}$ for some integer r. If r is the smallest such integer, we say that G is *nilpotent of class $r-1$*. A finite group is nilpotent if and only if it is the direct product of its Sylow subgroups (see [36, Theorem 2.3.5]). In particular, all p-groups are nilpotent.

Proposition 3.8 *Let G be a nilpotent group, and let H be a subgroup of G. If $H_2G_3 = G_2$, then $H_i = G_i$ for all $i \geqslant 2$.*

Proof: We prove first that $H_rG_{r+1} = G_r$ for all $r \geqslant 2$. We prove this equality using induction on r: it is true when $r = 2$, by assumption. Assume that $r > 2$ and that $H_sG_{s+1} = G_s$ whenever $2 \leqslant s < r$. Clearly $H_rG_{r+1} \leqslant G_r$, and so we must prove that $G_r \leqslant H_rG_{r+1}$. We need two preliminary results. Our first result asserts that whenever K, L and M are subgroups of G such that L is normal in G and such that $[K, M, L] \leqslant G_{r+1}$ then

$$[KL, M] \leqslant [K, M][L, M]G_{r+1}. \tag{3.10}$$

(Note that the right-hand side of (3.10) is indeed a subgroup, for $[L, M] \leqslant L$ as L is normal, and so $[[K, M], [L, M]] \leqslant [K, M, L] \leqslant G_{r+1}$.) To prove this result, observe that the subgroup $[KL, M]$ is generated by elements of the form $[xy, z]$ where $x \in K$, $y \in L$ and $z \in M$. But (3.2) expresses this commutator as the product of three commutators, the first in $[K, M]$, the second in G_{r+1} and the third in $[L, M]$ and so the result follows. We will use this result three times: with $K = H_{r-1}$, $L = G_r$ and $M = G$; with $K = H_2$, $L = G_3$ and $M = H_{r-2}$ and with $K = H_{r-1}$, $L = G_r$ and $M = H$. In all three cases, the condition $[K, M, L] \leqslant G_{r+1}$ follows by the fact that $H_i \leqslant G_i$ for all positive integers i and by Proposition 3.6.

The second result we require asserts that

$$[H_{r-1}, G] \leqslant [G, H_{r-2}, H][H, G, H_{r-2}]G_{r+1}. \tag{3.11}$$

(The right-hand side of (3.11) is a subgroup, since the subgroups $[G, H_{r-2}, H]$ and $[H, G, H_{r-2}]$ lie in G_r and so commute modulo G_{r+1}.) To see why (3.11) holds, first note that H_{r-1} is generated by elements of the form $[h', h]$ where $h \in H$ and $h' \in H_{r-2}$. We may use (3.2) and (3.5) to express a typical element of $[H_{r-1}, G]$ as a product of various elements of G_{r+1} and elements of the form $[h', h, g]$ where $h \in H$, $h' \in H_{r-2}$ and $g \in G$. Here we use the fact that

$$[H_{r-1}, G, H_{r-1}] \leqslant [G_{r-1}, G, G_{r-1}] \leqslant G_{2r-1} \leqslant G_{r+1}$$

by Proposition 3.6. So our second result holds if we can show that elements of the form $[h', h, g]$ all lie in $[G, H_{r-2}, H][H, G, H_{r-2}]G_{r+1}$. But Equation (3.7) shows that $[h', h, g]$ is equal to the product of a conjugate of an element of $[G, H_{r-2}, H]$ and a conjugate of an element of $[H, G, H_{r-2}]$. Since both $[G, H_{r-2}, H]$ and $[H, G, H_{r-2}]$ are contained in G_r, and since $[G_r, G] = G_{r+1}$, these conjugates differ from the original elements by members of G_{r+1}. Hence the result follows.

Using these two results, we may establish our inductive step:

$$
\begin{aligned}
G_r &= [G_{r-1}, G] \\
&= [H_{r-1}G_r, G] \text{ (by the inductive hypothesis)} \\
&= [H_{r-1}, G]G_{r+1} \text{ (by (3.10))} \\
&\leqslant [G, H_{r-2}, H][H, G, H_{r-2}]G_{r+1} \text{ (by (3.11))} \\
&\leqslant [G_{r-1}, H][G_2, H_{r-2}]G_{r+1} \text{ (using } H_i \leqslant G_i \text{ and Proposition 3.6)}
\end{aligned}
$$

$$= [H_{r-1}G_r, H][H_2G_3, H_{r-2}]G_{r+1} \text{ (by the inductive hypothesis)}$$

$$= [H_{r-1}, H][G_r, H][H_2, H_{r-2}][G_3, H_{r-2}]G_{r+1} \text{ (by (3.10))}$$

$$= H_rG_{r+1} \text{ (by Proposition 3.6).}$$

So, by induction on r, we find that $G_r = H_rG_{r+1}$ for $r \geqslant 2$ as required.

Next we prove that $G_i = H_iG_{i+k}$ for $k \geqslant 1$ and $i \geqslant 2$. This has just been proved when $k = 1$. Moreover, if $k > 1$ and $G_i = H_iG_{i+k-1}$ then

$$G_i = H_iG_{i+k-1} = H_iH_{i+k-1}G_{i+k} = H_iG_{i+k},$$

the second equality using the case $r = i+k-1$ of the fact that $G_r = H_rG_{r+1}$ for $r \geqslant 2$. So the claim follows by induction on k.

We are assuming that G is nilpotent, and so $G_{i+k} = \{1\}$ for all sufficiently large k. Hence the previous paragraph implies that $G_i = H_i$ for $i \geqslant 2$, as required.

3.3 The Frattini subgroup

Let G be a group. The *Frattini subgroup* $\Phi(G)$ of G is the characteristic subgroup of G defined as the intersection of all the maximal subgroups of G. This section investigates the Frattini subgroup of a group, with an eye to using the results in Section 4.2.

Lemma 3.9 *Let G be a finite group, and let X be a subset of G. Then $\Phi(G)$ and X together generate G if and only if X generates G.*

We may express this lemma more informally by saying that '$\Phi(G)$ is the set of non-generators of G'.

Proof: Clearly $G = \langle X \rangle$ implies that $G = \langle X \cup \Phi(G) \rangle$. For the converse, suppose that $G \neq \langle X \rangle$. Then $\langle X \rangle$ is contained in some maximal subgroup M of G. Since $\Phi(G)$ is the intersection of all maximal subgroups of G, we have that $\Phi(G) \leqslant M$. But now $\langle X \cup \Phi(G) \rangle \leqslant M < G$, and so $G \neq \langle X \cup \Phi(G) \rangle$.

When G is a p-group, the maximal subgroups of G have an especially simple form. This is a consequence of the next lemma.

Lemma 3.10 *Let G be a finite group, and let H be a subgroup of G. Suppose that H is a p-group, and that p divides the index of H in G. Then H is strictly contained in its normaliser $N_G(H)$. Moreover, there exists a subgroup K of G such that H has index p in K.*

Proof: There is a natural action of the subgroup H on the set Ω of cosets of H in G. A coset is fixed by H under this action if and only if it is contained in $N_G(H)$. So the index of H in $N_G(H)$ is equal to the number of trivial orbits of H acting on Ω. Now, p divides $|\Omega|$. Since H is a p-group, p divides the length of any non-trivial orbit in Ω. Hence p divides the number of trivial orbits. So p divides the index of H in $N_G(H)$ and thus H is strictly contained in $N_G(H)$. Let $x \in N_G(H)$ be such that xH has order p in $N_G(H)/H$. The last assertion of the lemma now follows by taking K to be the group generated by H and x.

Corollary 3.11 *Let G be a finite p-group. Then every maximal subgroup M of G is normal and has index p in G.*

Lemma 3.12 *Let G be a finite p-group. Then $G/\Phi(G)$ is an elementary abelian p-group of order p^d, where d is the minimal number of generators of G. Moreover $\Phi(G) = G^p G'$, where G^p is the subgroup generated by the set $\{x^p : x \in G\}$.*

Proof: By Corollary 3.11, every maximal subgroup M of G is normal and has index p in G. It follows that every maximal subgroup contains $G^p G'$, and so $G^p G' \leqslant \Phi(G)$ and $G/\Phi(G)$ is an elementary abelian group. Since G is generated by d elements, so is $G/\Phi(G)$ and so $|G/\Phi(G)| \leqslant p^d$. If $|G/\Phi(G)| < p^d$ then $G/\Phi(G)$ may be generated by $d-1$ elements, and so G may be generated by $\Phi(G)$ together with the inverse image of these elements under the natural map from G to $G/\Phi(G)$. By Lemma 3.9, this implies that G may be generated by $d-1$ elements. This contradiction shows that $|G/\Phi(G)| = p^d$.

We have seen that $\Phi(G) \geqslant G^p G'$. Since $G/(G^p G')$ is elementary abelian, the intersection of all maximal subgroups of $G/(G^p G')$ is trivial. The inverse images in G of these maximal subgroups under the natural homomorphism from G to $G/(G^p G')$ are maximal in G and their intersection is $G^p G'$. Thus $\Phi(G) \leqslant G^p G'$ and so $\Phi(G) = G^p G'$, as required.

Lemma 3.12 implies that the Frattini subgroup of a p-group is generated by the set of values of the words x^p and $[x, y]$ in G. (Thus $\Phi(G)$ is an example of a verbal subgroup; see [75].) This implies, in particular, that for any normal subgroup N of a p-group G, we have $\Phi(G/N) = \Phi(G)/N$.

We will use the next lemma in the proof of Pyber's theorem.

Lemma 3.13 *Let G be a finite p-group. Let B be the group of automorphisms of G that induce the identity automorphism on $G/\Phi(G)$. Then B is a p-group.*

Proof: Suppose that B is not a p-group. So there exists $x \in B$ such that x has order not a power of p. By replacing x by a suitable power of x, we may assume that x has order q, where q is a prime distinct from p.

Let $g \in G$. Since x induces the identity automorphism on $G/\Phi(G)$, we have that x permutes the elements in the coset $g\Phi(G)$. Each cycle of x has length either 1 or q. Since $g\Phi(G)$ has order a power of p (and since a power of p is not a multiple of q), not all these cycles can have length q. Hence x fixes an element in every coset $g\Phi(G)$ of $\Phi(G)$ in G. In particular the elements in G that are fixed by x generate G modulo $\Phi(G)$, and so generate G by Lemma 3.9. Since x fixes a generating set for G, we find that x is the identity automorphism. This contradicts the fact that x has order q, and so the lemma follows.

To finish this section, we prove (as an aside) a result due to Wielandt that gives a characterisation of nilpotency in terms of the Frattini subgroup.

Proposition 3.14 *Let G be a finite group. Then G is nilpotent if and only if $G' \leqslant \Phi(G)$.*

Proof: Suppose that G is nilpotent. Any proper subgroup H of a nilpotent group G is strictly contained in its normaliser. (To see this, let i be the largest integer such that G_i is not contained in H. Then $[G_i, H] \leqslant G_{i+1} \leqslant H$, and so $G_i \leqslant N_G(H)$.) In particular, any maximal subgroup M of G is normal. Since G/M is non-trivial and nilpotent, $(G/M)'$ is a proper subgroup of G/M. Hence, since M is maximal, $(G/M)'$ is trivial and therefore G/M is abelian. Thus $G' \leqslant M$ for any maximal subgroup M of G and so G' is contained in the intersection $\Phi(G)$ of all maximal subgroups of G.

To prove the converse, suppose that $G' \leqslant \Phi(G)$ (and so in particular every maximal subgroup of G is normal). Let P be a Sylow subgroup of G, and assume, seeking a contradiction, that $N_G(P)$ is a proper subgroup of G. Then $N_G(P) \leqslant M$ for some maximal subgroup M of G. Now $P \leqslant M$, and so P is a Sylow subgroup of the normal subgroup M. By the Frattini argument (see Gorenstein [36, page 12, Theorem 3.7]), $N_G(P)M = G$. But $N_G(P) \leqslant M$ and so we have our required contradiction. Thus every Sylow subgroup is normal in G and so G is nilpotent.

3.4 Linear algebra

This section contains some enumeration results from linear algebra, rehearses some of the standard material on alternating forms and draws out some of the immediate consequences for elementary abelian p-groups. Throughout this section, q will be a power of a prime number, and $\mathbb{F}_q$ will be the finite field with q elements.

Proposition 3.15 *Let V be a vector space of dimension d over the finite field $\mathbb{F}_q$. There are*

$$(q^d - 1)(q^d - q) \cdots (q^d - q^{k-1})$$

choices for a sequence $v_1, v_2, \ldots, v_k$ of linearly independent vectors in V. In particular,

$$|\mathrm{GL}(d, q)| = (q^d - 1)(q^d - q) \cdots (q^d - q^{d-1}) \leqslant q^{d^2}.$$

Proof: Assume that we have already chosen linearly independent vectors $v_1, v_2, \ldots, v_{k-1} \in V$. The result follows if we can show that there are $q^d - q^{k-1}$ choices for v_k. But v_k can be any vector of V that is not in the $(k-1)$-dimensional subspace U spanned by $v_1, v_2, \ldots, v_{k-1}$. There are q^d vectors in V and q^{k-1} vectors in U, and so there are $q^d - q^{k-1}$ possibilities for v_k and the first statement of the proposition follows.

The second statement of the proposition follows from the first statement together with the observation that a $d \times d$ matrix with entries in $\mathbb{F}_q$ is invertible if and only if its rows are linearly independent vectors in $(\mathbb{F}_q)^d$. The final inequality follows from the fact that $q^d - q^i < q^d$, but this bound can be seen directly by observing that a $d \times d$ matrix is specified by choosing its d^2 entries.

Proposition 3.16 *Let V be a vector space over $\mathbb{F}_q$ of dimension d. For $0 \leqslant k \leqslant d$, let $n_{d,k}$ be the number of subspaces of V of dimension k. Then*

$$n_{d,k} = \frac{(q^d - 1)(q^d - q) \cdots (q^d - q^{k-1})}{(q^k - 1)(q^k - q) \cdots (q^k - q^{k-1})}. \qquad (3.12)$$

Moreover,

$$q^{k(d-k)} \leqslant n_{d,k} \leqslant q^{k(d-k+1)}. \qquad (3.13)$$

Proof: A subspace of V is determined by choosing a basis for it. By Proposition 3.15, a basis $v_1, v_2, \ldots, v_k$ for a k-dimensional subspace may be chosen in $(q^d - 1)(q^d - q) \cdots (q^d - q^{k-1})$ ways. Each subspace U has $(q^k - 1)(q^k - q) \cdots (q^k - q^{k-1})$ bases (again by Proposition 3.15), and so the total number $n_{d,k}$ of subspaces satisfies (3.12), as required. The bounds for $n_{d,k}$ given in (3.13) are obtained by noting that for $0 \leqslant i < k \leqslant d$,

$$q^{d-k} \leqslant \frac{q^d - q^i}{q^k - q^i} \leqslant q^{d-k+1}$$

and so the proposition is proved.

Since an elementary abelian p-group may be thought of as a vector space over $\mathbb{F}_p$, we have the following corollary to Proposition 3.16.

Corollary 3.17 *Let p be a prime number, and let P be an elementary abelian group of order p^d. Then P has exactly $n_{d,k}$ subgroups of order p^k, where*

$$n_{d,k} = \frac{(p^d - 1)(p^d - p) \cdots (p^d - p^{k-1})}{(p^k - 1)(p^k - p) \cdots (p^k - p^{k-1})}.$$

We now turn to some standard results concerning alternating forms. Let V and W be vector spaces over a field F. Just as for abelian groups, a bilinear map $\phi : V \times V \to W$ is said to be *alternating* if $\phi(v, v) = 0$ for all $v \in V$. Note that $\phi(u, v) = -\phi(v, u)$ for any vectors $u, v \in V$ when ϕ is an alternating map: to see this, expand $\phi(u + v, u + v)$. If W has dimension 1 over F, we say that ϕ is an *alternating form* on V.

Let ϕ be an alternating form on V. For a subspace U of V, we define $U^\perp$ by

$$U^\perp = \{v \in V : \phi(u, v) = 0 \text{ for all } u \in U\}.$$

The *radical* R of an alternating form on V is the subspace defined by $R = V^\perp$, so

$$R = \{u \in V : \phi(u, v) = 0 \text{ for all } v \in V\}.$$

If the radical R of an alternating form ϕ is trivial, we say that ϕ is *non-degenerate*.

We say that a basis $u_1, u_2, \ldots, u_r, v_1, v_2, \ldots, v_r$ of V is *symplectic* (with respect to the alternating form ϕ) if for all $i, j \in \{1, 2, \ldots, r\}$,

$$\phi(u_i, u_j) = 0,$$

$$\phi(v_i, v_j) = 0 \text{ and}$$

$$\phi(u_i, v_j) = \begin{cases} 1 & \text{when } i = j, \\ 0 & \text{otherwise.} \end{cases}$$

Proposition 3.18 *Let V be a finite-dimensional vector space over a field F, and let $\phi : V \times V \to F$ be a non-degenerate alternating form on V. Then there exists a basis $u_1, u_2, \ldots, u_r, v_1, v_2, \ldots, v_r$ of V that is symplectic with respect to ϕ. In particular, the dimension of V must be even.*

This is a standard result in linear algebra. Here is a sketch of the proof. A symplectic basis for V may be built up as follows. Firstly choose any non-zero vector $u_1 \in V$. Since ϕ is non-degenerate, the linear map $\pi : V \to F$ defined by

$\pi(v) = \phi(u_1, v)$ is not zero, and so we may choose $v_1 \in V$ such that $\pi(v_1) = \phi(u_1, v_1) = 1$. It is now not difficult to show that vectors u_1 and v_1 are linearly independent, and so they span a subspace U of dimension 2. The subspace $U^\perp$ is a complement to U in V, and the restriction of ϕ to $U^\perp$ is again a non-degenerate alternating form. A symplectic basis $u_2, u_3, \ldots, u_r, v_2, v_3, \ldots, v_r$ for $U^\perp$ is chosen, and the resulting basis $u_1, u_2, \ldots, u_r, v_1, v_2, \ldots, v_r$ is a symplectic basis for V. It is not difficult to show that every symplectic basis may be constructed in this way.

Let V be a vector space of dimension $2r$ over a finite field $\mathbb{F}_q$, and let ϕ be a non-degenerate alternating form on V. We write $\mathrm{Sp}(2r, q)$ for the group of all linear transformations of V that preserve the alternating form, so

$$\mathrm{Sp}(2r, q) = \{g \in \mathrm{GL}(V) \mid \phi(u, v) = \phi(gu, gv) \text{ for all } u, v \in V\}.$$

Proposition 3.19 *Let V be a vector space of dimension $2r$ over a finite field $\mathbb{F}_q$, and let ϕ be a non-degenerate alternating form on V. Then the number of bases for V that are symplectic with respect to ϕ is*

$$(q^{2r} - 1)q^{2r-1}(q^{2r-2} - 1)q^{2r-3} \cdots (q^2 - 1)q = q^{r^2}(q^{2r} - 1)$$

$$(q^{2r-2} - 1) \cdots (q^2 - 1).$$

Moreover, this is the order of $\mathrm{Sp}(2r, q)$.

Proof: We prove the result by induction on the dimension $2r$ of V. When $r = 0$ the result is trivial. Assume that r is positive and, as an inductive hypothesis, assume the result to be true for vector spaces of smaller dimension than V. Consider the construction of a symplectic basis given above. There are clearly $q^{2r} - 1$ choices for the non-zero vector u_1. The kernel of the map $\pi : V \to F$ has codimension 1 and therefore contains q^{2r-1} vectors. The possible choices for the vector v_1 make up a particular coset of the kernel, and so there are q^{2r-1} choices for v_1. So the number of symplectic bases for V is $(q^{2r} - 1)q^{2r-1}$ times the number of symplectic bases for $U^\perp$. Our inductive hypothesis now implies that the number of symplectic bases of V agrees with the formula above. The statement follows by induction on r. The final statement of the proposition now follows fairly easily. For let $u_1, u_2, \ldots, u_r, v_1, v_2, \ldots, v_r$ be a fixed symplectic basis of V. Then it is easy to see that an (invertible) linear transformation g lies in $\mathrm{Sp}(2r, q)$ if and only if the image of $u_1, u_2, \ldots, u_r, v_1, v_2, \ldots, v_r$ under g is a symplectic basis. So there is a bijection between the elements of $\mathrm{Sp}(2r, q)$ and the set of symplectic bases of V, given by associating the element $g \in G$ with the image of $u_1, u_2, \ldots, u_r, v_1, v_2, \ldots, v_r$ under g.

4

Enumerating p-groups: a lower bound

Let p be a fixed prime number. For any positive integer m, $f(p^m)$ is the number of (isomorphism classes of) groups of order p^m. This chapter will show that

$$f(p^m) \geqslant p^{\frac{2}{27} m^2 (m-6)}.$$

This bound, and the proof we shall give, are due to Graham Higman [45]. The chapter is divided into two sections. The first section investigates a specific class of groups that we will need in our enumeration. In fact, this class is the set of relatively free groups in a certain variety of p-groups. Let G be a group from this class. We will show that the isomorphism classes of certain quotients of G are in one-to-one correspondence with the orbits of Aut G on the set X of subgroups of the Frattini subgroup of G. In the second section, we give an upper bound on the length of an orbit of Aut G acting on X. This enables us to show that a large number of isomorphism classes of groups occur as quotients of the group G, and this will provide the lower bound we need.

4.1 Relatively free groups

Let r be a positive integer. Let F_r be a free group of rank r, generated by $x_1, x_2, \ldots, x_r$. Let G_r be the quotient of F_r by the subgroup N generated by all words of the form x^{p^2}, $[x, y]^p$ and $[x, y, z]$. Note that all words of the form $[x^p, y]$ lie in N, by Lemma 3.3. The group G_r is the relatively free group in the variety of p-groups of Φ-class 2; see Hanna Neumann [75] for an introduction to varieties of groups in general. A p-group G is said to have Φ-class 2 if there exists a central elementary abelian subgroup H of G such

23

that G/H is elementary abelian. We identify the elements x_i with their images in G_r. Using this identification, $x_1, x_2, \ldots, x_r$ generate G_r.

Lemma 4.1 *Let H be a group of Φ-class 2, and let $y_1, y_2, \ldots, y_r \in H$. There is a homomorphism $\phi : G_r \to H$ such that $\phi(x_i) = y_i$ for $i \in \{1, 2, \ldots, r\}$.*

Proof: Since F_r is free, there is a unique homomorphism $\psi : F_r \to H$ such that $\psi(x_i) = y_i$ for all $i \in \{1, 2, \ldots, r\}$. By our restriction on H, we have $N \leqslant \ker \psi$ and so ψ induces a homomorphism $\phi : G_r \to H$ such that $\phi(x_i) = y_i$ for $i \in \{1, 2, \ldots, r\}$, as required.

Lemma 4.2 *The group G_r is a p-group. The Frattini subgroup $\Phi(G_r)$ of G_r is central of order $p^{\frac{1}{2}r(r+1)}$ and index p^r. Moreover, any automorphism $\alpha \in \operatorname{Aut} G_r$ that induces the identity mapping on $G_r/\Phi(G_r)$ fixes $\Phi(G_r)$ pointwise.*

Proof: Since any commutator or p^{th} power in G_r is central of order p, we have that $G_r^p G_r'$ is a central elementary abelian p-group. Since $G_r/G_r^p G_r'$ is also an elementary abelian p-group, G_r is a p-group.

By Lemma 3.12, $\Phi(G_r) = G_r^p G_r'$, and so $\Phi(G_r)$ is central. It is not difficult to show that $\Phi(G_r)$ is generated by the elements x_i^p for $i \in \{1, 2, \ldots, r\}$ and $[x_j, x_i]$ where $1 \leqslant i < j \leqslant r$. These elements form a minimal generating set of $\Phi(G_r)$; we can see this as follows. Suppose there exist $a_i \in \{0, 1, \ldots, p-1\}$ for $i = 1, 2, \ldots, r$ and $b_{i,j} \in \{0, 1, \ldots, p-1\}$ for $1 \leqslant i < j \leqslant r$ such that

$$\prod_{i=1}^{r} \left(x_i^p \right)^{a_i} \prod_{1 \leqslant i < j \leqslant r} [x_j, x_i]^{b_{i,j}} = 1. \tag{4.1}$$

Let H be the cyclic group of order p^2, generated by an element h. Since H has Φ-class 2, Lemma 4.1 implies that for all k such that $1 \leqslant k \leqslant r$, there exists a homomorphism $\psi_k : G_r \to H$ such that

$$\psi_k(x_i) = \begin{cases} 1 & \text{if } i \neq k, \\ h & \text{if } i = k. \end{cases}$$

Under ψ_k, Equation (4.1) becomes $(h^p)^{a_k} = 1$, and so $a_k = 0$. Hence $a_1 = a_2 = \cdots = a_r = 0$. We may show that $b_{i,j} = 0$ for all i and j by using a similar argument, where the homomorphism ϕ maps x_i and x_j to the generating elements

$$\begin{pmatrix} 1 & 1 & 0 \\ 0 & 1 & 0 \\ 0 & 0 & 1 \end{pmatrix} \quad \text{and} \quad \begin{pmatrix} 1 & 0 & 0 \\ 0 & 1 & 1 \\ 0 & 0 & 1 \end{pmatrix}$$

of the group H of 3×3 upper unitriangular matrices over $\mathbb{F}_p$ and where ϕ maps all other generators to 1. Hence the generating set of $\Phi(G_r)$ is minimal, and $|\Phi(G_r)| = p^{r + \frac{1}{2}r(r-1)}$.

Now $G_r / \Phi(G_r)$ is elementary abelian, and has the images of $x_1, x_2, \ldots, x_r$ as a minimal generating set. Therefore $\Phi(G_r)$ has index p^r.

Let $\alpha \in \operatorname{Aut} G_r$ be an automorphism that induces the identity mapping on G / G_r. So there exist $h_1, h_2, \ldots, h_r \in \Phi(G_r)$ such that $\alpha(x_i) = x_i h_i$ for $i = 1, 2, \ldots, r$. Since $\Phi(G_r)$ is central and of exponent p,

$$\alpha(x_i^p) = \alpha(x_i)^p = (x_i h_i)^p = x_i^p h_i^p = x_i^p.$$

Also, since $\Phi(G_r)$ is central,

$$\alpha([x_j, x_i]) = [\alpha(x_j), \alpha(x_i)] = [x_j h_j, x_i h_i] = [x_j, x_i].$$

Hence α fixes $\Phi(G_r)$ pointwise, as required.

Lemma 4.3 *Let $N_1, N_2 \leqslant \Phi(G_r)$. Then $G_r / N_1 \cong G_r / N_2$ if and only if there exists $\alpha \in \operatorname{Aut} G_r$ such that $\alpha(N_1) = N_2$.*

Proof: Note that quotients by N_1 and N_2 make sense, since $\Phi(G_r)$ is contained in the centre of G_r.

An element $\alpha \in \operatorname{Aut} G_r$ mapping N_1 to N_2 induces an isomorphism α' from G_r / N_1 to G_r / N_2. We need to show the converse.

Let $\alpha' : G_r / N_1 \to G_r / N_2$ be an isomorphism. Let $y_1, y_2, \ldots, y_r \in G_r$ be such that $\alpha'(x_i N_1) = y_i N_2$. Since G_r has Φ-class 2, Lemma 4.1 implies that there exists a homomorphism $\alpha : G_r \to G_r$ such that $\alpha(x_i) = y_i$. Now, since α' is an isomorphism, $y_1, y_2, \ldots, y_r$ and N_2 together generate G_r. Since $N_2 \leqslant \Phi(G_r)$, the elements $y_1, y_2, \ldots, y_r$ and $\Phi(G_r)$ together generate G_r, and so Lemma 3.9 tells us that $y_1, y_2, \ldots, y_r$ generate G_r. But the elements y_i are contained in the image of α, and so α is surjective. Since G_r is finite, this implies that $\alpha \in \operatorname{Aut} G_r$. Finally, we need to show that $\alpha(N_1) = N_2$. The definition of α shows that $\alpha(x)N_2 = \alpha'(xN_1)$ when x is one of the generators x_i, and so $\alpha(x)N_2 = \alpha'(xN_1)$ for all $x \in G_r$. In other words, the following diagram commutes, where the vertical maps are natural:

$$
\begin{array}{ccc}
G_r & \xrightarrow{\alpha} & G_r \\
\downarrow & & \downarrow \\
G_r / N_1 & \xrightarrow{\alpha'} & G_r / N_2
\end{array}
$$

It is easy to check from this diagram that $\alpha(N_1) = N_2$, and so the lemma follows.

4.2 Proof of the lower bound

This section contains a proof of Graham Higman's lower bound for the number of isomorphism classes of p-groups of a given order.

Proposition 4.4 *Let r be a positive integer, and let s be an integer such that $1 \leqslant s \leqslant \frac{1}{2}r(r+1)$. Then there are at least $p^{\frac{1}{2}rs(r+1)-r^2-s^2}$ isomorphism classes of groups of order p^{r+s}.*

Proof: Let G_r be the group defined in Section 4.1. Let X be the set of subgroups $N \leqslant \Phi(G_r)$ of index p^s in $\Phi(G_r)$. Each subgroup $N \in X$ gives rise to a group G_r/N of order p^{r+s}. Moreover, by Lemma 4.3 the set of isomorphism classes of groups that arise in this way is in one-to-one correspondence with the set of orbits of Aut G_r acting in the natural way on X.

Let θ be the natural homomorphism from Aut G_r to Aut $(G_r/\Phi(G_r))$. By Lemma 4.2, any automorphism $\alpha \in \ker \theta$ fixes $\Phi(G_r)$ pointwise and so acts trivially on X. Thus $\ker \theta$ is contained in the stabiliser of every element of X, and so the length of any orbit of Aut G_r acting on X is at most $|\text{Aut } G_r|/|\ker \theta| \leqslant |\text{Aut}(G_r/\Phi(G_r))|$.

We may regard $G_r/\Phi(G_r)$ as a vector space over $\mathbb{F}_p$, and group automorphisms of $G_r/\Phi(G_r)$ correspond exactly with invertible linear transformations in this setting. Hence $|\text{Aut}(G_r/\Phi(G_r))| = |\text{GL}(r, p)| \leqslant p^{r^2}$ by Proposition 3.15. So every orbit on X has length at most p^{r^2}. By Corollary 3.17, $|X| \geqslant p^{(\frac{1}{2}r(r+1)-s)s}$. Hence there are at least $p^{(\frac{1}{2}r(r+1)-s)s}/p^{r^2}$ orbits of Aut G_r on X. Thus there are at least $p^{\frac{1}{2}rs(r+1)-r^2-s^2}$ isomorphism classes of groups of order p^{r+s}, as required.

Theorem 4.5 *The number $f(p^m)$ of p-groups of order p^m is at least*

$$p^{\frac{2}{27}m^2(m-6)}.$$

Proof: The theorem is trivially true in the case when $m \leqslant 6$. When $m > 6$, define the integer s by

$$s = \begin{cases} \frac{1}{3}m & \text{if } m \equiv 0 \bmod 3, \\ \frac{1}{3}(m+2) & \text{if } m \equiv 1 \bmod 3, \\ \frac{1}{3}(m+1) & \text{if } m \equiv 2 \bmod 3, \end{cases}$$

and define $r = m - s$. Then, by Proposition 4.4,

$$f(p^m) \geqslant p^{\frac{1}{2}rs(r+1)-r^2-s^2} \geqslant p^{\frac{2}{27}m^2(m-6)},$$

as required.

The proof of Theorem 4.5 gives an indication as to why the constant $2/27$ appears in the group enumeration function: Proposition 4.4 yields approximately $p^{\frac{1}{2}a^2bm^3}$ groups having a Frattini subgroup of index p^{am} and order p^{bm}, and $2/27$ is the maximum (at $a = 2/3$) of the function $\frac{1}{2}a^2b$ subject to $a + b = 1$.

5

Enumerating p-groups: upper bounds

Our aim in this chapter is to prove a good upper bound on the number $f(p^m)$ of isomorphism classes of groups of order p^m. We will establish that

$$f(p^m) \leqslant p^{\frac{2}{27}m^3 + O(m^{5/2})}.$$

The leading term is the best possible, by the results of Chapter 4. This bound, but with the larger error term of $O(m^{8/3})$, is due to Charles Sims [86]. The modification of Sims' argument that leads to the improved error term is due to Mike Newman and Craig Seeley [77].

Before proving the upper bound above, we give an argument that produces a weaker upper bound, but which is much more elementary than the Sims approach.

5.1 An elementary upper bound

The upper bound for $f(p^m)$ we give in this section is due to Graham Higman [45]. A better upper bound, using more complex techniques, was also given by Higman in the same paper, but he remarks of the bound we will give 'It seems to me surprising that an upper bound obtained so simply should be as near the truth as [Theorem 4.5] shows that this must be'.

The strategy of the proof is to show that any group of order p^m has a presentation of a restricted type, and then to give an upper bound for the number of such presentations. This is also the basic strategy used to prove the Sims bound, but Sims uses a class of smaller presentations, thus improving the upper bound but at the expense of having to use more sophisticated arguments.

28

Theorem 5.1 *Let p be a prime number and let m be an integer such that $m \geqslant 1$. Then*

$$f(p^m) \leqslant p^{\frac{1}{6}(m^3 - m)}.$$

Proof: Let G be a group of order p^m, and let $G = G(0) > G(1) > \cdots > G(m-1) > G(m) = \{1\}$ be a chief series for G. Thus $G(i)$ is a normal subgroup of G of index p^i in G.

For $1 \leqslant i \leqslant m$, let $g_i \in G(i-1) \setminus G(i)$. Then every element $g \in G$ may be written uniquely in the form

$$g = g_1^{\alpha_1} g_2^{\alpha_2} \cdots g_m^{\alpha_m} \tag{5.1}$$

where $\alpha_1, \alpha_2, \ldots, \alpha_m \in \{0, 1, \ldots, p-1\}$. Moreover, for $i \geqslant 1$ we have that $g \in G(i)$ if and only if $\alpha_1 = \alpha_2 = \cdots = \alpha_i = 0$.

We have that $g_i^p \in G(i)$ since $G(i-1)/G(i)$ has order p. Hence for $1 \leqslant i \leqslant m$ there exist $\beta_{i,i+1}, \beta_{i,i+2}, \ldots, \beta_{i,m} \in \{0, 1, \ldots, p-1\}$ such that

$$g_i^p = g_{i+1}^{\beta_{i,i+1}} g_{i+2}^{\beta_{i,i+2}} \cdots g_m^{\beta_{i,m}}. \tag{5.2}$$

Now $G(j-1)/G(j)$ is central in $G/G(j)$ since it has order p and is a normal subgroup of the p-group $G/G(j)$. Hence the commutator $[g_j, g_i]$ lies in $G(j)$ for all integers i and j such that $1 \leqslant i < j \leqslant m$. Therefore we may write

$$[g_j, g_i] = g_{j+1}^{\gamma_{i,j,j+1}} g_{j+2}^{\gamma_{i,j,j+2}} \cdots g_m^{\gamma_{i,j,m}} \tag{5.3}$$

for some elements $\gamma_{i,j,k} \in \{0, 1, \ldots, p-1\}$.

It is not difficult to verify that the generators $g_1, g_2, \ldots, g_m$, the relations (5.2) for $1 \leqslant i \leqslant m$ and the relations (5.3) for $1 \leqslant i < j \leqslant m$ form a presentation for G. (This is a so-called power-commutator presentation for G.) To verify the relations do form a presentation, it is only necessary to verify that the product of two words in normal form (5.1) can be brought into normal form using the relations given. This can be done by one of the collection processes used in computational p-group theory. See Sims [87] for a detailed discussion of such processes. One method goes roughly as follows. Given the product of two words in normal form, all occurrences of the generator g_1 can be moved to the far left, using the relations (5.3) with $i = 1$. The resulting power of g_1 can then be made to lie in the correct range by using (5.2) with $i = 1$. This process results in a word of the form $g_1^{\alpha_1} g$, where g is a word in $g_2, g_3, \ldots, g_m$. The process is then repeated with g and the generator g_2. After m applications of this process, the result is a word in normal form.

As the relations (5.2) and (5.3) form a presentation for G, the isomorphism class of G is determined by the values of the elements $\beta_{i,k}$ where $1 \leqslant i < k \leqslant m$ and $\gamma_{i,j,k}$ where $1 \leqslant i < j < k \leqslant m$. There are at most p choices for the values of each of these $\frac{1}{6}(m^3 - m)$ elements, and so the number of isomorphism classes of groups of order p^m is at most $p^{\frac{1}{6}(m^3 - m)}$, as required.

5.2 An overview of the Sims approach

We need to improve the leading term of the upper bound given in Section 5.1. This is achieved by reducing the number of relations in our presentation for a group G of order p^m, by choosing the set of generators for the group more carefully. In the bound of Section 5.1, the power relations are specified by the elements $\beta_{i,k}$ where $1 \leqslant i < k \leqslant m$. Since there are at most $p^{\frac{1}{2}m(m-1)}$ choices for the elements $\beta_{i,k}$, the power relations do not affect the leading term of the enumeration function, and so we must concentrate on choosing the presentation of a group G in such a way that the number of commutator relations we need is reduced as much as possible. To this end, the role of the chief series of G in the previous section will be taken by the lower central series $G_1, G_2, \ldots$ of G. This will allow us to make more detailed use of the commutator structure of the group.

We define the *Sims rank* $s(G)$ of a p-group G to be the smallest non-negative integer s such that there exists $H \leqslant G$ with s generators such that $G_2 = H_2 G_3$. (Recall that, by Proposition 3.8, this condition on H implies that $H_i = G_i$ for $i \geqslant 2$.) We will show that when the Sims rank of a p-group is large, there are comparatively few possibilities for G/G_3. When the Sims rank is small, there are comparatively few possibilities for G once G/G_3 is fixed. This trade-off will be used to prove the upper bound on the number of groups of order p^m that we are seeking.

The proof is divided into three parts, each contained in one of the next three sections. Firstly, we will concentrate on bounding the number of possibilities for G/G_3 when the Sims rank $s(G)$ is fixed. It is the number of possibilities for the commutator structure of G/G_3 that is important, and we consider a 'linearised' version of the problem of counting the possible commutator structures that can arise. Secondly, we use our analysis of this linearised problem to show that we may always choose a presentation for G to have a special form, that has fewer commutator relations when compared with Higman's approach above. Finally, we give an upper bound for the number of groups of order p^m by bounding the number of choices for the presentations of this special form.

5.3 'Linearising' the problem

Let V and W be vector spaces over some field. Recall from Chapter 3 that a map $[\,,\,]: V \times V \to W$ is alternating if it is bilinear and if $[x, x] = 0$ for all $x \in V$. Let U_1 and U_2 be subspaces of V. We define $[U_1, U_2]$ to be the subspace of W spanned by all vectors of the form $[x, y]$ where $x \in U_1$ and $y \in U_2$.

A p-group G naturally gives rise to an alternating map as follows. There is a natural function $\phi: G \times G \to G_2/G_3\Phi(G_2)$, where $(x, y) \in G \times G$ is mapped to $[x, y]G_3\Phi(G_2)$. The facts that $[G_2, G_1] = G_3$ and that $[g^p, h] = [g, h]^p$ mod G_3 for all $g, h \in G$ imply that multiplying x and y by elements of $\Phi(G)$ does not change $\phi(x, y)$, and so ϕ induces a map $[\,,\,]: (G/\Phi(G)) \times (G/\Phi(G)) \to G_2/G_3\Phi(G_2)$. Now, both $G/\Phi(G)$ and $G_2/G_3\Phi(G_2)$ are elementary abelian p-groups, and so we may think of them as vector spaces V and W over $\mathbb{F}_p$. It is not difficult to verify that $[\,,\,]: V \times V \to W$ is bilinear and that $[x, x] = 0$ for all $x \in V$. Moreover, the Sims rank of G is s if and only if s is the minimal dimension of a subspace U of V such that $[U, U] = W$. (This follows since a subgroup H of G has the property that $H_2G_3 = G_2$ if and only if the image U of H under the natural homomorphism from G to V has the property that $[U, U] = W$.) The aim of this section is to give a bound on the number of possibilities for the map $[\,,\,]$ and some structural information about this map, if we know the Sims rank of G.

Proposition 5.2 *Let $[\,,\,]: V \times V \to W$ be an alternating map, and suppose that $[V, V] = W$. Suppose $\dim W > 0$. Let U be a subspace of V of minimal dimension subject to the property that $[U, U] = W$. Then U has a basis $x_1, x_2, \ldots, x_s$ such that if we define $V_i = \langle x_1, x_2, \ldots, x_i \rangle$ then for $1 \leqslant i \leqslant s - 1$*

$$[x_i, x_{i+1}] \notin [V_i, V_i].$$

Proof: Define $s = \dim U$. We show by induction that there exists a basis $x_1, x_2, \ldots, x_s$ of U satisfying the proposition, the inductive hypothesis being that there exists a (linearly independent) set $x_1, x_2, \ldots, x_k \in U$ such that

$$[x_i, x_{i+1}] \notin [V_i, V_i] \text{ for all } i \in \{1, 2, \ldots, k - 1\}. \tag{5.4}$$

The hypothesis is clearly true when $k = 2$, for since $\dim W > 0$ and $[U, U] = W$ we have that $\dim U \geqslant 2$ and there exist $x_1, x_2 \in U$ such that $[x_1, x_2] \neq 0$.

To prove the inductive step, suppose that $2 < k \leqslant s$ and there exists a linearly independent set $y_1, y_2, \ldots, y_{k-1} \in U$ such that

$$[y_i, y_{i+1}] \notin [\langle y_1, y_2, \ldots, y_i \rangle, \langle y_1, y_2, \ldots, y_i \rangle] \text{ for all } i \in \{1, 2, \ldots, k - 2\}.$$

For any $i \in \{1, 2, \ldots, k-1\}$, define $U_i = \langle y_1, y_2, \ldots, y_i \rangle$. We consider four cases:

- There exists $x \in U \setminus U_{k-1}$ such that $[y_{k-1}, x] \notin [U_{k-1}, U_{k-1}]$. In this case, setting $x_i = y_i$ for $1 \leqslant i \leqslant k-1$ and setting $x_k = x$, we see that (5.4) holds.
- $[y_{k-1}, x] \in [U_{k-1}, U_{k-1}]$ for all $x \in U \setminus U_{k-1}$, but there exists $x \in U \setminus U_{k-1}$ such that $[y_j, x] \notin [U_{k-1}, U_{k-1}]$ for some $j \in \{1, 2, \ldots, k-2\}$. Set $x_i = y_i$ for all $i \in \{1, 2, \ldots, k-2\}$, set $x_{k-1} = y_{k-1} + y_j$ and set $x_k = x$. Note that $V_i = U_i = \langle x_1, x_2, \ldots, x_i \rangle$ whenever $i \leqslant k-1$. Hence $[V_i, V_i] = [U_i, U_i]$ whenever $i \leqslant k-1$ and so (5.4) clearly holds when $i < k-2$. But (5.4) also holds when $i = k-2$:

$$[x_{k-2}, x_{k-1}] = [y_{k-2}, y_{k-1}] + [y_{k-2}, y_j] \notin [U_{k-2}, U_{k-2}]$$

since $[y_{k-2}, y_{k-1}] \notin [U_{k-2}, U_{k-2}]$ and $[y_{k-2}, y_j] \in [U_{k-2}, U_{k-2}]$. Furthermore, (5.4) holds when $i = k-1$:

$$[x_{k-1}, x_k] = [y_{k-1}, x] + [y_j, x] \notin [U_{k-1}, U_{k-1}]$$

since $[y_{k-1}, x] \in [U_{k-1}, U_{k-1}]$ and $[y_j, x] \notin [U_{k-1}, U_{k-1}]$. Hence (5.4) holds in this case.

- Suppose that the above two possibilities do not occur. So we have that $[U_{k-1}, U] = [U_{k-1}, U_{k-1}]$ and there exist $x, y \in U \setminus U_{k-1}$ such that $[x, y] \notin [U_{k-1}, U_{k-1}]$. Suppose further that

$$[y_{k-2}, y_{k-1}] + [y_{k-2}, x] \in [U_{k-2}, U_{k-2}].$$

This implies that $[y_{k-2}, x] \notin [U_{k-2}, U_{k-2}]$. Set $x_i = y_i$ for all $i \in \{1, 2, \ldots, k-2\}$, set $x_{k-1} = x$ and $x_k = y$. So $V_i = U_i$ for $1 \leqslant i \leqslant k-2$. Now, (5.4) clearly holds when $i < k-2$, and also in the case $i = k-2$ since $[y_{k-2}, x] \notin [U_{k-2}, U_{k-2}]$. To see that the case $i = k-1$ also holds, we observe that

$$[V_{k-1}, V_{k-1}] \subseteq [U_{k-2}, U_{k-2}] + [U_{k-2}, x_{k-1}] \subseteq [U_{k-1}, U_{k-1}].$$

Since $[x, y] \notin [U_{k-1}, U_{k-1}]$, (5.4) holds in the case when $i = k-1$.

- Finally, we suppose that $[U_{k-1}, U] = [U_{k-1}, U_{k-1}]$ and there exist $x, y \in U \setminus U_{k-1}$ such that $[x, y] \notin [U_{k-1}, U_{k-1}]$ and such that

$$[y_{k-2}, y_{k-1}] + [y_{k-2}, x] \notin [U_{k-2}, U_{k-2}].$$

Set $x_i = y_i$ for all $i \in \{1, 2, \ldots, k-2\}$, set $x_{k-1} = y_{k-1} + x$ and set $x_k = y$. Again, it is clear that (5.4) holds when $i < k-2$. Since $V_{k-2} = U_{k-2}$ and $[x_{k-2}, x_{k-1}] = [y_{k-2}, y_{k-1}] + [y_{k-2}, x] \notin [U_{k-2}, U_{k-2}]$, (5.4) holds when $i = k-2$. Finally,

$$[V_{k-1}, V_{k-1}] \subseteq [U_{k-2}, U_{k-2}] + [U_{k-2}, y_{k-1} + x] \subseteq [U_{k-1}, U_{k-1}],$$

and so the fact that $[x_{k-1}, x_k] = [y_{k-1}, y] + [x, y]$ implies that (5.4) holds when $i = k - 1$.

The proposition now follows by induction on k.

Corollary 5.3 *Let V and W be vector spaces over some field, and let $[,]$: $V \times V \to W$ be an alternating map such that $[V, V] = W$. Then there exists a subspace U of V such that $[U, U] = W$ and $\dim U \leqslant \dim W + 1$.*

Proof: The corollary is trivial when $\dim W = 0$, so assume that $\dim W > 0$. Let U be a subspace of V of minimal dimension subject to the property that $[U, U] = W$. Proposition 5.2 shows that U has a basis $x_1, x_2, \ldots, x_s$ such that the $s - 1$ vectors $[x_1, x_2], [x_2, x_3], \ldots, [x_{s-1}, x_s]$ form a linearly independent set in W. Hence $\dim U - 1 \leqslant \dim W$ and so $\dim U \leqslant \dim W + 1$.

Corollary 5.4 *Let V and W be vector spaces over some field, and let $[,]$: $V \times V \to W$ be an alternating map such that $[V, V] = W$. Let s be the smallest integer such that there exists an s-dimensional subspace U of V such that $[U, U] = W$. Then for any subspace K of V,*

$$\dim[K, K] \leqslant \dim K + \dim W + 1 - s.$$

Proof: The map $[,]$ induces an alternating map $[,]' : V \times V \to W/[K, K]$. Since $[V, V]' = W/[K, K]$, Corollary 5.3 implies that there is a subspace L of V such that $\dim L \leqslant \dim(W/[K, K]) + 1$ and $[L, L]' = W/[K, K]$. But now, $[K + L, K + L] = W$. Thus

$$s \leqslant \dim(K + L) \leqslant \dim K + \dim L \leqslant \dim K + \dim W - \dim[K, K] + 1$$

and the corollary is proved.

We have established some information about the structure of alternating maps arising from p-groups with Sims rank s as defined on p. 30. The following proposition uses this information to provide two upper bounds on the number of such maps. Each bound is good in certain situations, but both bounds must be used together to provide the approximations we require. The idea of using bounds on the number of alternating maps in this way is the contribution of Newman and Seeley.

Proposition 5.5 *Let V and W be vector spaces over $\mathbb{F}_p$ of dimensions r_1 and r_2 respectively, and let s be a positive integer. Let $N_p(r_1, r_2, s) \in \mathbb{R}$ be such that $p^{N_p(r_1, r_2, s)}$ is the number of alternating maps $[,] : V \times V \to W$ for which* *there exists an s-dimensional subspace U of V with the property that $[U, U] = W$, but no $(s-1)$-dimensional subspace U_0 exists such that $[U_0, U_0] = W$. Then*

$$N_p(r_1, r_2, s) \leqslant \tfrac{1}{2}r_1^2(r_2 - (s-1)) + O((r_1 + r_2)^{8/3}) \text{ and} \qquad (5.5)$$

$$N_p(r_1, r_2, s) \leqslant \tfrac{1}{2}r_1^2(r_2 - (s-1)) + \tfrac{1}{2}r_1 r_2(s-1) + O((r_1 + r_2)^{5/2}). \qquad (5.6)$$

Proof: We begin by proving the bound (5.5). Define $f = \lfloor r_1^{2/3} \rfloor$ and define $g = \lceil r_1/f \rceil$. Let $x_1, x_2, \dots, x_{r_1}$ be a basis of V. Note that the map $[\,,]$ is determined by $[x_k, x_\ell]$ where $1 \leqslant k < \ell \leqslant r_1$. Since there are at most p^{r_2} choices for each vector $[x_k, x_\ell]$, we have (trivially) that the number of maps of the form we are looking for is at most $p^{\binom{r_1}{2} r_2}$. When $s < 2r_1^{2/3} + 2$, this bound implies (5.5) and so we may assume that $s \geqslant 2r_1^{2/3} + 2$.

Define subspaces $V_1, V_2, \dots, V_g$ of V by

$$V_i = \langle x_{(i-1)f+1}, x_{(i-1)f+2}, \dots, x_{(i-1)f+f} \rangle$$

for $i \in \{1, 2, \dots, g-1\}$ and

$$V_g = \langle x_{(g-1)f+1}, x_{(g-1)f+2}, \dots, x_{r_1} \rangle.$$

For all integers i and j such that $1 \leqslant i < j \leqslant g$, $\dim(V_i + V_j) \leqslant \dim V_i + \dim V_j \leqslant 2f$ and so Corollary 5.4 implies that

$$\dim[V_i + V_j, V_i + V_j] \leqslant 2f + r_2 + 1 - s < r_2,$$

the last inequality coming from the fact that $s \geqslant 2r_1^{2/3} + 2$. For all integers i and j such that $1 \leqslant i < j \leqslant g$, let W_{ij} be a $(2f + r_2 + 1 - s)$-dimensional subspace of W containing $[V_i + V_j, V_i + V_j]$.

We may now obtain the bound (5.5) as follows. Proposition 3.16 shows that the number of subspaces in a vector space of dimension r_2 over $\mathbb{F}_p$ is at most $p^{r_2^2}$ (the bound clearly holds when $d = 0$, and the upper bound of the proposition implies this bound whenever $d \geqslant 1$). So there are at most $p^{r_2^2\binom{g}{2}}$ choices for the subspaces W_{ij}. Once the subspaces W_{ij} have been fixed, there are at most p^{2f+r_2+1-s} choices for each image $[x_k, x_\ell]$, since $x_k, x_\ell \in V_i + V_j$ for some integers i and j such that $1 \leqslant i < j \leqslant g$ and then $[x_k, x_\ell] \in W_{ij}$. Hence, once the subspaces W_{ij} have been fixed, there are at most $p^{\binom{r_1}{2}(2f+r_2+1-s)}$ choices for the map $[\,,]$. Now, $f = r_1^{2/3} + O(1)$ and $g = r_1^{1/3} + O(1)$. Hence $p^{N_p(r_1, r_2, s)}$ is bounded above by

$$p^{\binom{r_1}{2}(2f + r_2 + 1 - s) + r_2^2\binom{g}{2}} = p^{\binom{r_1}{2}(r_2 - (s-1)) + O((r_1 + r_2)^{8/3})}.$$

This establishes the bound (5.5).

We now establish the bound (5.6). We will use the same techniques as in the proof of the bound (5.5), except our choice of f will differ and our counting argument changes slightly. When $s \leqslant 2r_1^{1/2} + 2$, the trivial upper bound of $p^{\binom{r_1}{2} r_2}$ implies the bound (5.6). So we may assume that $s > 2r_1^{1/2} + 2$. Let $f = \lfloor r_1^{1/2} \rfloor$ and define $g = \lceil r_1/f \rceil$. We choose a basis $x_1, x_2, \ldots, x_{r_1}$ and define subspaces V_i and W_{ij} just as before. Note that W_{ij} is well-defined, since $r_2 - s + 1 + 2f < r_2$.

By Proposition 3.16, the number of d-dimensional subspaces of W is bounded above by $p^{(r_2-d+1)d}$. Hence the number of possibilities for each subspace W_{ij} is at most $p^{(r_2-(r_2-s+1+2f)+1)(r_2-s+1+2f)} = p^{(s-2f)(r_2-s+1+2f)}$. Just as before, there are $\binom{g}{2}$ subspaces W_{ij}, and once these subspaces have been chosen there are at most $p^{r_2-s+1+2f}$ choices for each of the $\binom{r_1}{2}$ images $[x_k, x_\ell]$ with $1 \leqslant k < \ell \leqslant r_1$. So $p^{N_p(r_1,r_2,s)}$ is bounded above by p to the power

$$\binom{g}{2}(s-2f)(r_2-s+1+2f) + \binom{r_1}{2}(r_2-s+1+2f).$$

But $f = r_1^{1/2} + O(1)$ and $g = r_1^{1/2} + O(1)$, and so the bound (5.6) follows.

5.4 A small set of relations

In this section we show that we may always find a presentation for a p-group G of a restricted type. Our eventual enumeration in the next section will depend on the fact that there are comparatively few possibilities for presentations of this sort.

We begin by choosing our generating set for G and finding a collection of relations this set satisfies. We then prove that we have a presentation for G.

Let G be a group of order p^m. Let $G = G_1 \geqslant G_2 \geqslant \cdots \geqslant G_c \geqslant G_{c+1} = \{1\}$ be the lower central series of G. Define integers $r_1, r_2, \ldots, r_c$ by setting r_i to be the rank of G_i/G_{i+1}. Define $V = G/\Phi(G)$ and $W = G_2/G_3\Phi(G_2)$. The characterisation of the Frattini subgroup of a p-group given in Lemma 3.12 implies that $\dim V = r_1$ and $\dim W = r_2$. We choose a basis $\{x_i\}$ for V and a basis $\{y_j\}$ for W as follows. As before, the process of forming commutators in G induces an alternating bilinear map $[\,,\,] : V \times V \to W$. Let s be the Sims rank of G. Let U be a subspace of V of minimal dimension such that $[U, U] = W$; then U has dimension s. By Corollary 5.3 we have that $s \leqslant r_2 + 1$. By Proposition 5.2, there exists a basis $x_1, x_2, \ldots, x_s$ of U such that for all $i \in \{1, 2, \ldots, s-1\}$, $[x_i, x_{i+1}] \notin [V_i, V_i]$, where we define $V_i = \langle x_1, x_2, \ldots, x_i \rangle$. Define

$$W_i = [V_i, V_i] \quad \text{and} \quad d_i = \dim W_i.$$

Note that $0 = d_1 < d_2 < \cdots < d_s = r_2$. Define $d_0 = 0$. Choose a basis $y_1, y_2, \ldots, y_{r_2}$ for W such that $W_i = \langle y_1, y_2, \ldots, y_{d_i} \rangle$ for all $i \in \{2, 3, \ldots, s\}$. Now

$$W_i = \langle W_{i-1}, [x_1, x_i], [x_2, x_i], \ldots, [x_{i-1}, x_i] \rangle$$

and so we may in addition choose the vectors y_j to have the following property: there exist integers $f(1), f(2), \ldots, f(r_2)$ such that

$$y_j = [x_{f(j)}, x_i] \text{ where } 1 \leqslant f(j) < i \text{ whenever } d_{i-1} < j \leqslant d_i. \qquad (5.7)$$

Finally, we extend $x_1, x_2, \ldots, x_s$ to a basis $x_1, x_2, \ldots, x_{r_1}$ of V.

Let $g_{11}, g_{12}, \ldots, g_{1r_1}$ be representatives of $x_1, x_2, \ldots, x_{r_1}$ in G. Define $g_{21}, g_{22}, \ldots, g_{2r_2}$ by setting $g_{2j} = [g_{1f(j)}, g_{1i}]$ whenever $d_{i-1} < j \leqslant d_i$. Note that g_{2j} is a representative of y_j in G. Define H to be the subgroup of G generated by $g_{11}, g_{12}, \ldots, g_{1s}$. Now H_2, the second term in the lower central series of H, contains all the elements g_{2j}. Since $y_1, y_2, \ldots, y_{r_2}$ generate $G_2/G_3\Phi(G_2)$, we have that $H_2 G_3 = G_2$ and so Proposition 3.8 shows that $H_i = G_i$ for all $i \geqslant 2$.

For $i \in \{3, 4, \ldots, c\}$, let $g_{i1}, g_{i2}, \ldots, g_{ir_i} \in G$ be chosen so that G_i/G_{i+1} is generated by $g_{i1}G_{i+1}, g_{i2}G_{i+1}, \ldots, g_{ir_i}G_{i+1}$. Since $G_i = H_i$ when $i \geqslant 2$, we may choose (by Proposition 3.7) $g_{ij} = [g_{(i-1)k}, g_{1\ell}]$ for some k and ℓ where $1 \leqslant k \leqslant r_{i-1}$ and $1 \leqslant \ell \leqslant s$. In fact, we claim that when $i = 3$ we can choose k such that $d_{\ell-1} < k$. To show that we may do this, it suffices to prove that every commutator in the set $\{[g_{2k}, g_{1\ell}] \mid 1 \leqslant \ell \leqslant s, 1 \leqslant k \leqslant r_2\}$ lies in the subgroup L generated by G_4 and the set $\{[g_{2k}, g_{1\ell}] \mid 1 \leqslant \ell \leqslant s$ and $d_{\ell-1} < k \leqslant r_2\}$. Let a be an integer such that $3 \leqslant a \leqslant s$ and assume (as an inductive hypothesis) that we have shown that $[g_{2k}, g_{1\ell}] \in L$ whenever $1 \leqslant \ell < a$ and $1 \leqslant k \leqslant r_2$. Since $d_0 = d_1 = 0$, the hypothesis is true when $a = 3$. We show that $[g_{2k}, g_{1a}] \in L$ whenever $1 \leqslant k \leqslant r_2$. When $k > d_{a-1}$, the commutator $[g_{2k}, g_{1a}]$ lies in L by the definition of L so we may assume that $1 \leqslant k \leqslant d_{a-1}$. But in this case (5.7) implies that $g_{2k} = [g_{1f(k)}, g_{1i}]$ where $1 \leqslant f(k) < i \leqslant a - 1$, and so

$$[g_{2k}, g_{1a}] = [g_{1f(k)}, g_{1i}, g_{1a}]$$
$$= [g_{1i}, g_{1a}, g_{1f(k)}]^{-1}[g_{1a}, g_{1f(k)}, g_{1i}]^{-1} \bmod G_4.$$

Now, modulo G_3 the commutators $[g_{1i}, g_{1a}]$ and $[g_{1a}, g_{1f(k)}]$ are equal to a product of the elements g_{2j}. Since $f(k) < a$ and $i < a$ our inductive hypothesis implies that the expression above is an element of L modulo G_4 and hence, since L contains G_4, we find that $[g_{2k}, g_{1a}] \in L$, as required. Hence, by induction on a, our claim follows.

We have now chosen our generating set for G. We now exhibit a collection of relations that these elements satisfy.

For $1 \leqslant i \leqslant c$ and $1 \leqslant j \leqslant r_i$, let $a(i, j)$ be the smallest positive integer such that $g_{ij}^{p^{a(i,j)}} \in \langle G_{i+1}, g_{i1}, g_{i2}, \ldots, g_{i(j-1)} \rangle$. Note that every element $g \in G$ may be written uniquely in the form

$$g = g_{11}^{e(1,1)} g_{12}^{e(1,2)} \cdots g_{1r_1}^{e(1,r_1)} g_{21}^{e(2,1)} \cdots g_{cr_c}^{e(c,r_c)}$$

where $0 \leqslant e(i, j) < p^{a(i,j)}$. In the following, we will abbreviate this product to

$$\prod g_{ij}^{e(i,j)}, \quad 1 \leqslant i \leqslant c, \ 1 \leqslant j \leqslant r_i.$$

Note that $g \in G_i$ if and only if $e(u, v) = 0$ whenever $u < i$. This implies that

$$|G_i| = p^{\sum_{u=i}^{c} \sum_{v=1}^{r_u} a(u,v)}.$$

In particular, since $|G| = p^m$,

$$\sum_{u=1}^{c} \sum_{v=1}^{r_u} a(u, v) = m. \tag{5.8}$$

Note that (5.8) and the fact that the integers $a(u, v)$ are positive implies that for any $i \in \{1, 2, \ldots, c\}$,

$$\sum_{u=i+1}^{c} \sum_{v=1}^{r_u} a(u, v) \leqslant m - r_1 - r_2 - \cdots - r_i. \tag{5.9}$$

For all integers i and j such that $1 \leqslant i \leqslant c$ and $1 \leqslant j \leqslant r_i$, define integers $b(i, j, u, v)$ by

$$g_{ij}^{p^{a(i,j)}} = \prod g_{uv}^{b(i,j,u,v)}, \quad 1 \leqslant u \leqslant c, \ 1 \leqslant v \leqslant r_u. \tag{5.10}$$

Similarly, for all i, j and k such that $1 \leqslant i \leqslant c$, $1 \leqslant j \leqslant r_i$ and $1 \leqslant k \leqslant r_1$, define the integers $c(i, j, k, u, v)$ by

$$[g_{ij}, g_{1k}] = \prod g_{uv}^{c(i,j,k,u,v)}, 1 \leqslant u \leqslant c, \ 1 \leqslant v \leqslant r_u. \tag{5.11}$$

Note that $b(i, j, u, v) = 0$ whenever $u < i$ and whenever $u = i$ and $v \geqslant j$. Moreover, $c(i, j, k, u, v) = 0$ whenever $u \leqslant i$.

We have defined a generating set for G and a collection of relations that these generators satisfy. We now show that a subset of these relations (together with the integers p, c, s, d_i and r_i) suffice to define the isomorphism class of G uniquely.

Theorem 5.6 *The isomorphism class of the group G is determined by the integers p, c, s, $r_1, r_2, \ldots, r_c$, $d_0, d_1, \ldots, d_s$ together with the integers*

$$
\begin{array}{ll}
a(i,j) & 1 \leqslant i \leqslant c,\ 1 \leqslant j \leqslant r_i; \\
b(i,j,u,v) & 1 \leqslant i \leqslant c,\ 1 \leqslant j \leqslant r_i,\ i \leqslant u \leqslant c,\ 1 \leqslant v \leqslant r_u; \\
c(1,j,k,u,v) & 1 \leqslant j < k \leqslant r_1,\ 2 \leqslant u \leqslant c,\ 1 \leqslant v \leqslant r_u; \\
c(2,j,k,u,v) & 1 \leqslant k \leqslant s,\ d_{k-1} < j \leqslant r_2,\ 3 \leqslant u \leqslant c,\ 1 \leqslant v \leqslant r_u; \\
c(i,j,k,u,v) & 3 \leqslant i \leqslant c,\ 1 \leqslant j \leqslant r_i,\ 1 \leqslant k \leqslant s,\ i < u \leqslant c,\ 1 \leqslant v \leqslant r_u.
\end{array}
$$

Proof: We will show that any group G of nilpotency class at most c generated by the set of elements $\{g_{ij} \mid 1 \leqslant i \leqslant c, 1 \leqslant j \leqslant r_i\}$ and satisfying the relations (5.10) when $1 \leqslant i \leqslant c$, $1 \leqslant j \leqslant r_i$ and the relations (5.11) when $i = 1$ and $1 \leqslant j < k \leqslant r_1$; when $i = 2$, $1 \leqslant k \leqslant s$ and $d_{k-1} < j \leqslant r_2$; and when $3 \leqslant i \leqslant c$, $1 \leqslant j \leqslant r_i$ and $1 \leqslant k \leqslant s$ has order at most p^m. This is sufficient to prove the theorem, for suppose L and M are groups that give rise to the same collection of integers. Since L and M each possess a generating set that satisfies the above relations, both L and M are isomorphic to quotients of G. Moreover, L and M have order p^m, by (5.8). If G has order at most p^m, we have that L and M must in fact be isomorphic to G, and so L and M are isomorphic.

To show that G has order at most p^m, let $H \leqslant G$ be the subgroup generated by $\{g_{11}, g_{12}, \ldots, g_{1s}\}$. Define subgroups $H_2, H_3, \ldots, H_c$ by

$$
H_i = \langle \{g_{uv} \mid i \leqslant u \leqslant c,\ 1 \leqslant v \leqslant r_u\} \rangle
$$

and define $H_{c+1} = \{1\}$. For $u \geqslant 2$, the relations we are given include those defining each g_{uv} as a commutator and so we may write each element g_{uv} as a commutator of length u in $g_{11}, g_{12}, \ldots, g_{1s}$. Hence

$$
G \geqslant H \geqslant H_2 \geqslant H_3 \geqslant \cdots \geqslant H_c \geqslant H_{c+1} = \{1\}.
$$

Moreover, since G has nilpotency class at most c and the generators of H_c may be represented as commutators of length c, we have that H_c is central in G and so in particular H_c is normal in H.

As the notation suggests, the subgroups H_i are the terms in the lower central series for H. However, we do not prove this. We do, however, wish to show that H_i/H_{i+1} is central in H/H_{i+1} for all i such that $i \geqslant 2$. Suppose firstly that $i \geqslant 3$ and we have shown that H_{i+1} is normal in H. (This is certainly true when $i = c - 1$, since H_c is normal even in G.) The relations (5.11) that we are given include those that imply that H_i/H_{i+1} is centralised by $g_{11}H_{i+1}, g_{12}H_{i+1}, \ldots, g_{1s}H_{i+1}$, and so H_i/H_{i+1} is central in H/H_{i+1}; in particular H_i is normal in H. Induction on $c - i$ now shows that H_i/H_{i+1} is central in H/H_{i+1} for $i \geqslant 3$. It remains to show that H_2/H_3 is central in H/H_3.

Assume, as an inductive hypothesis, that

$$H_2/H_3 \text{ is centralised by } g_{11}H_3, g_{12}H_3, \ldots, g_{1(a-1)}H_3.$$

This hypothesis is true when $2 \leqslant a \leqslant 3$, for since $d_0 = d_1 = 0$ it is implied by the relations (5.11). For our inductive step, it is sufficient to show that $[g_{2\ell}, g_{1a}] \in H_3$ for $1 \leqslant \ell \leqslant r_2$. This is implied by our commutator relations when $\ell > d_{a-1}$, and so we may assume that $1 \leqslant \ell \leqslant d_{a-1}$. One of the given relations (5.11) is the equality

$$g_{2\ell} = [g_{1x}, g_{1y}]$$

where $1 \leqslant x < y < a$, by our choice (5.7) of the generators g_{2j}. We claim that for any $i \in \{1, 2, \ldots, r_2\}$, $[g_{2i}, [g_{1y}, g_{1a}]] \in H_3$. When $i > d_{a-1}$, this follows from our given commutator relations. But when $i \leqslant d_{a-1}$, the commutator relation (5.7) shows that $g_{2i} = [g_{1u}, g_{1v}]$ where $1 \leqslant u < v \leqslant a - 1$. By our inductive hypothesis, $g_{1u}H_3$ and $g_{1v}H_3$ centralise H_2/H_3, and so $[g_{1u}, g_{1v}]H_3$ centralises H_2/H_3. Since one of the given commutator relations expresses $[g_{1y}, g_{1a}]$ as a product of elements in H_2, this implies that $[g_{2i}, [g_{1y}, g_{1a}]] \in H_3$ for all values of i and so our claim follows. Now

$$g_{1a}^{-1} g_{2\ell} g_{1a} = g_{1a}^{-1}[g_{1x}, g_{1y}]g_{1a}$$

$$= [g_{1a}^{-1} g_{1x} g_{1a}, g_{1a}^{-1} g_{1y} g_{1a}]$$

$$= [g_{1x}[g_{1x}, g_{1a}], g_{1y}[g_{1y}, g_{1a}]].$$

Since $[g_{1x}, g_{1a}] \in H_2$, this element commutes with g_{1x} and g_{1y} modulo H_3 by our inductive hypothesis and commutes with $[g_{1y}, g_{1a}]$ modulo H_3 by the claim we proved above. Hence,

$$g_{1a}^{-1} g_{2\ell} g_{1a} = [g_{1x}, g_{1y}[g_{1y}, g_{1a}]] \bmod H_3.$$

But $[g_{1y}, g_{1a}] \in H_2$, and so commutes modulo H_3 with g_{1x} and g_{1y} by our inductive hypothesis. Therefore

$$g_{1a}^{-1} g_{2\ell} g_{1a} = [g_{1x}, g_{1y}] = g_{2\ell} \bmod H_3,$$

so g_{1a} centralises H_2 modulo H_3, as required. Thus, by induction on a, H_2/H_3 is central in H/H_3.

The given commutator relations directly imply that H/H_2 is abelian, that H is normal in G and that G/H is abelian, hence the quotients of consecutive terms in the sequence

$$G \geqslant H \geqslant H_2 \geqslant H_3 \geqslant \cdots \geqslant H_c \geqslant 1$$

are abelian. But now the relations (5.10) imply that

$$|H_i/H_{i+1}| \leqslant p^{\sum_{j=1}^{r_i} a(i,j)},$$

$$|H/H_2| \leqslant p^{\sum_{j=1}^{s} a(1,j)} \text{ and}$$

$$|G/H| \leqslant p^{\sum_{j=s+1}^{r_1} a(1,j)}.$$

Hence $|G| \leqslant p^{\sum_{i=1}^{c}\sum_{j=1}^{r_i} a(i,j)} = p^m$, and so the theorem follows.

5.5 Proof of the upper bound

This section completes the enumeration of p-groups by providing an upper bound on the number of isomorphism classes of groups of order p^m.

Theorem 5.7 *Let p be a prime number. Then*

$$f(p^m) \leqslant p^{\frac{2}{27}m^3 + O(m^{5/2})}.$$

Theorems 4.5 and 5.7 combine to produce the following theorem.

Theorem 5.8 *Let p be a prime number. Then*

$$f(p^m) = p^{\frac{2}{27}m^3 + O(m^{5/2})}.$$

Before embarking on proving Theorem 5.7, we give the following elementary lemma.

Lemma 5.9 *Let n be a positive integer. Then there are exactly 2^{n-1} ordered partitions of n.*

Proof: An ordered partition of n is produced by taking the expression

$$(1+1+1+\cdots+1) = n$$

and grouping the terms of this sum by replacing some of the '+' signs by ')+('. There are $n-1$ plus signs in this expression, and so there are 2^{n-1} ways of adding brackets. So the number of ordered partitions of n is 2^{n-1}.

Corollary 5.10 *Let n be a positive integer. The number $p(n)$ of (unordered) partitions of n is at most 2^{n-1}.*

Proof of Theorem 5.7: Let a prime p and a positive integer m be fixed. The previous section showed that the isomorphism class of a group of order p^m is determined by a certain collection of integers, namely the integers $c, s, r_i, d_i, a(i, j), b(i, j, u, v)$ and $c(i, j, k, u, v)$. We will provide an upper bound on the number of choices for these integers. We start by showing that there are relatively few choices for the integers $c, s, r_i, d_i, a(i, j)$ and $b(i, j, u, v)$, and so we need to concentrate on finding a good upper bound for the number of choices for the integers $c(i, j, k, u, v)$ to provide a good upper bound on the number of p-groups of order p^m.

The sum $\sigma = r_1 + r_2 + \cdots + r_c$ is such that $1 \leqslant \sigma \leqslant m$. The integers $r_1, r_2, \ldots, r_c$ form an ordered partition of the integer σ with c parts, and so once σ is fixed the number of choices for c and $r_1, r_2, \ldots, r_c$ is at most equal to the number of ordered partitions of σ. Lemma 5.9 shows that the number of ordered partitions of σ is $2^{\sigma-1}$. Since $1 \leqslant \sigma \leqslant m$, there are at most $\sum_{\sigma=1}^{m} 2^{\sigma-1}$ choices for c and $r_1, r_2, \ldots, r_c$, which is less than 2^m. The integers $d_2 - d_1, d_3 - d_2, \ldots, d_s - d_{s-1}$ form an ordered partition of r_2, since the integers d_i are strictly increasing. Moreover, since $d_0 = d_1 = 0$, the integers d_i and s are determined by this partition. So, again using Lemma 5.9, the number of choices for s and $d_0, d_1, \ldots, d_s$ is at most $2^{r_2-1} \leqslant 2^{m-1}$. The integers $a(i, j)$ are positive and sum to m, and so the number of choices for $a(i, j)$ is at most the number of ordered partitions of m. Hence there are at most 2^{m-1} choices for the integers $a(i, j)$. Finally, to count the number of choices for the integers $b(i, j, u, v)$, note that $0 \leqslant b(i, j, u, v) < p^{a(u,v)}$ and so there are at most $p^{a(u,v)}$ choices for each integer $b(i, j, u, v)$. When i and j are fixed, the number of choices for the integers $b(i, j, u, v)$ is at most

$$\prod_{u=i}^{c} \prod_{v=1}^{r_u} p^{a(u,v)} = p^{\sum_{u=i}^{c} \sum_{v=1}^{r_u} a(u,v)} \leqslant p^{\sum_{u=1}^{c} \sum_{v=1}^{r_u} a(u,v)} = p^m.$$

Thus there are at most p^m choices for the integers $b(i, j, u, v)$ when i and j are fixed. But there are at most m choices for the integers i and j, for once i has been chosen there are r_i choices for j, and $\sum_{i=1}^{c} r_i \leqslant m$. So there are at most p^{m^2} choices for the integers $b(i, j, u, v)$.

In summary, the number of choices for the integers $c, s, r_i, d_i, a(i, j)$ and $b(i, j, u, v)$ is bounded above by

$$2^m 2^{m-1} 2^{m-1} p^{m^2}$$

and this is much less than our error term of $p^{O(m^{5/2})}$. Hence to prove the upper bound we require, it suffices to provide a good upper bound on the number of choices for the integers $c(i, j, k, u, v)$ once the integers $s, c, r_i, d_i, a(i, j)$ and $b(i, j, u, v)$ have been chosen.

We need to choose the integers $c(i, j, k, u, v)$ when $i = 1$ and $1 \leqslant j <$ $k \leqslant r_1$; when $i = 2$, $1 \leqslant k \leqslant s$ and $d_{k-1} < j \leqslant r_2$; when $3 \leqslant i \leqslant c$, $1 \leqslant j \leqslant$ r_i and $1 \leqslant k \leqslant s$. In all these cases, $i < u \leqslant c$ and $1 \leqslant v \leqslant r_u$. For each choice of i, j, k, u, v, we know that $0 \leqslant c(i, j, k, u, v) \leqslant p^{a(u,v)} - 1$. Now, the integers $c(1, j, k, 2, v) \bmod p$ determine the map $[,] : V \times V \to W$ defined in Section 5.4, and so we may choose the integers $c(i, j, k, 2, v)$ subject to the additional condition that the minimum dimension of a subspace U of V such that $[U, U] = W$ is s.

In the notation of Proposition 5.5, there are $p^{N_p(r_1, r_2, s)}$ choices for a map $[,] : V \times V \to W$ such that the minimum dimension of a subspace U of V satisfying $[U, U] = V$ is s. Once this map has been chosen, the values $c(1, j, k, 2, v) \bmod p$ are determined, and so the number of choices for the integers $c(1, j, k, 2, v)$ is at most

$$p^{N_p(r_1, r_2, s)} \prod_{j=1}^{r_1} \prod_{k=j+1}^{r_1} \prod_{v=1}^{r_2} p^{a(2,v)-1} = p^{N_p(r_1, r_2, s) + \binom{r_1}{2}\left(\left(\sum_{v=1}^{r_2} a(2,v)\right) - r_2\right)}, \qquad (5.12)$$

since there are $\binom{r_1}{2}$ choices for j and k. Similarly, the number of choices for the integers $c(1, j, k, u, v)$ where $u \geqslant 3$ is at most

$$\prod_{j=1}^{r_1} \prod_{k=j+1}^{r_1} \prod_{u=3}^{c} \prod_{v=1}^{r_u} p^{a(u,v)} = p^{\binom{r_1}{2} \sum_{u=3}^{c} \sum_{v=1}^{r_u} a(u,v)}. \qquad (5.13)$$

So, from (5.12) and (5.13), the number of choices for the integers $c(1, j, k, u, v)$ is at most

$$p^{N_p(r_1, r_2, s) + \binom{r_1}{2}\left[\left(\sum_{u=2}^{c} \sum_{v=1}^{r_u} a(u,v)\right) - r_2\right]} \leqslant p^{N_p(r_1, r_2, s) + \binom{r_1}{2}(m - r_1 - r_2)},$$

the final inequality following from (5.9) with $i = 1$.

We now give an upper bound on the number of choices for the integers $c(i, j, k, u, v)$ when $i > 1$. When i, j and k are fixed, the number of choices for the integers $c(i, j, k, u, v)$ is at most

$$\prod_{u=i+1}^{c} \prod_{v=1}^{r_u} p^{a(u,v)} \leqslant p^{m - r_1 - r_2 - \cdots - r_i},$$

by (5.9). When $i \geqslant 3$, there are sr_i choices for j and k and so there are at most

$$p^{sr_i(m - r_1 - r_2 - \cdots - r_i)}$$

choices for the integers $c(i, j, k, u, v)$. When $i = 2$, there are $r_2 - d_{k-1}$ choices for j once k is fixed. Since the sequence $d_1, d_2, \ldots, d_s$ is strictly increasing and non-negative, we have that $d_{k-1} \geqslant k-2$ whenever $k \geqslant 2$. Hence, since $d_0 = 0$, we find that the number of choices for j and k is

$$sr_2 - \sum_{k=1}^{s} d_{k-1} = sr_2 - \sum_{k=2}^{s} d_{k-1}$$

$$\leqslant sr_2 - \sum_{k=2}^{s}(k-2)$$

$$= sr_2 - \binom{s-1}{2},$$

and so there are at most

$$p^{(sr_2 - \binom{s-1}{2})(m-r_1-r_2)}$$

choices for the integers $c(2, j, k, u, v)$.

Thus the number of choices for the integers $c(i, j, k, u, v)$ is at most p to the power M, where

$$M = N_p(r_1, r_2, s) + \binom{r_1}{2}(m - r_1 - r_2) + \left(sr_2 - \binom{s-1}{2}\right)(m - r_1 - r_2)$$

$$+ \sum_{i=3}^{c} sr_i(m - r_1 - r_2 - \cdots - r_i).$$

Now $r_i(m - r_1 - \cdots - r_i) \leqslant \sum_{j=1}^{r_i}(m - r_1 - \cdots r_{i-1} - j)$ and so

$$\sum_{i=3}^{c} sr_i(m - r_1 - r_2 - \cdots - r_i) \leqslant s\sum_{i=3}^{c}\sum_{j=1}^{r_i}(m - r_1 - \cdots - r_{i-1} - j)$$

$$= s\sum_{k=1}^{r_3+r_4+\cdots+r_c}(m - r_1 - r_2 - k)$$

$$\leqslant s\sum_{k=1}^{m-r_1-r_2}(m - r_1 - r_2 - k)$$

$$= s\binom{m - r_1 - r_2}{2}.$$

So we find that

$$M \leqslant N_p(r_1, r_2, s) + \frac{1}{2}r_1^2(m - r_1 - r_2)$$

$$+ \left((s-1)r_2 - \frac{1}{2}(s-1)^2\right)(m - r_1 - r_2) \tag{5.14}$$

$$+ \frac{1}{2}(s-1)(m - r_1 - r_2)^2 + O(m^2).$$

We need to show that this upper bound for M is at most $\frac{2}{27}m^3 + O(m^{5/2})$. We consider two cases, depending on the size of r_1. Suppose that $r_1 \geqslant \frac{6}{10}m$. By the upper bound (5.6),

$$M \leqslant \frac{1}{2}r_1^2(m - r_1 - (s-1)) + \frac{1}{2}r_1 r_2(s-1)$$
$$+ \left((s-1)r_2 - \frac{1}{2}(s-1)^2 \right)(m - r_1 - r_2)$$
$$+ \frac{1}{2}(s-1)(m - r_1 - r_2)^2 + O(m^{5/2}).$$

Setting $x = r_1/m$, $y = r_2/m$, $z = (m - r_1 - r_2)/m$ and $u = (s-1)/m$, we need to show that the function $A(x, y, z, u)$ defined by

$$\frac{1}{2}x^2(z + y - u) + \frac{1}{2}xyu + \left(uy - \frac{1}{2}u^2 \right)z + \frac{1}{2}uz^2$$

is at most $\frac{2}{27}$ when x, y, z, u are non-negative and satisfy $x + y + z = 1$, $u \leqslant y$ and $x \geqslant \frac{6}{10}$. This may be shown using standard techniques—see Lemma A.1 in Appendix A for details.

Now suppose that $r_1 \leqslant \frac{6}{10}m$. The bounds (5.5) and (5.14) combine to show that

$$M \leqslant \frac{1}{2}r_1^2(m - r_1 - (s-1)) + \left((s-1)r_2 - \frac{1}{2}(s-1)^2 \right)(m - r_1 - r_2)$$
$$+ \frac{1}{2}(s-1)(m - r_1 - r_2)^2 + O(m^{8/3}).$$

As before, define $x = r_1/m$, $y = r_2/m$, $z = (m - r_1 - r_2)/m$ and $u = (s-1)/m$ and let $B(x, y, z, u)$ be defined by

$$B(x, y, z, u) = \frac{1}{2}x^2(z + y - u) + \left(uy - \frac{1}{2}u^2 \right)z + \frac{1}{2}uz^2.$$

It is not difficult to show that the maximum value μ of $B(x, y, z, u)$ on the region defined by $x \geqslant 0$, $y \geqslant 0$, $z \geqslant 0$, $u \geqslant 0$, $x + y + z = 1$, $u \leqslant \min\{x, y\}$ and $x \leqslant \frac{6}{10}$ is strictly less than $\frac{2}{27}$ (see Lemma A.2 in Appendix A). But this implies that $M \leqslant \mu m^3 + O(m^{8/3}) \leqslant \frac{2}{27}m^3 + O(m^{5/2})$ when $r_1 \leqslant \frac{6}{10}m$. We have shown that $M \leqslant \frac{2}{27}m^3 + O(m^{5/2})$ whatever the value of r_1, and so the theorem follows.

III

Pyber's theorem

6

Some more preliminaries

This chapter contains material that we will use in the proof of Pyber's upper bound on the number of soluble groups of order n. Section 6.1 contains results on Hall subgroups and Sylow systems of soluble groups and Section 6.2 deals with the Fitting subgroup of a soluble group. Section 6.3 concerns itself with primitivity in permutation and linear groups, and also contains a result bounding the number of generators of a permutation group.

6.1 Hall subgroups and Sylow systems

Let G be a finite group. A subgroup H of G is said to be a *Hall subgroup* if $|H|$ and $|G : H|$ are coprime. Let π be a set of prime numbers. A subgroup H of G is a *Hall π-subgroup* if $|H|$ is a π number (a product of primes in π) and $|G : H|$ is a π' number (a product of primes not in π).

Theorem 6.1 *Let G be a soluble group of order $p_1^{\alpha_1} \cdots p_k^{\alpha_k}$. Let r be an integer such that $1 \leqslant r \leqslant k$ and let $\pi = \{p_1, p_2, \ldots, p_r\}$. Then:*

  (i) *The group G has a Hall π-subgroup (of order $p_1^{\alpha_1} \cdots p_r^{\alpha_r}$).*
 (ii) *Any two Hall π-subgroups are conjugate in G.*
(iii) *Any π-subgroup of G is contained in a Hall π-subgroup.*

Proof: The proof is by induction on $|G|$. The theorem is trivial when $|G| = 1$, so we may assume that $|G| > 1$ and that the theorem is true for all groups of order less than $|G|$. We must show that the group G satisfies properties (i), (ii) and (iii) of the theorem.

Let M be a minimal normal subgroup of G. Since G is soluble, $|M| = p^\alpha$ for some prime p dividing n. Applying the inductive hypothesis to G/M we find that G/M has a Hall π-subgroup K/M. Further, all such subgroups are conjugate in G/M and any π-subgroup of G/M is contained in a Hall π-subgroup of G/M. The proof is now divided into two cases.

The first case is when p belongs to π. In this case K is a Hall π-subgroup of G, and so property (i) of the theorem follows. If H is any Hall π-subgroup of G then H contains M (since HM is a π-group). Moreover, H/M is a Hall π-subgroup of G/M and so our inductive hypothesis implies that $(gM)^{-1}(H/M)(gM) = K/M$ for some $gM \in G/M$. But this means that $K = g^{-1}Hg$. Thus we have shown that property (ii) follows. Finally, suppose that L is a π-subgroup of G. Then LM/M is a π-subgroup of G/M and so is contained in a Hall π-subgroup H/M of G/M. But then H is a Hall π-subgroup of G, and $L \leqslant H$. Thus (iii) follows, and we have established the inductive step in this case.

The second case is when p does not belong to π. Then $|K/M|$ and $|M|$ are coprime. The Schur–Zassenhaus theorem implies that there is a complement H for M in K and any two complements are conjugate in K. (We will prove this in Chapter 7, Corollary 7.17.) But then $|H| = |K/M|$ and so H is a Hall π-subgroup of G, and so property (i) follows. Let H_1 be any Hall π-subgroup of G. To establish property (ii), it is enough to show that there exists a conjugate H_2 of H_1 that lies in K. (For then H_2 is a complement of M in K, and so the Schur–Zassenhaus theorem implies that H_2 and H are conjugate.) Now, $H_1 M/M$ is a Hall π-subgroup of G/M. By the inductive hypothesis there exists g_1 in G such that $(g_1 M)^{-1}(H_1 M/M)(g_1 M) = K/M$. Thus $g_1^{-1} H_1 g_1 \leqslant K$ and so property (ii) follows. It remains to prove property (iii). Let L be a π-subgroup of G. Now LM/M is contained in some Hall π-subgroup K_1/M of G/M (so $L \leqslant K_1$). Since all Hall π-subgroups of G/M are conjugate in G/M, K_1 and K are conjugate in G. So there exists a conjugate L_1 of L such that $L_1 \leqslant K$. We have that $K = HM$ and so

$$L_1 M = L_1 M \cap HM = (L_1 M \cap H)M.$$

Now L_1 and $L_1 M \cap H$ are complements for M in $L_1 M$ and therefore are conjugate by the Schur–Zassenhaus theorem. Thus L_1 is conjugate to a subgroup of H. Thus L is contained in a Hall π-subgroup of G (namely an appropriate conjugate of H). So we have established property (iii) in this case, and so the inductive step follows here also.

The theorem now follows by induction on $|G|$.

Let G be a group of order $p_1^{\alpha_1} p_2^{\alpha_2} \cdots p_k^{\alpha_k}$. A *Sylow system* in G is a family $P_1, P_2, \ldots, P_k$ where P_i is a Sylow p_i-subgroup and $P_i P_j = P_j P_i$ for all $i, j \in \{1, 2, \ldots, k\}$. Whenever $i \neq j$, we find that $P_i P_j$ is a subgroup and its order is $p_i^{\alpha_i} p_j^{\alpha_j}$. Further, if $\pi \subseteq \{p_1, p_2, \ldots, p_k\}$ then $\prod_{p_i \in \pi} P_i$ is a Hall π-subgroup of G.

Theorem 6.2 *Let G be a soluble group of order $p_1^{\alpha_1} p_2^{\alpha_2} \cdots p_k^{\alpha_k}$. Then*

(i) *G has a Sylow system;*
(ii) *any two Sylow systems are conjugate in G;*
(iii) *if $H \leqslant G$ and $Q_1, \ldots, Q_k$ is a Sylow system for H then there is a Sylow system $P_1, \ldots, P_k$ for G such that $Q_i = H \cap P_i$ for all $i \in \{1, 2, \ldots, k\}$.*

Proof: For each $i \in \{1, 2, \ldots, k\}$, define $\pi_i = \{p_1, p_2, \ldots, p_k\} \setminus \{p_i\}$. We say that a subgroup H is a *Hall p_i-complement* if H is a Hall π_i-subgroup. Let Σ be the set of all Sylow systems in G and let Σ_i be the set of all Hall p_i-complements in G for $1 \leqslant i \leqslant k$. We claim that there is a bijection between the set Σ of Sylow systems $P_1, P_2, \ldots, P_k$ of G and the set $\Sigma_1 \times \Sigma_2 \times \cdots \times \Sigma_k$ of sequences $G_1, G_2, \ldots, G_k$ where each G_i is a Hall p_i-complement. For, given a Sylow system $P_1, P_2, \ldots, P_k$, we may define $G_i = \prod_{j \neq i} P_j$ for all $i \in \{1, 2, \ldots, k\}$. The remark after the definition of a Sylow system shows that each G_i is a Hall p_i-complement. In the reverse direction, suppose that $G_1, G_2, \ldots, G_k$ are such that each G_j is a Hall p_i-complement. The index of G_j is $p_j^{\alpha_j}$ and hence (since these indices are coprime) we find that for any subset $S \subseteq \{1, 2, \ldots, k\}$ the index of $\bigcap_{j \in S} G_j$ is $\prod_{j \in S} p_j^{\alpha_j}$ and so $|\bigcap_{j \in S} G_j| = \prod_{j \notin S} p_j^{\alpha_j}$. Define $P_i = \bigcap_{j \neq i} G_j$. Then $|P_i| = p_i^{\alpha_i}$ and so P_i is a Sylow p_i-subgroup of G. Moreover, for $i \neq j$ we find that $\langle P_i, P_j \rangle = \bigcap_{\ell \notin \{i, j\}} G_\ell$, a group of order $p_i^{\alpha_i} p_j^{\alpha_j}$. Hence $P_i P_j = P_j P_i = \langle P_i, P_j \rangle$, and so the P_i form a Sylow system for G. We have therefore constructed a bijection from Σ to $\Sigma_1 \times \Sigma_2 \times \cdots \times \Sigma_k$ as required.

Theorem 6.1 shows that a sequence $G_1, G_2, \ldots, G_k$ of Hall p_i-complements exists, and so the above correspondence shows that G has a Sylow system and part (i) of the theorem follows.

The group G acts naturally on Σ and the sets Σ_i by conjugation. The bijection we constructed above respects conjugation, and so the sets Σ and $\Sigma_1 \times \Sigma_2 \times \cdots \times \Sigma_k$ are isomorphic as G-sets. To prove part (ii) of the theorem, we must show that G acts transitively on Σ, which is the same as showing that G acts transitively on $\Sigma_1 \times \Sigma_2 \times \cdots \times \Sigma_k$. By Theorem 6.1 part (ii), we see that the action of G on Σ_i is transitive. Further, for $i = 1, \ldots, k$

$$|\Sigma_i| = |G : N_G(G_i)| = p_i^{\beta_i}$$

for some β_i such that $\beta_i \leqslant \alpha_i$. Therefore the action of G on $\Sigma_1 \times \cdots \times \Sigma_k$ is transitive as the Σ_i are transitive G-spaces with orders that are pairwise coprime. This establishes part (ii) of the theorem.

Let H be a subgroup of G and $Q_1, \ldots, Q_k$ a Sylow system for H. Let $H_i = \prod_{j \neq i} Q_j$. So H_i is a Hall p_i-complement in H for $i = 1, \ldots, k$. For each i there exists a Hall p_i-complement G_i in G such that $H_i \leqslant G_i$ (by part (iii) of

Theorem 6.1). As seen previously, these p_i-complements determine a Sylow system $P_1, \ldots, P_k$ in G and it is not difficult to see that $H \cap P_i = Q_i$ for $1 \leqslant i \leqslant k$. Thus part (iii) of the theorem follows.

6.2 The Fitting subgroup

This section contains some basic results concerning the Fitting subgroup $F(G)$ of a group G. We stress that these results depend on G being finite. We prove results about $F(G)$ itself, and about the relationship between $F(G)$ and the Frattini subgroup $\Phi(G)$.

If P and Q are normal p-subgroups of a group G, then PQ is a normal p-subgroup containing both P and Q. If P and Q are maximal with respect to the property of being a normal p-subgroup of G, then $P = PQ = Q$ (the first equality by the maximality of P, the second by the maximality of Q). So there is a unique maximal normal p-subgroup of a finite group G, and we denote this subgroup by $O_p(G)$. (In particular, $O_p(G) = G$ in the case when G is a p-group.) We define the *Fitting subgroup* by

$$F(G) = \langle O_p(G) \mid p \text{ is a prime} \rangle.$$

If p and q are distinct primes then

$$[O_p(G), O_q(G)] \leqslant O_p(G) \cap O_q(G) = \{1\}.$$

Thus

$$F(G) = O_{p_1}(G) \times O_{p_2}(G) \times \cdots \times O_{p_k}(G)$$

where $p_1, \ldots, p_k$ are the distinct primes dividing $|G|$. In particular, since $F(G)$ is the direct product of its Sylow subgroups, $F(G)$ is nilpotent.

Proposition 6.3 *Let G be a finite group. Then $F(G)$ is the unique maximal nilpotent normal subgroup of G.*

Proof: If N is a nilpotent normal subgroup of G, then

$$N = Q_1 \times Q_2 \times \cdots \times Q_k$$

where Q_i is the Sylow p_i-subgroup of N. Since the Q_i are characteristic subgroups of N, they are normal subgroups of G. Therefore $Q_i \leqslant O_{p_i}(G)$. Thus $N \leqslant F(G)$, as required.

Theorem 6.4 *Let G be a finite group, and let S be the (unique) maximal soluble normal subgroup of $C_G(F(G))$. Then $S = Z(F(G))$.*

Proof: Define $C = C_G(F(G))$ and $Z = Z(F(G))$. Clearly $Z \leqslant S$.

Suppose, for a contradiction, that Z is strictly contained in S. Since S is normal in G, this implies that S/Z contains a minimal normal subgroup M/Z of G/Z. Since S/Z is soluble, M/Z is abelian and so $M' \leqslant Z$. Since $M \leqslant C$, we find that $[M', M] \leqslant [F(G), M] = 1$, and so M is nilpotent (of class at most 2). Thus $M \leqslant F(G)$ by Proposition 6.3. The fact that $M \leqslant C$ now implies that $M \leqslant Z$. This contradicts the fact that M/Z is a minimal normal subgroup of G/Z, and so the theorem follows.

Corollary 6.5 *If G is a finite soluble group, then $C_G(F(G)) = Z(F(G))$.*

The remainder of this section is concerned with proving the following theorem due to W. Gaschütz:

Theorem 6.6 *If G is finite, then $F(G/\Phi(G)) = F(G)/\Phi(G)$.*

In preparation for the proof of this theorem, we prove two lemmas. The second of these two lemmas, Lemma 6.8, is a generalisation of the Frattini argument (which is the special case when $N = K$).

Lemma 6.7 *Let N be a finite group with a normal subgroup K, and suppose that N/K is nilpotent. Then any two Sylow p-subgroups of N are conjugate by an element of K.*

Proof: Suppose that P_1 and P_2 are Sylow p-subgroups of N. Then P_1K/K and P_2K/K are Sylow p-subgroups of N/K. Since N/K is nilpotent it has a unique Sylow p-subgroup. So we have $P_1K/K = P_2K/K$. Thus $P_1K = P_2K$ and P_2 is a Sylow p-subgroup of P_1K. Hence $P_2 = y^{-1}P_1y$ for some y in P_1K. Now $y = xk$ for some x in P_1 and k in K. Therefore

$$P_2 = y^{-1}P_1y = k^{-1}P_1k.$$

Lemma 6.8 *Let G be a finite group with a normal subgroup N. Suppose K is a normal subgroup of N such that N/K is nilpotent. Then for any Sylow subgroup P of N*

$$G = N_G(P)K.$$

Proof: Let g be an element of G. Then $g^{-1}Pg$ is a Sylow subgroup of N. By Lemma 6.7 there exists $k \in K$ such that $g^{-1}Pg = k^{-1}Pk$. But then $gk^{-1} \in N_G(P)$ and so $g \in N_G(P)K$. Hence $G = N_G(P)K$, as required.

We now show that the statement of Theorem 6.6 makes sense:

Proposition 6.9 *Let G be a finite group, and let N be a normal subgroup of G. Suppose that N contains $\Phi(G)$ and that $N/\Phi(G)$ is nilpotent. Then N is nilpotent. In particular, $\Phi(G)$ is nilpotent and so $\Phi(G) \leqslant F(G)$.*

Proof: Let P be a Sylow p-subgroup of N. Lemma 6.8 (applied in the case when $K = \Phi(G)$) implies that $N_G(P)\Phi(G) = G$. But $\Phi(G)$ is a set of 'non-generators' for G by Lemma 3.9, and so $N_G(P) = G$. In particular, every Sylow subgroup of N is normal and so N is nilpotent.

Proof of Theorem 6.6: Let N be the subgroup of G containing $\Phi(G)$ with the property that $N/\Phi(G) = F(G/\Phi(G))$. To prove the theorem, it suffices to show that $F(G) = N$.

Now, $F(G)$ contains $\Phi(G)$, by Proposition 6.9. Moreover, $F(G)/\Phi(G)$ is a nilpotent normal subgroup of $G/\Phi(G)$ and so is contained in $F(G/\Phi(G))$ by Proposition 6.3. Hence $F(G) \leqslant N$. Furthermore, since $F(G/\Phi(G))$ is nilpotent Proposition 6.9 implies that N is nilpotent. Hence $N \leqslant F(G)$. So $N = F(G)$ and the theorem is proved.

6.3 Permutations and primitivity

This section begins by proving an elementary theorem that bounds the number of generators of a permutation group. After this, the section discusses primitivity in permutation and linear groups.

Let G be a permutation group acting on a set Ω. We will write the action of G on the left, so for $g \in G$ and $\alpha \in \Omega$ we write $g\alpha$ for the image of α under g. We assume throughout this section that all sets on which groups act are finite.

Theorem 6.10 *Let Ω be a set of order n, and let G be a permutation group acting on Ω. If G has r orbits on Ω, then G can be generated by $n - r$ elements. In particular, any permutation group of degree n can be generated by $n - 1$ elements.*

Proof: Let Ω be fixed. We prove the theorem by induction (on $n - r$). When $r = n$, the group G is trivial and so the result holds. Assume, as an inductive hypothesis, that any permutation group on Ω with more than r orbits satisfies the theorem. Let G have r orbits on Ω, and let ω lie in a non-trivial orbit Δ of G. We may write $\Delta = \Delta_1 \cup \Delta_2 \cup \cdots \cup \Delta_k$, where the sets Δ_i are orbits of the stabiliser G_ω of ω in G. We may assume (without loss of generality) that $\Delta_k = \{\omega\}$. Since Δ is a non-trivial orbit, $k > 1$. For $i \in \{1, 2, \ldots, k-1\}$, let $g_i \in G$ be such that $g_i \omega \in \Delta_i$. Now, G is generated by G_ω together with the $k - 1$ elements $g_1, g_2, \ldots, g_{k-1}$. But G_ω has at least $k + (r - 1)$ orbits on Ω, and so our inductive hypothesis implies that G_ω can be generated by $n - (r - k - 1)$ elements. Hence G may be generated by $n - r$ elements, and so the theorem follows.

In fact, it is possible to strengthen Theorem 6.10 further. The following theorem is due to McIver and Neumann, and is stated in their paper [67]. See Cameron, Solomon and Turull [14] for a sketch proof of this theorem.

Theorem 6.11 *Let Ω be a set of order n, and let G be a permutation group acting on Ω. When $n \neq 3$ then G can be generated by $\lfloor n/2 \rfloor$ elements.*

We omit the proof of this theorem (and remark that the current proof of the theorem uses the classification of finite simple groups and therefore is not elementary—it would be very interesting to find such a proof, or an efficient algorithm to produce such a small generating set).

We now turn to the notion of a primitive permutation group. Let G be a permutation group acting on a set Ω. The first step in trying to understand G is often to write Ω as the disjoint union $\Omega = \Omega_1 \cup \Omega_2 \cup \cdots \cup \Omega_r$, where the subsets Ω_i are the orbits of G on Ω. Then $G \leqslant G^{\Omega_1} \times G^{\Omega_2} \times \cdots \times G^{\Omega_r}$, where G^{Ω_i} is the permutation group induced by G on Ω_i. For many problems, it is now sufficient to consider the transitive groups G^{Ω_i}. The next step is often to decompose a transitive permutation group further, by analysing equivalence relations that the group preserves. In analogy to the above reduction to a direct product of transitive groups, this process will often result in a reduction to a wreath product of primitive groups. We now define these notions, and discuss some of their basic properties.

A *congruence* is a G-invariant equivalence relation on Ω. So an equivalence relation ρ is a congruence if for all $g \in G$ and $\alpha, \beta \in \Omega$

$$\alpha \sim_\rho \beta \text{ if and only if } g\alpha \sim_\rho g\beta.$$

An equivalence class of a congruence is known as a *congruence class*. A subset $\Delta \subseteq \Omega$ is a *block* if and only if for all $g \in G$ either $g\Delta = \Delta$ or $g\Delta \cap \Delta = \emptyset$. It is not difficult to show that $\Delta \subseteq \Omega$ is a block if and only if it is a congruence class.

The trivial equivalence relation (where $\alpha \sim \beta$ if and only if $\alpha = \beta$) and the universal equivalence relation (where $\alpha \sim \beta$ for all $\alpha, \beta \in \Omega$) are congruences for any permutation group G acting on Ω. (In the trivial case the blocks are singletons, and in the universal case there is a single block Ω.) We say that a transitive permutation group G is *primitive* if there are no other congruences, otherwise we say that G is *imprimitive*. Note that a primitive group is transitive by definition. The following lemma shows that more is true.

Lemma 6.12 *Let G be a primitive permutation group, acting on the set Ω. Let N be a non-trivial normal subgroup of G. Then N is transitive.*

Proof: Define an equivalence relation ρ on Ω by $\alpha \sim_\rho \beta$ if and only if there exists $h \in N$ such that $h\alpha = \beta$. In other words, $\alpha \sim_\rho \beta$ if and only if α and β lie in the same N-orbit. Let $g \in G$ and let $\alpha, \beta \in \Omega$ be such that $\alpha \sim_\rho \beta$. Let $h \in N$ be such that $h\alpha = \beta$. Since N is normal, $gh = h'g$ for some $h' \in N$. But

$$h'g\alpha = gh\alpha = g\beta$$

and so $g\alpha \sim_\rho g\beta$. Similarly, $g\alpha \sim_\rho g\beta$ implies that $\alpha \sim_\rho \beta$ and so ρ is a congruence; the orbits of N are the blocks of ρ. Since N is non-trivial, ρ is not the trivial equivalence relation. Since G is primitive, this implies that ρ is the universal equivalence relation. But the definition of ρ now shows that N is transitive and so the lemma follows.

Recall that a permutation group G is *regular* if it is transitive and the stabiliser G_α of a point $\alpha \in \Omega$ is the trivial group. It is easy to show (by observing that point stabilisers of a transitive group are conjugate) that any transitive abelian permutation group is regular.

Proposition 6.13 *Let G be a primitive soluble group acting faithfully on a set Ω. Let M be a minimal normal subgroup of G.*

(i) *The subgroup M acts regularly on Ω. In particular, we have that $|\Omega| = |M| = p^d$ for some prime number p and positive integer d.*

(ii) *Let $\alpha \in \Omega$, and let G_α be the stabiliser of α in G. Then G is the semidirect product of M by G_α. Moreover, the action of G_α on M by conjugation is faithful, and G_α is isomorphic to an irreducible soluble subgroup of $\mathrm{GL}(d, p)$.*

Proof: By Lemma 6.12, M acts transitively on Ω. Since G is soluble, M is an elementary abelian group of order p^d for some prime number p and some positive integer d. Since M is abelian and transitive, M is regular. This proves part (i).

Define $H = G_\alpha$. Since M is transitive, $G = MH$. Since M acts regularly, $M \cap H = \{1\}$. So G is the semidirect product of M by H. Suppose that

$h \in H$ centralises M. Then $hg\alpha = gh\alpha = g\alpha$ for all $g \in M$. But, since M is transitive, $g\alpha$ runs through all of Ω as g runs through M. Hence h is the identity permutation. Thus the action of H on M by conjugation is faithful. In particular, H may be identified with a subgroup of $\text{Aut}(M)$. Since M is elementary abelian of order p^d, we have that $\text{Aut}(M) \cong \text{GL}(d, p)$, and so H may be identified with a subgroup of $\text{GL}(d, p)$. Moreover, this subgroup is irreducible by our choice of M, since H-invariant 'subspaces' of M correspond to normal subgroups of G contained in M. Hence the proposition follows.

We round off this brief discussion of primitive soluble groups by making the following remark. Since the subgroup M acts regularly, we may identify M and Ω by identifying $g \in M$ with $g\alpha \in \Omega$. Now, for any $h \in H$ and $g \in M$, we have that

$$h(g\alpha) = hgh^{-1}(h\alpha) = hgh^{-1}\alpha,$$

and so the action of H on Ω is determined by the action of H on M by conjugation. This shows that the permutation group G is determined (up to isomorphism) by the action of H on M. (The conjugacy class of G in $\text{Sym}(\Omega)$ is determined by the conjugacy class of H in $\text{GL}(d, p)$.)

We now consider imprimitive groups. Let k be an integer such that $k \geqslant 2$, and let Δ be a set such that $|\Delta| \geqslant 2$. Let $\Omega = \Delta \times \{1, 2, \ldots, k\}$, so Ω is the union of k copies $\Delta_1, \Delta_2, \ldots, \Delta_k$ of the set Δ (where $\Delta_i = \Delta \times \{i\}$). Let ρ be the equivalence relation on Ω defined by $(\delta, i) \sim_\rho (\delta', i')$ if and only if $i = i'$. So the equivalence classes of ρ are the subsets Δ_i of Ω. We define an imprimitive permutation group on Ω that preserves ρ as follows. Let H be a permutation group on Δ, and let Q be a permutation group on $\{1, 2, \ldots, k\}$. Let B be the product of k copies of H. We may regard B as a permutation group on Ω by defining

$$(h_1, h_2, \ldots, h_k)(\delta, i) = (h_i\delta, i) \tag{6.1}$$

for all $(h_1, h_2, \ldots, h_k) \in B$ and all $(\delta, i) \in \Omega$. The group Q can naturally be regarded as a permutation group on Ω by defining

$$y(\delta, i) = (\delta, yi) \tag{6.2}$$

for all $(\delta, i) \in \Omega$ and all $y \in Q$. In $\text{Sym}(\Omega)$, it is easy to check that Q normalises B, and that $B \cap Q = \{1\}$, so BQ is a semidirect product of B by Q. Since both B and Q preserve the congruence ρ above, therefore so does the group BQ. This group is the *wreath product* $H \text{wr} Q$ of H by Q. Since ρ is a non-trivial congruence, $H \text{wr} Q$ is an imprimitive group.

In fact, we may just as easily define the wreath product as an abstract group, and then give an action on $\Delta \times \{1, 2, \ldots, k\}$ to turn it into a permutation group. In this approach, we note that any $y \in Q$ is associated with the automorphism $(h_1, h_2, \ldots, h_k) \mapsto (h_{y^{-1}1}, h_{y^{-1}2}, \ldots, h_{y^{-1}k})$ of B. So we may define $H \operatorname{wr} Q$ as the semidirect product of B by Q, arising from the resulting natural homomorphism from Q to $\operatorname{Aut} B$. More concretely, we may define $H \operatorname{wr} Q$ as the set of expressions of the form $(h_1, h_2, \ldots, h_k) y$, with $h_i \in H$ and $y \in Q$. The multiplication in $H \operatorname{wr} Q$ is then defined by

$$(h_1, h_2, \ldots, h_k) y (h'_1, h'_2, \ldots, h'_k) y' = (h_1 h_{y^{-1}1}, h_2 h_{y^{-1}2}, \ldots, h_k h_{y^{-1}k}) y y'.$$

Once the wreath product is defined as an abstract group, the imprimitive action of $H \operatorname{wr} Q$ on $\Delta \times \{1, 2, \ldots, k\}$ may be defined by (6.1) and (6.2).

If G_1, G_2, G_3 are permutation groups acting on sets $\Omega_1, \Omega_2, \Omega_3$ respectively, then we may form the wreath product $(G_1 \operatorname{wr} G_2) \operatorname{wr} G_3$ on $(\Omega_1 \times \Omega_2) \times \Omega_3$ and the wreath product $G_1 \operatorname{wr} (G_2 \operatorname{wr} G_3)$ on $\Omega_1 \times (\Omega_2 \times \Omega_3)$. The natural bijection between $(\Omega_1 \times \Omega_2) \times \Omega_3$ and $\Omega_1 \times (\Omega_2 \times \Omega_3)$ shows that $(G_1 \operatorname{wr} G_2) \operatorname{wr} G_3$ and $G_1 \operatorname{wr} (G_2 \operatorname{wr} G_3)$ are isomorphic as permutation groups, and so we may unambiguously write both these groups as $G_1 \operatorname{wr} G_2 \operatorname{wr} G_3$. Similarly, if $G_1, G_2, \ldots, G_r$ are permutation groups acting on $\Omega_1, \Omega_2, \ldots, \Omega_r$ respectively, the group $G_1 \operatorname{wr} G_2 \operatorname{wr} \cdots \operatorname{wr} G_r$ acting on $\Omega_1 \times \Omega_2 \times \cdots \times \Omega_r$ is unambiguously defined.

The importance of the wreath product construction comes from the following result.

Proposition 6.14 *Let G be a transitive permutation group on the set Ω. Let $\rho_0 < \rho_1 < \ldots < \rho_r$ be a maximal chain in the lattice of congruences on Ω (we define a partial order on the set of congruences by $\rho \leqslant \rho'$ whenever every congruence class of ρ is contained in a congruence class of ρ'). Let $\Delta_0 \subseteq \Delta_1 \subseteq \cdots \subseteq \Delta_r$ be such that Δ_i is a congruence class of ρ_i. For all $i \in \{1, 2, \ldots, r\}$, let Ω_i be the set of congruence classes of ρ_{i-1} contained in Δ_i and let G_i be the group $(G_{\Delta_i})^{\Omega_i}$ of permutations of Ω_i induced from the setwise stabiliser G_{Δ_i} of Δ_i. Then the groups G_i are primitive, and we may identify Ω with $\Omega_1 \times \Omega_2 \times \cdots \times \Omega_r$ in such a way that*

$$G \leqslant G_1 \operatorname{wr} G_2 \operatorname{wr} \cdots \operatorname{wr} G_r.$$

Proof: We first note that since the chain of congruences is maximal, we have that ρ_0 is the trivial congruence and ρ_r is the universal congruence on Ω. In particular, this implies that $\Delta_r = \Omega$.

It is not difficult to show that the congruences ρ preserved by G_i are in one-to-one correspondence with those congruences ρ' preserved by G such that $\rho_{i-1} \leqslant \rho' \leqslant \rho_i$. (Here, a block B of ρ corresponds to the block $\cup_{\Gamma \in B} \Gamma$ of ρ'. The remaining blocks of ρ' are the translates of this one block by elements of G.) Since the chain $\rho_0 < \rho_1 < \ldots < \rho_r$ is maximal, no congruence lies strictly between ρ_{i-1} and ρ_i, and so G_i is primitive.

To prove that G can be embedded in the wreath product, we use induction on the length r of the maximal chain of congruences. Assume that the result is true for maximal chains of congruences of smaller length. Let $\Delta = \Delta_{r-1}$ and $H = (G_\Delta)^\Delta$. Since G is a transitive permutation group on Ω, we have that H is a transitive permutation group on Δ. The restrictions of $\rho_0, \rho_1, \ldots, \rho_{r-1}$ to Δ form a maximal chain of equivalence relations preserved by H. So our inductive hypothesis implies that Δ may be identified with $\Omega_1 \times \Omega_2 \times \cdots \times \Omega_{r-1}$ in such a way that $H \leqslant G_1 \mathrm{wr}\, G_2 \mathrm{wr} \cdots \mathrm{wr}\, G_{r-1}$. To prove the proposition, it is therefore sufficient to show that Ω may be identified with $\Delta \times \Omega_r$ in such a way that $G \leqslant H \mathrm{wr}\, G_r$.

Let $\Omega_r = \{\Gamma_1, \Gamma_2, \ldots, \Gamma_k\}$. Since $\Delta_r = \Omega$, we have that Ω_r is the set of all blocks of ρ_{r-1} and so Ω is the disjoint union of the sets Γ_i. Since G is transitive, any block Γ_i of ρ_{r-1} is a translate of Δ. So for $i \in \{1, 2, \ldots, k\}$ we may choose $x_i \in G$ such that $x_i \Delta = \Gamma_i$. We define a bijection $f : \Omega \to \Delta \times \Omega_r$ by

$$f(\alpha) = (x_i^{-1}\alpha, \Gamma_i) \text{ for all } \alpha \in \Gamma_i.$$

It is not too difficult to verify that the map taking $g \in G$ to $fgf^{-1} \in \mathrm{Sym}(\Delta \times \Omega_r)$ embeds G into $H \mathrm{wr}\, G_r$. Indeed, let $g \in G$ and let $y \in G_r$ be the permutation of Ω_r induced by g. Let $i \in \{1, 2, \ldots, k\}$. Then there exists $j \in \{1, 2, \ldots, k\}$ such that $g^{-1}\Gamma_i = \Gamma_j$. Now, $x_i^{-1}gx_j$ preserves Δ and so $x_i^{-1}gx_j \in G_\Delta$. Let h_i be the element of H induced by $x_i^{-1}gx_j$. Then it is not difficult to check that

$$fgf^{-1} = (h_1, h_2, \ldots, h_k)y \in H \mathrm{wr}\, G_r,$$

and so the proposition follows.

We now turn to an application of wreath products to the theory of linear groups. A wreath product may be regarded as a linear group as follows. Let W be a vector space over a field F and let k be a positive integer, where $k \geqslant 2$. Let $H \leqslant \mathrm{GL}(W)$ and $Q \leqslant \mathrm{Sym}(k)$. Then the direct sum kW of k copies of W may be made into a $H \mathrm{wr}\, Q$-module in a natural way as follows. Let $(w_1, w_2, \ldots, w_k) \in kW$. Let $(h_1, h_2, \ldots, h_k)y \in H \mathrm{wr}\, Q$, where $h_i \in H$ and $y \in Q$. Then we define

$$(h_1, h_2, \ldots, h_k)y(w_1, w_2, \ldots, w_k) = (h_1 w_{y^{-1}1}, h_2 w_{y^{-1}2}, \ldots, h_k w_{y^{-1}k}).$$

It is an easy exercise to show that kW is an irreducible $H \mathrm{wr}\, Q$-module if and only if H is non-trivial, H acts irreducibly on W and Q is transitive.

Let V be a vector space over a field F, and let $G \leqslant \mathrm{GL}(V)$ be irreducible. We say that G is *imprimitive* if for some integer k such that $k \geqslant 2$ there exists a decomposition $V = V_1 \oplus V_2 \oplus \cdots \oplus V_k$ of V such that G permutes the direct summands. More precisely, for all $i \in \{1, 2, \ldots, k\}$ and $g \in G$ we have that $gV_i = V_j$ for some $j \in \{1, 2, \ldots, k\}$. If no such decomposition exists, we say that G is *primitive*. The irreducible wreath product construction above gives an example of an imprimitive group (where we define the subspaces V_i to be the k natural images of W in kW).

The structural theory for primitive linear groups G is more complicated than that for permutation groups, and we will spend Chapters 9 and 13 developing some of the appropriate theory when G is soluble. However, the role of wreath products in reduction from the irreducible to the primitive case is essentially the same for linear groups as for permutation groups, as can be seen by the following proposition.

Proposition 6.15 *Let G be a group acting faithfully and irreducibly on a vector space V. Suppose G is imprimitive, so there exists an integer $k \geqslant 2$ and a decomposition $V = V_1 \oplus V_2 \oplus \cdots \oplus V_k$ of V such that G permutes the direct summands. Let $Q \leqslant \mathrm{Sym}(k)$ be the permutation group induced by the action of G on the set of summands V_i. Define $W = V_1$, and define $H \leqslant \mathrm{GL}(W)$ to be the group induced by the stabiliser of W in G. Then Q is transitive and H is irreducible. Moreover, G is isomorphic as a linear group to a subgroup of $H \mathrm{wr}\, Q$ (where we regard $H \mathrm{wr}\, Q$ as a linear group as above).*

Proof: Since G is irreducible, Q is transitive (for otherwise the sum $\oplus V_i$, where i runs over a non-trivial orbit of Q, is a proper G-submodule of V). For $i \in \{1, 2, \ldots, k\}$ choose $x_i \in G$ so that $x_i W = V_i$.

To prove that H is irreducible, assume for a contradiction that U is a proper H-submodule. The subspace $x_1 U \oplus x_2 U \oplus \cdots \oplus x_k U$ of V is proper; we show that it is a G-submodule. Let $g \in G$ and $i \in \{1, 2, \ldots, k\}$. Now, $gx_i W = gV_i = V_j = x_j W$ for some $j \in \{1, 2, \ldots, k\}$, and so $x_j^{-1} gx_i$ stabilises W. By the definition of H, we know that $x_j^{-1} gx_i$ acts like an element of H when restricted to W, and so $x_j^{-1} gx_i U = U$. This means that $gx_i U = x_j U$. So $x_1 U \oplus x_2 U \oplus \cdots \oplus x_k U$ is a proper G-submodule of V, which contradicts the fact that G is irreducible. Hence H is irreducible.

The remainder of the proof is now very similar to the end of the proof of Proposition 6.14. We define a bijective linear map $f : V \to kW$ by

$$f(v_1 + v_2 + \cdots + v_k) = (x_1^{-1} v_1, x_2^{-1} v_2, \ldots, x_k^{-1} v_k).$$

for all vectors $v_1, v_2, \ldots, v_k$ in $V_1, V_2, \ldots, V_k$ respectively. Just as before, it is possible to check that the map $g \mapsto fgf^{-1}$ maps G into $H \mathrm{wr}\, Q$, and so the proposition follows. (Define $y \in \mathrm{Sym}(k)$ to be the permutation induced by g acting on the summands V_i. For $i \in \{1, 2, \ldots, k\}$, define $h_i \in \mathrm{GL}(W)$ to be the element induced by the restriction of $x_i^{-1} g x_j$ to W, where $j \in \{1, 2, \ldots, k\}$ is defined by $g^{-1} V_i = V_j$. This definition makes sense, and $h_i \in H$. Then $fgf^{-1} = (h_1, h_2, \ldots, h_k) y \in H \mathrm{wr}\, Q$.)

We remark that if the decomposition $V = V_1 \oplus V_2 \oplus \cdots \oplus V_k$ in Proposition 6.15 is chosen so that k is as large as possible, then H will in fact be a primitive subgroup of $\mathrm{GL}(W)$. For let $W = W_1 \oplus W_2 \oplus \cdots \oplus W_\ell$ be a decomposition of W such that $\ell \geqslant 2$ and H permutes the direct summands W_i. It is not difficult to check that G permutes the direct summands of the decomposition $\oplus_{1 \leqslant i \leqslant k} (x_i W_1 \oplus x_i W_2 \oplus \cdots \oplus x_i W_\ell)$ of V, contradicting the maximality of k.

7
Group extensions and cohomology

The main aim of this chapter is to explore the connection between group extensions and the second cohomology group. Towards the end of the chapter we shall illustrate the use of the second cohomology group in the enumeration of soluble A-groups. (An *A-group* is a group whose nilpotent subgroups are abelian. It is easy to see that a finite group is an A-group if and only if all its Sylow subgroups are abelian.)

7.1 Group extensions

Let E be a group and let M be a normal subgroup of E. We write $G = E/M$ and call E an extension of M by G. More formally, by an *extension* of a group M by a group G we shall mean a short exact sequence

$$1 \longrightarrow M \xrightarrow{i} E \xrightarrow{\rho} G \longrightarrow 1. \tag{7.1}$$

So the homomorphism $i : M \longrightarrow E$ is injective, the map $\rho : E \longrightarrow G$ is surjective and $\ker \rho = i(M)$.

Two extensions E and E' of M by G are said to be *equivalent* if there exists a homomorphism $\phi : E \to E'$ such that the following diagram commutes. In the diagram id stands for the identity map.

$$
\begin{array}{ccccccccc}
1 & \longrightarrow & M & \longrightarrow & E & \longrightarrow & G & \longrightarrow & 1 \\
 & & \text{id} \downarrow & & \phi \downarrow & & \text{id} \downarrow & & \\
1 & \longrightarrow & M & \longrightarrow & E' & \longrightarrow & G & \longrightarrow & 1
\end{array}
\tag{7.2}
$$

Note 7.1 *The map ϕ has to be bijective and so equivalent extensions are also isomorphic.*

60

For the rest of this section we shall restrict ourselves to the case where M is an abelian group. Note that we shall be writing M multiplicatively.

We identify M with $i(M)$ and regard M as an abelian normal subgroup of E. Now E acts on M by conjugation and M is abelian, so M is contained in the kernel of the action of E. Hence E/M acts on M in a natural way. As G is isomorphic to E/M, there is a natural action of G on M.

Note 7.2 *If E and E' are equivalent extensions, then they determine the same action of G on M.*

Note 7.3 *If M is a $\mathbb{Z}G$-module then there is always an extension E realising the action of G on M, namely the semidirect product of M by G arising from the action of G on M.*

Now given a $\mathbb{Z}G$-module M we would like to determine extensions E of M by G realising the action. Let E be such an extension. Let $\{s_g \mid g \in G\}$ be a transversal for M in E such that $\rho(s_g) = g$. Assume for convenience that $s_1 = 1$. Let $g \in G$ and let $m \in M$. The action of g on m is written as $g \cdot m$ and is given by $g \cdot m = s_g m s_g^{-1}$.

Since ρ is a homomorphism, for any $g, h \in G$, the elements $s_g s_h$ and s_{gh} have the same image gh under ρ. So

$$s_g s_h = f(g, h) s_{gh}$$

for some unique element $f(g, h)$ in M. Thus $f : G \times G \to M$ and since $s_1 = 1$, we get that

$$f(g, 1) = 1 = f(1, h) \tag{7.3}$$

for all g, h in G. The associative law in E also imposes certain conditions on f. For

$$(s_g s_h) s_k = f(g, h) s_{gh} s_k = f(g, h) f(gh, k) s_{ghk}$$

and

$$s_g (s_h s_k) = s_g (f(h, k) s_{hk}) = s_g f(h, k) s_g^{-1} s_g s_{hk} = (g \cdot f(h, k)) f(g, hk) s_{ghk} .$$

Consequently we must have

$$f(g, h) f(gh, k) = (g \cdot f(h, k)) f(g, hk) \tag{7.4}$$

for all $g, h, k \in G$. We say that a function $f : G \times G \to M$ satisfying (7.3) and (7.4) is a *factor set*.

Let $Z^2(G, M)$ denote the set of all factor sets from $G \times G$ to M. We can define an addition on $Z^2(G, M)$ as follows. Let f and f' be in $Z^2(G, M)$. Define

$$(f + f')(g, h) = f(g, h)f'(g, h)$$

for all g, h in G. It can be shown easily that under this addition $Z^2(G, M)$ is an abelian group.

So given a $\mathbb{Z}G$-module M, an extension E of M by G realising the module action, and a transversal $\{s_g \mid g \in G\}$ for M in E with the property that $s_1 = 1$ we get an element f of $Z^2(G, M)$. However, this element is dependent on our choice of transversal for M in E.

Suppose we choose a different transversal $\{s'_g \mid g \in G\}$ for M in E with the property that $s'_1 = 1$. Then for any $g \in G$, there exists a unique element $a(g)$ in M such that $s'_g = a(g)s_g$. Thus we get a function $a : G \to M$ such that $a(1) = 1$. Further, corresponding to the new transversal we get a new factor set f' in $Z^2(G, M)$.

Now for any g, h in G,

$$
\begin{aligned}
f'(g, h) &= s'_g s'_h {s'_{gh}}^{-1} \\
&= a(g)s_g\, a(h)s_h \left(a(gh)s_{gh}\right)^{-1} \\
&= a(g)\left(s_g a(h)s_g^{-1}\right) s_g s_h s_{gh}^{-1} a(gh)^{-1} \\
&= a(g)\, (g \cdot a(h))\, f(g, h)\, a(gh)^{-1} \\
&= (g \cdot a(h))\, a(gh)^{-1} a(g)\, f(g, h).
\end{aligned}
\tag{7.5}
$$

For any function $a : G \to M$ with $a(1) = 1$, define $\delta a : G \times G \to M$ as follows. For any $g, h \in G$,

$$\delta a(g, h) = (g \cdot a(h))\, a(gh)^{-1} a(g). \tag{7.6}$$

Define $B^2(G, M)$ as follows: $B^2(G, M) = \{\delta a \mid a : G \to M \text{ and } a(1) = 1\}$. It is not difficult to check that $\delta a \in Z^2(G, M)$. Thus $B^2(G, M) \subseteq Z^2(G, M)$.

Let a, b be functions from G to M satisfying $a(1) = 1$ and $b(1) = 1$. For any $g \in G$, define $(a + b)(g) = a(g)b(g)$. This definition of addition makes the set $\{a : G \to M \mid a(1) = 1\}$ into a group. It is easy to show that δ is a homomorphism from the above group to $Z^2(G, M)$. Hence the image $B^2(G, M)$ of δ is a subgroup of $Z^2(G, M)$.

Define $H^2(G, M) = Z^2(G, M)/B^2(G, M)$. This abelian group $H^2(G, M)$ is called the 2-*dimensional cohomology group* of M with respect to G. Now

(7.5) tells us that $f' - f = \delta a$. So $f' - f \in B^2(G, M)$ or $f + B^2(G, M) = f' + B^2(G, M)$. Thus f and f' determine the same element of $H^2(G, M)$. Hence given a $\mathbb{Z}G$-module M and an extension E of M by G realising the action, E determines a unique element of $H^2(G, M)$. We are in a position to state the following theorem.

Theorem 7.4 *Two extensions of M by G are equivalent if and only if they determine the same action of G on M and the same element of $H^2(G, M)$.*

Proof: We have already mentioned in Note 7.2 that equivalent extensions give rise to the same action and it also follows easily from the definition of equivalence that they determine the same element of $H^2(G, M)$.

Let E and E' be two extensions of M by G that realise the same action of G on M and also determine the same element of $H^2(G, M)$. We wish to construct a homomorphism ϕ from E to E' such that diagram (7.2) commutes.

Let E determine an element $f + B^2(G, M)$ of $H^2(G, M)$. We shall show that multiplication in E is completely determined by the action of G on M and the element $f + B^2(G, M)$ and use this to define ϕ.

Formally, let the extension E be given as in (7.1). As we had done earlier we shall identify M with $i(M)$ and regard M as an abelian normal subgroup of E. Let $\{s_g \mid g \in G\}$ be a transversal for M in E with $s_1 = 1$ and $\rho(s_g) = g$.

Every element of E can be represented uniquely as ms_g for some $m \in M$ and $g \in G$. Multiplication in E is determined by the data we have collected, because

$$(ms_g)(m's_h) = m(s_g m' s_g^{-1})(s_g s_h)$$

$$= \{m(g \cdot m')f(g, h)\} s_{gh}.$$

Now E' determines the same action of G on M and the same element $f + B^2(G, M)$ of $H^2(G, M)$. Let $T' = \{s'_g \mid g \in G\}$ be a transversal for M in E' with $s'_1 = 1$. Every element of E' can be represented uniquely as ms'_g for some $m \in M$ and $g \in G$. Define $f'(g, h) = s'_g s'_h s'^{-1}_{gh}$ for all g, h in G.

Since E and E' give rise to the same element of the group $H^2(G, M)$, we must have $f' + B^2(G, M) = f + B^2(G, M)$. Thus $f' - f = \delta a$ for some $\delta a \in B^2(G, M)$. Define $\phi : E \to E'$ as $\phi(ms_g) = ma(g)^{-1} s'_g$. We can show without difficulty that ϕ is a homomorphism such that diagram (7.2) commutes. This establishes the equivalence of E and E'.

Theorem 7.5 *Let M be a $\mathbb{Z}G$-module and let $\mathcal{M}$ be the set of extensions of M by G realising the action of G on M. Then*

(i) *There is a bijection between the set of equivalence classes of extensions in $\mathcal{M}$ and elements of $H^2(G, M)$.*

(ii) *The number of isomorphism classes of extensions contained in $\mathcal{M}$ is at most $|H^2(G, M)|$.*

Proof: We can define a map from the set of equivalence classes of extensions in $\mathcal{M}$ to $H^2(G, M)$ by associating the equivalence class of E with $f + B^2(G, M)$, where $f + B^2(G, M)$ is the unique element of $H^2(G, M)$ determined by E. Theorem 7.4 shows that this map is well-defined and injective. We only have to show that it is also surjective.

Let $f' + B^2(G, M)$ be an element of $H^2(G, M)$. Let $E' = M \times G$ and define a multiplication in E' as follows:

$$(m, g)\,(m', h) = (m\,(g \cdot m')\,f'(g, h),\, gh)$$

for all m, m' in M and g, h in G.

It is easy to check that E' is a group under this multiplication and is an extension of M by G in $\mathcal{M}$ determining the element $f' + B^2(G, M)$ of $H^2(G, M)$. Hence our map is surjective. The second part of the theorem follows by Note 7.1 and part (i).

We have explored previously extensions of an abelian group M by an arbitrary group G. We now consider the special case when G is abelian and focus on extensions of M by G that are abelian. Of course, M must be a trivial $\mathbb{Z}G$-module for abelian extensions to arise.

By Theorem 7.5 we may regard $H^2(G, M)$ as the group of equivalence classes of extensions of M by G. We define $\mathrm{Ext}(G, M)$ to be the subset of $H^2(G, M)$ consisting of equivalence classes of abelian extensions.

Note 7.6 *An element $f + B^2(G, M)$ of $H^2(G, M)$ belongs to $\mathrm{Ext}(G, M)$ if and only if $f(g, h) = f(h, g)$ for all $h, g \in G$. In other words, an extension E is abelian if and only if a factor set f determined by it satisfies the symmetry condition given above. The condition also clearly ensures that $\mathrm{Ext}(G, M) \leqslant H^2(G, M)$.*

The second cohomology group is just one of an infinite family of groups $H^n(G, M)$, defined for any positive integer n. We will make this generalisation in the next section, but we finish this section by describing the group $H^1(G, M)$ and its role in the classification of certain subgroups in the semidirect product of M by G.

Let G be a group and let M be a $\mathbb{Z}G$-module. As above, we write the action of $g \in G$ on $m \in M$ as $g \cdot m$. Define $Z^1(G, M)$ to be the set of all functions $f : G \to M$ such that $f(1) = 1$ and such that

$$f(g_1 g_2) = (g_1 \cdot f(g_2)) f(g_1)$$

for all $g_1, g_2 \in G$. Such functions are known as *derivations* from G to M. If $f, f' \in Z^1(G, M)$, we define the function $f + f'$ from G to M by

$$(f + f')(g) = f(g) f'(g)$$

for all $g \in G$. It is easy to check that $f + f'$ is again a derivation, and this definition of addition makes $Z^1(G, M)$ into an abelian group.

Let $m \in M$. Then the function $\delta_m : G \to M$ defined by

$$\delta_m(g) = (g \cdot m^{-1}) m$$

for all $g \in G$ is a derivation, since clearly $\delta_m(1) = 1$ and since

$$\delta_m(g_1 g_2) = (g_1 g_2 \cdot m^{-1}) m = (g_1 \cdot ((g_2 \cdot m^{-1}) m m^{-1})) m$$
$$= (g_1 \cdot ((g_2 \cdot m^{-1}) m))(g_1 \cdot m^{-1}) m = (g_1 \cdot \delta_m(g_2)) \delta_m(g_1).$$

The functions δ_m are known as *inner derivations*. (Let E be an extension of M by G. The derivation δ_m is related to the inner automorphism of E obtained by conjugation by $x \in E$, where $xM = g^{-1}$. Indeed, if E is a semidirect product of M by G then $\delta_m(g) = [g^{-1}, m]$.) We write the set of inner derivations from G to M as $B^1(G, M)$. It is not difficult to show that $B^1(G, M)$ is a subgroup of $Z^1(G, M)$ (by showing that $\delta_{m_1 m_2} = \delta_{m_1} + \delta_{m_2}$ for all $m_1, m_2 \in M$). We define $H^1(G, M)$ to be the quotient $Z^1(G, M)/B^1(G, M)$.

Let E be the semidirect product $E = M \rtimes G$ of M by G realising the action of G on M. Recall that a subgroup X of E is a *complement to M* if $E = MX$ and $M \cap X = \{1\}$. So G, thought of as a subgroup of E, is a complement to M in E. We will now show that the groups $Z^1(G, M)$ and $H^1(G, M)$ are intimately associated with the complements of M in E.

Proposition 7.7 *Let G be a group, let M be a $\mathbb{Z}G$-module and let $E = M \rtimes G$. There is a bijection between the set $Z^1(G, M)$ of derivations from G to M and the set of complements of M in E.*

Proof: Let X be a complement to M in E. We will define a derivation f from G to M as follows. The natural homomorphism from E onto G induces an isomorphism $\phi : X \to G$. Let $\psi : G \to X$ be the inverse of this isomorphism.

Thinking of G as a subgroup of E, we note that ψ is the unique map from G to X such that $\psi(g) \in Mg$ for all $g \in G$. We define $f: G \to M$ by

$$f(g) = g(\psi(g))^{-1}$$

for all $g \in G$. Note that $\psi(g) = mg$ for some $m \in M$, and so $g(\psi(g))^{-1} = g(mg)^{-1} = m^{-1} \in M$. Thus f is well-defined. We find that $f \in Z^1(G, M)$ for, since $\psi(1) = 1$, we have that $f(1) = 1$ and also

$$
\begin{aligned}
f(g_1 g_2) &= g_1 g_2 (\psi(g_1 g_2))^{-1} \\
&= g_1 g_2 (\psi(g_2))^{-1} (\psi(g_1))^{-1} \\
&= g_1 g_2 (\psi(g_2))^{-1} g_1^{-1} g_1 (\psi(g_1))^{-1} \\
&= (g_1 \cdot f(g_2)) f(g_1).
\end{aligned}
$$

Conversely, given $f \in Z^1(G, M)$ we may define

$$X = \{f(g)^{-1} g : g \in G\}.$$

We find that X is a subgroup, since

$$
\begin{aligned}
f(g_1)^{-1} g_1 f(g_2)^{-1} g_2 &= f(g_1)^{-1} g_1 f(g_2)^{-1} g_1^{-1} g_1 g_2 \\
&= (g_1 \cdot f(g_2) f(g_1))^{-1} g_1 g_2 \\
&= f(g_1 g_2)^{-1} g_1 g_2.
\end{aligned}
$$

Moreover, since $f(g)^{-1} \in M$ the elements of X form a complete set of coset representatives for M in E and so X is a complement for M in E.

It is not difficult to check that the two correspondences defined above are inverses of each other, and so the proposition is established.

Theorem 7.8 *Let G be a group, let M be a $\mathbb{Z}G$-module and let $E = M \rtimes G$. There is a bijection between the set $H^1(G, M)$ and the conjugacy classes of complements to M in E.*

Proof: Let X_1 and X_2 be complements to M in E, and suppose that X_1 and X_2 are conjugate. So $X_1 = h X_2 h^{-1}$ for some $h \in E$. For $i \in \{1, 2\}$, let $\psi_i : G \to X_i$ be the unique isomorphism such that $\psi_i(g) \in Mg$ for all $g \in G$ and let $f_i \in Z^1(G, M)$ be the derivation corresponding to X_i. We must show that $f_1 - f_2 \in B^1(G, M)$.

Since X_2 is a complement to M in E, we may write the element h defined above in the form $h = m_0^{-1} x_2$ for some $m_0 \in M$ and $x_2 \in X_2$. But then $X_1 = m_0^{-1} x_2 X_2 x_2^{-1} m_0 = m_0^{-1} X_2 m_0$. So the isomorphism $\lambda : X_2 \to X_1$ given by

$\lambda(x) = m_0^{-1}xm_0$ for all $x \in X_2$ is well-defined. Since $m_0 \in M$, we have that $\lambda(x) \in Mx$ for all $x \in X_2$. But then both ψ_1 and $\lambda\psi_2$ are isomorphisms from G to X_1 preserving cosets of M, and so $\psi_1 = \lambda\psi_2$:

$$\psi_1(g) = m_0^{-1}\psi_2(g)m_0$$

for all $g \in G$. Thus for any $g \in G$,

$$
\begin{aligned}
(f_1 - f_2)(g) &= g\psi_1(g)^{-1}\psi_2(g)g^{-1} \\
&= gm_0^{-1}\psi_2(g)^{-1}m_0\psi_2(g)g^{-1} \\
&= g\psi_2(g)^{-1}[\psi_2(g)^{-1}, m_0]\psi_2(g)g^{-1} \\
&= [\psi_2(g)^{-1}, m_0],
\end{aligned}
$$

the last equality following since $\psi_2(g)g^{-1}$ and $[\psi_2(g)^{-1}, m_0]$ lie in the abelian group M and so commute. But $\psi_2(g) = gm$ for some $m \in M$, and so

$$
\begin{aligned}
[\psi_2(g)^{-1}, m_0] &= [m^{-1}g^{-1}, m_0] = (g[m^{-1}, m_0]g^{-1})[g^{-1}, m_0] \\
&= [g^{-1}, m_0] = (g \cdot m_0^{-1})m_0 = \delta_{m_0}(g).
\end{aligned}
$$

Thus $f_1 - f_2 = \delta_{m_0} \in B^1(G, M)$. Reversing this argument shows that whenever $f_1 - f_2 = \delta_{m_0} \in B^1(G, M)$ we have that $X_1 = m_0^{-1}X_2m_0$. Hence the theorem is proved.

7.2 Cohomology

In the previous section we defined the 1- and 2-dimensional cohomology groups. As mentioned earlier, these groups are just two of an infinite family $H^n(G, M)$, defined for any positive integer n. We shall describe these groups here. For the rest of this section M will be written additively.

Let M be a $\mathbb{Z}G$-module. Define $C^n(G, M)$ to be the set of all functions f from $G \times G \times \cdots \times G$ (with n factors G) to M subject to the condition

$$f(g_1, \ldots, g_n) = 0 \tag{7.7}$$

whenever there exists $i \in \{1, 2, \ldots, n\}$ such that $g_i = 1$. The elements of $C^n(G, M)$ are called *n-dimensional (normalised) cochains*. We define an addition on $C^n(G, M)$ as follows. Let f, f' be in $C^n(G, M)$. Define

$$(f + f')(g_1, \ldots, g_n) = f(g_1, \ldots, g_n) + f'(g_1, \ldots, g_n).$$

Under this operation $C^n(G, M)$ is an abelian group.

For each cochain f in $C^n(G, M)$ we can define a function $\delta_n f$ of $n+1$ variables as follows. Let $g_i \in G$ for $i = 1, \ldots, n+1$. Define

$$\delta_n f(g_1, \ldots, g_{n+1}) = g_1 \cdot f(g_2, \ldots, g_{n+1})$$

$$+ \sum_{i=1}^{n} (-1)^i f(g_1, \ldots, g_{i-1}, g_i g_{i+1}, g_{i+2}, \ldots, g_{n+1})$$

$$+ (-1)^{n+1} f(g_1, \ldots, g_n).$$

It is easy to check that $\delta_n f$ belongs to $C^{n+1}(G, M)$. Further, it can be shown that the map $\delta_n : C^n(G, M) \to C^{n+1}(G, M)$ is a homomorphism. This map δ_n is called the *coboundary operator* and $\delta_n f$ is called the *coboundary* of f.

Define $Z^n(G, M)$ as the kernel of δ_n and $B^n(G, M)$ as the image of δ_{n-1}. Elements of $Z^n(G, M)$ are called *n-dimensional cocycles* and elements of $B^n(G, M)$ are called *n-dimensional coboundaries*.

It is not difficult to check that $\delta_n \delta_{n-1} = 0$. Thus the image of δ_{n-1} is contained in the kernel of δ_n. So $B^n(G, M) \leqslant Z^n(G, M)$. We define the *n-dimensional cohomology group* $H^n(G, M)$ to be $Z^n(G, M)/B^n(G, M)$.

Note 7.9 *Equations (7.3) and (7.4) rewritten additively show that factor sets are nothing but 2-dimensional cocycles. Again (7.6) rewritten additively shows that δa is a 2-dimensional coboundary. Thus our previous definitions of $Z^2(G, M)$, $B^2(G, M)$ and $H^2(G, M)$ agree with our current definitions. Similarly, our previous definitions of $Z^1(G, M)$, $B^1(G, M)$ and $H^1(G, M)$ agree with our current definitions.*

There is another approach used to define the *n*-dimensional cohomology group which uses the concept of 'free resolutions'. The rest of this section is devoted to describing this method. Our main source of reference for the material presented here is [89]. The advantages of using this approach over the previous one will be evident when we consider the restriction and transfer maps in the next section.

As in the previous case we are given a group G and a $\mathbb{Z}G$-module M. We can regard $\mathbb{Z}$ as a $\mathbb{Z}G$-module by setting $(\sum_{g \in G} n_g g)n = (\sum_{g \in G} n_g)n$. Under this action $\mathbb{Z}$ is called the *trivial $\mathbb{Z}G$-module*. Before we present the definition of a free resolution let us dwell briefly on the notion of a 'free module'.

Let R be any ring. (We assume that all rings have a multiplicative identity.) Let A be an R-module. Let X be a subset of A satisfying the property that any map from X into any R-module M is *uniquely extendable* to an R-module homomorphism from A to M. Then we say that A is a *free R-module freely generated by X*.

We now present a concrete construction of a free $\mathbb{Z}G$-module on a set X, so showing that free $\mathbb{Z}G$-modules always exist. Let us denote by R the group ring $\mathbb{Z}G$ and let X be any non-empty set. Consider the set (of formal linear combinations)

$$RX = \left\{ \sum_{i=1}^{n} r_i x_i \mid r_i \in R, x_i \in X, n \geqslant 0 \right\}.$$

Each element of RX has a (unique) expression as $\sum_{x \in X} r_x x$ where $r_x = 0$ except for finitely many $x \in X$.

We can define an addition between elements in RX by setting

$$\sum_{x \in X} r_x x + \sum_{x \in X} s_x x = \sum_{x \in X} (r_x + s_x) x.$$

Clearly under this addition RX is an abelian group. We can make it a left R-module by defining $r(\sum_{x \in X} r_x x) = \sum_{x \in X} (rr_x)x$.

It is easy to see that the module RX has the property that any map from X into an R-module M extends uniquely to an R-module homomorphism between RX and M. Thus RX is a free R-module freely generated by X.

Note that if X is taken as the empty set then the free module generated by X is the zero module.

Any two free R-modules freely generated by X are isomorphic, as the identity map on X extends to an isomorphism between any two such modules. Moreover, it is easy to show that any R-module is a homomorphic image of some free R-module.

It is easy to see that a free R-module RX has the following 'projective' property. Let M and N be R-modules, and let $\alpha : RX \to N$ and $\beta : M \to N$ be homomorphisms. Then provided that $\operatorname{im} \alpha \subseteq \operatorname{im} \beta$ there exists a homomorphism $\pi : RX \to M$ such that $\beta\pi = \alpha$.

We say that a sequence $\mathcal{X}$ consisting of free $\mathbb{Z}G$-modules X_i and $\mathbb{Z}G$-module homomorphisms d_i is a *free resolution* of the trivial $\mathbb{Z}G$-module $\mathbb{Z}$ if the sequence

$$\mathcal{X}: \quad \cdots \longrightarrow X_n \xrightarrow{d_n} X_{n-1} \xrightarrow{d_{n-1}} \cdots \longrightarrow X_1 \xrightarrow{d_1} X_0 \xrightarrow{d_0} \mathbb{Z} \longrightarrow 0$$

is exact, that is, $\operatorname{im} d_n = \ker d_{n-1}$ for all $n \geqslant 1$ and $\operatorname{im} d_0 = \mathbb{Z}$.

Note 7.10 *Free resolutions exist. We can construct a sequence of free modules and module-homomorphisms to get a free resolution by the following method. The trivial $\mathbb{Z}G$-module $\mathbb{Z}$ is the homomorphic image of some free $\mathbb{Z}G$-module, say X_0. Let $X_0/K_1 \cong \mathbb{Z}$. Then there exists a free $\mathbb{Z}G$-module X_1 such that K_1 is a homomorphic image of X_1. So there is a submodule K_2 such that*

$X_1/K_2 \cong K_1$. *Proceeding similarly we get a sequence of free $\mathbb{Z}G$-modules and $\mathbb{Z}G$-module homomorphisms such that the sequence*

$$\cdots \longrightarrow X_n \cdots \longrightarrow X_1 \longrightarrow X_0 \longrightarrow \mathbb{Z} \longrightarrow 0$$

is exact.

Keeping to the notation established in the paragraph preceding Note 7.10 we find that the $\mathbb{Z}G$-module homomorphisms d_i from X_i to X_{i-1} induce abelian group homomorphisms d_i^* from $\mathrm{Hom}_{\mathbb{Z}G}(X_{i-1}, M)$ to $\mathrm{Hom}_{\mathbb{Z}G}(X_i, M)$. For any $f \in \mathrm{Hom}_{\mathbb{Z}G}(X_{i-1}, M)$ and $x \in X_i$, we have $d_i^*(f)(x) = f(d_i(x))$. Thus $d_i^*(f) = f\, d_i$. Since the sequence $\mathcal{X}$ is exact, we find that $d_i\, d_{i+1} = 0$ and so $d_{i+1}^* d_i^* = 0$. Consequently, im d_i^* is contained in ker d_{i+1}^*.

We define the *n-dimensional cohomology group* $H^n(G, M)$ to be the quotient ker $d_{n+1}^*/\mathrm{im}\, d_n^*$. This definition seems to depend on the free resolution that we have chosen but this is not the case: this is the content of Theorem 7.11 below. We have now given two definitions of a cohomology group. That these definitions are compatible can be shown using the concept of 'standard bar resolutions': see the discussion after the proof of Theorem 7.11.

Theorem 7.11 *The definition of an n-dimensional cohomology group does not depend on the free resolution $\mathcal{X}$.*

Proof: Let $\mathcal{X}$ and $\overline{\mathcal{X}}$ be two free resolutions of the trivial $\mathbb{Z}G$-module $\mathbb{Z}$. We first show that there exists a *morphism* $\pi : \mathcal{X} \to \overline{\mathcal{X}}$; in other words, a sequence $\pi_0, \pi_1, \ldots$ of $\mathbb{Z}G$-module homomorphisms such that the following diagram commutes:

$$
\begin{array}{ccccccccccccc}
\mathcal{X} : \cdots & \xrightarrow{d_{n+1}} & X_n & \xrightarrow{d_n} & X_{n-1} & \xrightarrow{d_{n-1}} & \cdots & \xrightarrow{d_1} & X_0 & \xrightarrow{d_0} & \mathbb{Z} & \longrightarrow & 0 \\
 & & \downarrow{\pi_n} & & \downarrow{\pi_{n-1}} & & & & \downarrow{\pi_0} & & \downarrow{\mathrm{id}} & & \\
\overline{\mathcal{X}} : \cdots & \xrightarrow{\overline{d}_{n+1}} & \overline{X}_n & \xrightarrow{\overline{d}_n} & \overline{X}_{n-1} & \xrightarrow{\overline{d}_{n-1}} & \cdots & \xrightarrow{\overline{d}_1} & \overline{X}_0 & \xrightarrow{\overline{d}_0} & \mathbb{Z} & \longrightarrow & 0
\end{array}
$$

To see why π exists, note first that im $\overline{d}_0 = \mathrm{im}\, d_0$, and so the fact that X_0 is free implies that there exists a map π_0 such that $d_0 = \overline{d}_0 \pi_0$. Assume, as an inductive hypothesis, that $\pi_0, \pi_1, \ldots, \pi_{n-1}$ have been chosen so that $\overline{d}_i \pi_i = \pi_{i-1} d_i$ for $0 \leqslant i < n$ (where we set $\pi_{-1} = \mathrm{id}$). Then $\overline{d}_{n-1} \pi_{n-1} d_n = \pi_{n-2} d_{n-1} d_n = 0$ and so im $\pi_{n-1} d_n \subseteq \ker \overline{d}_{n-1} = \mathrm{im}\, \overline{d}_n$. So, since X_n is free, there exists a homomorphism π_n such that $\overline{d}_n \pi_n = \pi_{n-1} d_n$. Hence, by induction, a morphism $\pi : \mathcal{X} \to \overline{\mathcal{X}}$ exists.

For any non-negative integer n, we find that π induces a homomorphism $\pi_n^* : \mathrm{Hom}_{\mathbb{Z}G}(\overline{X}_n, M) \to \mathrm{Hom}_{\mathbb{Z}G}(X_n, M)$ where $\pi_n^*(f) = f\pi_n$. It is not difficult to verify that $\pi_n^*(\ker \overline{d}_{n+1}^*) \subseteq \ker d_{n+1}^*$ and $\pi_n^*(\mathrm{im}\, \overline{d}_n^*) \subseteq \mathrm{im}\, d_n^*$, and so π also induces a homomorphism $\pi_n' : \ker \overline{d}_{n+1}^*/\mathrm{im}\, \overline{d}_n^* \to \ker d_{n+1}^*/\mathrm{im}\, d_n^*$.

Similarly, there exists a morphism $\overline{\pi} : \overline{\mathcal{X}} \to \mathcal{X}$ that induces maps $\overline{\pi}'_n :$ $\ker d^*_{n+1}/\operatorname{im} d^*_n \to \ker \overline{d}^*_{n+1}/\operatorname{im} \overline{d}^*_n$. We aim to show that the maps π'_n are isomorphisms, by showing that $\overline{\pi}'_n = (\pi'_n)^{-1}$. This is sufficient to prove the theorem.

Consider the morphism $\rho : \mathcal{X} \to \mathcal{X}$ defined by $\rho = \overline{\pi}\pi$. So $\rho_n = \overline{\pi}_n\pi_n$, $\rho^*_n = \pi^*_n\overline{\pi}^*_n$ and $\rho'_n = \pi'_n\overline{\pi}'_n$ for $n \geqslant 0$. We claim that ρ is 'homotopic to the identity', namely that there exist homomorphisms $\sigma_n : X_n \to X_{n+1}$ such that $\rho_n - \operatorname{id} = d_{n+1}\sigma_n + \sigma_{n-1}d_n$ for $n \geqslant 0$ (where we set $\sigma_{-1} = 0$). To see this, first note that the following diagram commutes:

$$\cdots \xrightarrow{d_{n+1}} X_n \xrightarrow{d_n} X_{N-1} \xrightarrow{d_{n-1}} \cdots \xrightarrow{d_1} X_0 \xrightarrow{d_0} \mathbb{Z} \longrightarrow 0$$
$$\downarrow \rho_n-\mathrm{id} \qquad \downarrow \rho_{n-1}-\mathrm{id} \qquad\qquad \downarrow \rho_0-\mathrm{id} \qquad \downarrow 0$$
$$\cdots \xrightarrow{d_{n+1}} X_n \xrightarrow{d_n} X_{n-1} \xrightarrow{d_{n-1}} \cdots \xrightarrow{d_1} X_0 \xrightarrow{d_0} \mathbb{Z} \longrightarrow 0$$

In particular, $d_0(\rho_0 - \operatorname{id}) = 0$, and so $\operatorname{im}(\rho_0 - \operatorname{id}) \subseteq \ker d_0 = \operatorname{im} d_1$. Since X_0 is free, there exists a homomorphism σ_0 such that $d_1\sigma_0 = \rho_0 - \operatorname{id}$, as required. Assume, as an inductive hypothesis, that $\sigma_0, \sigma_1, \ldots, \sigma_n$ have been chosen. Now,

$$d_{n+1}(\rho_{n+1} - \operatorname{id} - \sigma_n d_{n+1}) = (\rho_n - \operatorname{id})d_{n+1} - d_{n+1}\sigma_n d_{n+1}$$
$$= (d_{n+1}\sigma_n + \sigma_{n-1}d_n)d_{n+1} - d_{n+1}\sigma_n d_{n+1}$$
$$= 0$$

and so $\operatorname{im}(\rho_{n+1} - \operatorname{id} - \sigma_n d_{n+1}) \subseteq \ker d_{n+1} = \operatorname{im} d_{n+2}$. Since X_{n+1} is free, there exists a map σ_{n+1} such that $d_{n+2}\sigma_{n+1} = \rho_{n+1} - \operatorname{id} - \sigma_n d_{n+1}$, as required. So the maps $\sigma_0, \sigma_1, \ldots$ exist by induction, and our claim follows.

To prove that the homomorphisms ρ'_n are identity maps, we need to show that $(\rho^*_n - \operatorname{id})f \in \operatorname{im} d^*_n$ whenever $f \in \ker d^*_{n+1}$. But

$$(\rho^*_n - \operatorname{id})f = f(\rho_n - \operatorname{id})$$
$$= f(d_{n+1}\sigma_n + \sigma_{n-1}d_n)$$
$$= f\sigma_{n-1}d_n \text{ since } f \in \ker d^*_{n+1}$$
$$\in \operatorname{im} d^*_n.$$

Hence $\rho'_n = \pi'_n\overline{\pi}'_n = \operatorname{id}$.

A similar argument based on the morphism $\overline{\rho} : \overline{\mathcal{X}} \to \overline{\mathcal{X}}$ defined by $\overline{\rho} = \pi\overline{\pi}$ shows that $\overline{\pi}'_n\pi'_n = \operatorname{id}$. Thus $\overline{\pi}'_n = (\pi'_n)^{-1}$ and so π'_n is an isomorphism between $\ker d^*_{n+1}/\operatorname{im} d^*_n$ and $\ker \overline{d}^*_{n+1}/\operatorname{im} \overline{d}^*_n$. This proves the theorem.

The *standard bar resolution* of $\mathbb{Z}$ is defined as follows. Let n be any positive integer. Consider the collection of all symbols $[x_1|x_2|\cdots|x_n]$ where x_i are non-identity elements of G, and let B_n be the free $\mathbb{Z}G$-module generated freely by these symbols. We include the case when the x_i are identity elements by setting $[x_1|x_2|\cdots|x_n] = 0$ if any x_i is the identity. For $n = 0$, let B_0 be the free $\mathbb{Z}G$-module generated freely by a single symbol [ ]. Note that $B_0 \cong \mathbb{Z}G$. Therefore there exists a surjective $\mathbb{Z}G$-module homomorphism ∂_0 from B_0 to $\mathbb{Z}$ which corresponds to the mapping $\epsilon : \mathbb{Z}G \to \mathbb{Z}$ given by $\epsilon(\sum_{g \in G} n_g g) = \sum_{g \in G} n_g$. For all positive integers n, define $\partial_n : B_n \to B_{n-1}$ to be the $\mathbb{Z}G$-module homomorphism such that

$$\partial_n[x_1|x_2|\cdots|x_n] = x_1 \cdot [x_2|x_3|\cdots|x_n]$$

$$+ \sum_{i=1}^{n} (-1)^i [x_1|\cdots|x_i x_{i+1}|\cdots|x_n]$$

$$+ (-1)^n [x_1|\cdots|x_{n-1}]$$

for all $x_1, x_2, \ldots, x_n \in G \setminus \{1\}$.

It is not difficult to check that the sequence $\mathcal{B}$ consisting of the free $\mathbb{Z}G$-modules B_n and $\mathbb{Z}G$-module homomorphisms ∂_n defines a free resolution of $\mathbb{Z}$. It is this free resolution that is called the standard bar resolution. Since B_n is the free $\mathbb{Z}G$-module on the set of all symbols, we find that an element f in $\mathrm{Hom}_{\mathbb{Z}G}(B_n, M)$ is determined by its value at these $[x_1|\cdots|x_n]$. Conversely, we may assign arbitrary values in M to these symbols to produce an element f in $\mathrm{Hom}_{\mathbb{Z}G}(B_n, M)$. Thus we find that $\mathrm{Hom}_{\mathbb{Z}G}(B_n, M)$ is isomorphic to the additive group $C^n(G, M)$ defined above. Using this we can show that the n-dimensional cohomology group arising from the free resolution $\mathcal{B}$ is isomorphic to the quotient $Z^n(G, M)/B^n(G, M)$, and so the two definitions of $H^n(G, M)$ are compatible.

For the rest of this chapter $H^n(G, M)$ will be defined via free resolutions. To end this section we prove a result regarding the cohomology group of a sum of $\mathbb{Z}G$-modules.

Proposition 7.12 *Let A and B be $\mathbb{Z}G$-modules. Then $H^n(G, A + B) \cong H^n(G, A) + H^n(G, B)$.*

Proof: Let $\{X_n\}$ be a free resolution of the trivial $\mathbb{Z}G$-module $\mathbb{Z}$. Then for all n,

$$\mathrm{Hom}_{\mathbb{Z}G}(X_n, A + B) \cong \mathrm{Hom}_{\mathbb{Z}G}(X_n, A) + \mathrm{Hom}_{\mathbb{Z}G}(X_n, B)$$

as abelian groups. Let θ represent the above isomorphism and π_A and π_B the respective projections. It is easy to check that $\pi_A(\theta) \, d_n^* = d_n^* \, \pi_A(\theta)$ and similarly $\pi_B(\theta) \, d_n^* = d_n^* \, \pi_B(\theta)$. Consequently, θ induces an isomorphism from $H^n(G, A + B)$ onto $H^n(G, A) + H^n(G, B)$.

7.3 Restriction and transfer

Let M be a $\mathbb{Z}G$-module and let H be a subgroup of G. Then M can also be regarded as a $\mathbb{Z}H$-module by restricting the module action to H, so we can define $H^n(H, M)$. It is natural to ask if there is any connection between $H^n(H, M)$ and $H^n(G, M)$. The answer lies in two homomorphisms, the restriction map and the transfer, which connect the two.

Before we define these maps we need to note several facts. The first is that if X is a free $\mathbb{Z}G$-module then it is also a free $\mathbb{Z}H$-module. So if we consider a free resolution $\mathcal{X}$ of the trivial $\mathbb{Z}G$-module $\mathbb{Z}$, the free $\mathbb{Z}G$-modules X_i in the resolution are free $\mathbb{Z}H$-modules. Further, the $\mathbb{Z}G$-module homomorphisms d_i in the resolution are also trivially $\mathbb{Z}H$-module homomorphisms. Thus $\mathcal{X}$ is also a free resolution of the trivial $\mathbb{Z}H$-module $\mathbb{Z}$. Consequently, we can study simultaneously, cohomology groups of G and H using a single free resolution $\mathcal{X}$ of the trivial $\mathbb{Z}G$-module $\mathbb{Z}$. Let us assume that $\mathcal{X}$ is a fixed free resolution for the trivial $\mathbb{Z}G$-module $\mathbb{Z}$.

We define the *restriction map* $\rho^* : H^n(G, M) \to H^n(H, M)$ as follows. Let $f \in \operatorname{Hom}_{\mathbb{Z}G}(X_n, M)$. Then f is also a $\mathbb{Z}H$-module homomorphism from the $\mathbb{Z}H$-module X_n to the $\mathbb{Z}H$-module M. We define ρf to be the element f viewed as a member of $\operatorname{Hom}_{\mathbb{Z}H}(X_n, M)$, and so we have a map $\rho : \operatorname{Hom}_{\mathbb{Z}G}(X_n, M) \to \operatorname{Hom}_{\mathbb{Z}H}(X_n, M)$. The map ρ is a homomorphism which satisfies $\rho\, d_n^* = d_n^* \rho$. So ρ induces a homomorphism $\rho^* : H^n(G, M) \to H^n(H, M)$; this map is the restriction map from $H^n(G, M)$ to $H^n(H, M)$.

We define the *transfer map* $\tau^* : H^n(H, M) \to H^n(G, M)$ (also known as *induction*, or *corestriction*) as follows. Let H be a subgroup of G such that $|G : H| = k$. Let M be a $\mathbb{Z}G$-module and let $\{t_1, \ldots, t_k\}$ be a right transversal for H in G. Define $\tau : \operatorname{Hom}_{\mathbb{Z}H}(X_n, M) \to \operatorname{Hom}_{\mathbb{Z}G}(X_n, M)$ as follows. For any $f \in \operatorname{Hom}_{\mathbb{Z}H}(X_n, M)$ and $x \in X_n$, let

$$(\tau f)(x) = \sum_{i=1}^{k} t_i^{-1} \cdot f(t_i \cdot x).$$

It can be shown that τ is independent of the choice of transversal and is a homomorphism satisfying $\tau\, d_n^* = d_n^* \tau$. So τ induces a homomorphism $\tau^* : H^n(H, M) \to H^n(G, M)$; this is the transfer map from $H^n(H, M)$ to $H^n(G, M)$.

With the definitions and notations established above, the following theorem of Gaschütz and Eckmann holds:

Theorem 7.13 *Let H be a subgroup of finite index k in a group G. Then, for any element $\hat{f}$ of $H^n(G, M)$, we have*

$$\tau^*(\rho^*\hat{f}) = k\hat{f} \ .$$

Proof: Let $\hat{f} = f + \text{im } d_n^*$ for some $f \in \ker d_{n+1}^*$. Then $\tau^*(\rho^*\hat{f}) = \tau(\rho f) + \text{im } d_n^*$.

Now for any $x \in X_n$,

$$\tau(\rho f)(x) = \sum_{i=1}^{k} t_i^{-1} \cdot \rho f(t_i \cdot x)$$

$$= \sum_{i=1}^{k} t_i^{-1} \cdot f(t_i \cdot x)$$

$$= \sum_{i=1}^{k} t_i^{-1} \cdot (t_i \cdot f(x)) \quad \text{since } f \in \text{Hom}_{\mathbb{Z}G}(X_n, M)$$

$$= \sum_{i=1}^{k} f(x)$$

$$= k f(x).$$

Consequently $\tau^*(\rho^*\hat{f}) = k\hat{f}$.

Keeping to the notation of Theorem 7.13 we have the following corollaries.

Corollary 7.14 *Every element $\hat{f}$ of $H^n(G, M)$ that is contained in $\ker \rho^*$ satisfies $k\hat{f} = 0$.*

Corollary 7.15 *If G has finite order then every element $\hat{f}$ of $H^n(G, M)$ has finite order dividing $|G|$.*

Proof: Let H be the identity subgroup of G. Now

$$\cdots \to 0 \to 0 \to 0 \to \mathbb{Z} \to \mathbb{Z} \to 0$$

is a free resolution $\{X_i\}$ of the trivial $\mathbb{Z}H$-module. So $\text{Hom}_{\mathbb{Z}H}(X_n, M) = \text{Hom}_{\mathbb{Z}H}(0, M) = \{0\}$, and hence $H^n(H, M) = \{0\}$, for any positive integer n. Therefore by Theorem 7.13 we get that $|G|\hat{f} = 0$ for all $\hat{f} \in H^n(G, M)$.

Corollary 7.16 *Let G be a finite group and let P be a Sylow p-subgroup of G for some prime p. Then $H^n(G, M)_{(p)}$, the p-primary component of $H^n(G, M)$, is isomorphic to a subgroup of $H^n(P, M)$.*

Proof: Since P is a Sylow p-subgroup of G, we have that $\gcd(|G:P|, p) = 1$. Further, by Corollary 7.14, for all $\hat{f}$ in $\ker \rho^*$, we have $(|G:P|)\hat{f} = 0$. So $\ker \rho^*$ has no p-elements other than the identity. Consequently, ρ^* restricted to $H^n(G, M)_{(p)}$ is an injective homomorphism from $H^n(G, M)_{(p)}$ to $H^n(P, M)$.

The following corollary is due to Schur.

Corollary 7.17 *Let G be a group and let M be a finite $\mathbb{Z}G$-module. Suppose that $\gcd(|G|, |M|) = 1$. Let E be an extension of M by G realising the action of G on M. Then E is equivalent to a semidirect product of M by G (and so M has a complement in E). Moreover, all the complements to M in E are conjugate.*

Proof: Corollary 7.15 implies that every element of $H^n(G, M)$ has order dividing $|G|$. But every element of $H^n(G, M)$ has order dividing $|M|$ (since $H^n(G, M)$ is a quotient of a group $Z^n(G, M)$ of functions to M under point-wise addition). This implies that every element of $H^n(G, M)$ has order dividing $\gcd(|G|, |M|)$. Since $\gcd(|G|, |M|) = 1$ we find that $H^n(G, M) = 0$. In particular, since $H^2(G, M) = \{0\}$, Theorem 7.5 implies that every extension E of M by G is equivalent to the semidirect product of M by G. Since $H^1(G, M) = \{0\}$, Theorem 7.8 implies that all the complements to M in E are conjugate. So the corollary follows.

We remark that the conditions of Corollary 7.17 may be weakened to the case when E is an extension of a group M by a group G with $\gcd(|G|, |M|) = 1$ and where one of G and M is soluble. Indeed, the proof of Corollary 7.17 is the main part of the proof of this more general result. By the Feit–Thompson Theorem [33], we can then drop the assumption that one of G and M is soluble.

7.4 The McIver and Neumann bound

In this section we shall discuss a bound (Theorem 7.18) proved by McIver and Neumann [67] on the number of isomorphism classes of finite A-groups. We shall, however, restrict ourselves to presenting the proof of the enumeration for soluble A-groups. Let $f_{A, \text{sol}}(n)$ denote the number of soluble A-groups of order n up to isomorphism. Recall that, for any positive integer n, the integer $\lambda = \lambda(n)$ is defined to be the number of prime divisors of n including multiplicities.

Theorem 7.18 *We have that $f_{A, \text{sol}}(n) \leqslant n^{\lambda+1}$.*

Proof: Let G be a soluble A-group of order n and let $n = \prod_{i=1}^k p_i^{\alpha_i}$ be the prime factorisation of n. Let M be a minimal normal subgroup. Note that since G is soluble, M is elementary abelian. Define $q = |M|$, where $q = p^\alpha$ and $p = p_i$ for some $i \in \{1, \ldots, k\}$. Let $H = G/M$. Since M is a minimal normal subgroup of G, we find that it is an irreducible $\mathbb{F}_p H$-module.

Thus G is determined by a quadruple (q, H, M, E) where q is a p-power divisor of n for some prime p, H is a soluble A-group of order n/q, M is an irreducible $\mathbb{F}_p H$-module of order q and E is a cohomology class corresponding to an extension of M by H whose Sylow p-subgroups are abelian.

Given q, the number of possibilities for H is $f_{A,\mathrm{sol}}(n/q)$ and as inductive hypothesis we assume that

$$f_{A,\mathrm{sol}}(n/q) \leqslant \left(\frac{n}{q}\right)^{\lambda(n/q)+1} = \left(\frac{n}{q}\right)^{\lambda-\alpha+1}.$$

Up to isomorphism, there are at most $|H|$ irreducible $\mathbb{F}_p H$-modules. This is because of the following. Since an irreducible module is cyclic, it is a homomorphic image of $\mathbb{F}_p H$ (as a module). By the Jordan–Hölder theorem, the number of simple composition factors of $\mathbb{F}_p H$ is, up to isomorphism, at most $\dim \mathbb{F}_p H$, which is $|H|$. Thus once we have made our choice of H there are at most n/q choices for M.

Let $H^2(H, M)$ denote the second cohomology group which counts the equivalence classes of extensions of M by H and let $H_A^2(H, M)$ denote the subset of $H^2(H, M)$ corresponding to those extensions which are A-groups. Let Q be a Sylow p-subgroup of H. If Q does not centralise M then $H_A^2(H, M) = \emptyset$. So assume that Q centralises M. Keeping to the notation established in Section 7.1, it is not difficult to see that an element $f + B^2(H, M)$ of $H^2(H, M)$ belongs to $H_A^2(H, M)$ if and only if $f(g, h) = f(h, g)$ for all h, g that are elements of any Sylow p-subgroup of H. In other words, an extension E of M by H is an A-group if and only if a factor set f determined by E satisfies the condition given above. The condition also clearly ensures that $H_A^2(H, M) \leqslant H^2(H, M)$.

Consider the standard bar resolution for the trivial $\mathbb{Z}H$-module $\mathbb{Z}$. Since $\mathrm{Hom}_{\mathbb{Z}H}(B_n, M)$ is a group of functions into an abelian group of exponent p, we find that $\mathrm{Hom}_{\mathbb{Z}H}(B_n, M)$ is abelian of exponent p. Since a function in $\mathrm{Hom}_{\mathbb{Z}H}(B_n, M)$ is determined by its values on a (finite) free generating set of B_n, we find that $\mathrm{Hom}_{\mathbb{Z}H}(B_n, M)$ is a finite p-group, and so $H^2(H, M)$ is a finite p-group. By Corollary 7.16, the restriction map induces an injection from $H^2(H, M)$ to $H^2(Q, M)$. Moreover, this map takes an element of $H_A^2(H, M)$ to an element of $H_A^2(Q, M)$, and so $|H_A^2(H, M)| \leqslant |H_A^2(Q, M)|$. But $H_A^2(Q, M)$ is simply the group $\mathrm{Ext}(Q, M)$.

Let i be such that $p = p_i$, and define $\beta = \alpha_i$. Then $|\mathrm{Ext}(Q, M)| \leqslant q^{\beta-\alpha}$. We can indicate why this is true as follows. Let d be the number of generators of Q. We may present Q as

$$Q = \langle x_1, \ldots, x_d \mid x_1^{m_1} = x_2^{m_2} = \ldots = x_d^{m_d} = 1 \rangle$$

in the category of abelian groups. An abelian extension P of M by Q can be presented in the form

$$P = \langle y_1, \ldots, y_d, M \mid y_1^{m_1} = z_1, \ldots, y_d^{m_d} = z_d \rangle$$

where $z_1, \ldots, z_d \in M$. Clearly there are at most $|M|$ possibilities for each of the z_i, and so we get that $|\mathrm{Ext}(Q, M)| \leqslant |M|^d \leqslant q^{\beta-\alpha} \leqslant q^{\lambda-\alpha}$.

Now if we put these estimates together, we see that the number of quadruples (q, H, M, E) as specified above is at most

$$\sum_q f_{A, \mathrm{sol}}(n/q) \left(\frac{n}{q} \right) q^{\lambda-\alpha}$$

where q runs over the prime-power divisors of n. So

$$f_{A, \mathrm{sol}}(n) \leqslant \sum_q \left(\frac{n}{q} \right)^{\lambda-\alpha+1} \left(\frac{n}{q} \right) q^{\lambda-\alpha} < n^{\lambda+1} \sum_q \frac{1}{q^2}$$

$$< \left(\frac{\pi^2}{6} - 1 \right) n^{\lambda+1} \tag{7.8}$$

$$\leqslant n^{\lambda+1},$$

and the theorem follows.

Note 7.19 *In fact, $f_A(n) \leqslant n^{\lambda+1}$. This can be shown by estimating the number of A-groups which have no non-trivial abelian normal subgroups. It was shown in [67] that the number of such A-groups of order n (up to isomorphism) is less than $\frac{1}{4} n^{\lambda+1}$. Combining this bound with (7.8) we get the required result. Later we shall see that in fact there are bounds of the form $f_A(n) \leqslant n^{a\mu+b}$, where $\mu = \mu(n)$. See Corollary 16.22.*

Note 7.20 *By counting groups that are extensions of an elementary abelian p-group by an elementary abelian q-group for certain fixed primes p and q, we will show in Theorem 18.4 that there exists $c > 0$ such that*

$$f_{A, \mathrm{sol}}(n) > n^{c\mu}$$

for infinitely many n, with arbitrarily large μ. The discussion at the end of Section 18.1 shows that we may take c to be very close to $3 - 2\sqrt{2}$.

8

Some representation theory

In Chapters 9 and 13 we will develop some of the structure theory of soluble subgroups of the general linear group. The aim of this chapter is to review the representation theory that we need for this task.

We assume that the reader is familiar with some of the basics of representation theory. In particular, we assume that the reader knows about such objects as algebras and modules, has understood the notions of indecomposable, irreducible and completely reducible modules and the connection between representations of a group and modules of the corresponding group algebra. All this material may be found in Chapter II of Curtis and Reiner [20]; see also Collins [18] or James and Liebeck [53]. We also assume that the reader has met the theory of semisimple algebras and some of the theory of induced modules. The first two sections of this chapter are intended as reminders of the structure theory of semisimple algebras and of Clifford's theorem respectively. We omit the proofs of many standard results in these two sections. For a full treatment, see Chapters IV and VII of Curtis and Reiner [20]. The final two sections contain proofs of two more advanced results that we require, namely the Skolem–Noether theorem and Wedderburn's theorem that a finite skew field is a field. For the most part, these sections follow the approach given in Cohn [17, Section 7.1].

8.1 Semisimple algebras

Let F be a field, and let R be a finite-dimensional F-algebra. From now on, we will assume that all vector spaces are finite-dimensional and refer to a finite-dimensional F-algebra simply as an F-algebra. We say that an ideal I of R is *nilpotent* if $I^k = \{0\}$ for some positive integer k. In other words, I is nilpotent if there exists a positive integer k such that

$$r_1 r_2 \cdots r_k = 0 \text{ for all } r_1, r_2, \ldots, r_k \in I.$$

We define the *radical* rad$R$ of R to be the sum of all nilpotent left ideals of R. The radical is, in fact, a nilpotent two-sided ideal of R and $R/(\text{rad}R)$ has zero radical (see [20, Theorem 24.4]).

An F-algebra R is called *semisimple* if rad$R = \{0\}$. By the statement at the end of the previous paragraph, $R/(\text{rad}R)$ is always semisimple. An F-algebra R is semisimple if and only if every R-module is completely reducible (see [20, (25.8)]). We now state the main structure theorem for semisimple algebras. Here, and in future, by an R-module we mean a left R-module. Recall that a non-trivial F-algebra R is simple if it has no two-sided ideals other than $\{0\}$ and R.

Theorem 8.1 *Let R be a semisimple F-algebra. Then*

(i) *There are finitely many isomorphism classes of irreducible R-modules. Let r be the number of such isomorphism classes and let $M_1, M_2, \ldots, M_r$ be irreducible R-modules that are pairwise non-isomorphic.*

(ii) *For any $i \in \{1, 2, \ldots, r\}$, define S_i to be the sum of all minimal left ideals of R that are isomorphic (as R-modules) to M_i. Then S_i is a simple F-algebra. Moreover, S_i is a two-sided ideal in R and*

$$R = S_1 \oplus S_2 \oplus \cdots \oplus S_r. \tag{8.1}$$

We call S_i the simple component of R corresponding to M_i.

(iii) *For all $i, j \in \{1, 2, \ldots, r\}$,*

$$S_i M_j = \begin{cases} \{0\} & \text{if } i \neq j, \\ M_j & \text{if } i = j. \end{cases}$$

Theorem 8.1 parts (i) and (ii) follow directly from Theorems 25.10 and 25.15 of Curtis and Reiner [20]. To show part (iii) of the theorem, observe that M_j is isomorphic to a minimal left ideal L of R contained in S_j. But now (8.1) implies that $S_i L = \{0\}$ whenever $i \neq j$ and so $S_i M_j = \{0\}$ when $i \neq j$. Since M_j is an irreducible R-module, $R M_j \neq \{0\}$ and so $S_j M_j \neq \{0\}$. But $S_j M_j$ is an R-submodule of M_j since S_j is a left ideal of R. Hence, since M_j is irreducible, $S_j M_j = M_j$.

This structure theorem allows us to prove the following proposition, which is part of Exercise 26.2 in Curtis and Reiner [20].

Proposition 8.2 *Let R be an F-algebra. If there exists a faithful irreducible R-module M, then R is simple.*

Proof: Suppose M is a faithful irreducible R-module. Since $\mathrm{rad}\,R$ is a left ideal of R, we find that $(\mathrm{rad}\,R)M$ is a submodule of M. Since M is irreducible, $(\mathrm{rad}\,R)M = \{0\}$ or $(\mathrm{rad}\,R)M = M$. But the second case cannot occur, since by the nilpotency of $\mathrm{rad}\,R$ we would then have that

$$M = (\mathrm{rad}\,R)M = (\mathrm{rad}\,R)^2 M = \cdots = \{0\}M = 0.$$

Thus $(\mathrm{rad}\,R)M = \{0\}$. This implies, since M is faithful, that $\mathrm{rad}\,R = \{0\}$ and so R is semisimple.

Define the integer r, the irreducible R-modules M_i and the simple components S_i of R as in Theorem 8.1. Suppose, for a contradiction, that $r > 1$. Now, Theorem 8.1 (i) implies that M is isomorphic to M_j for some $j \in \{1, 2, \ldots, r\}$. Let $i \in \{1, 2, \ldots, r\} \setminus \{j\}$. Then, by Theorem 8.1 (iii), $S_i M = \{0\}$. Since M is faithful, this implies that $S_i = \{0\}$, which is a contradiction. Hence $r = 1$ and R is simple, as required.

Theorem 8.1 reduces the structure theory of semi-simple F-algebras to the case when S is a simple F-algebra. The following theorem of Wedderburn examines this case. Recall that a skew field is an object satisfying the axioms of a field except that multiplication is not necessarily commutative.

Theorem 8.3 *Let S be a simple F-algebra. Then there exists a unique positive integer k such that S is isomorphic to the ring $\mathcal{M}_k(T)$ of $k \times k$ matrices over a skew field T. The isomorphism class of T is uniquely determined by S.*

This theorem appears as Theorem 26.4 in Curtis and Reiner [20].

Let S be a simple F-algebra; so $S \cong \mathcal{M}_k(T)$ for some skew field T and positive integer k. Let $I_k \in \mathcal{M}_k(T)$ be the $k \times k$ identity matrix. It is easy to prove that the centre of $\mathcal{M}_k(T)$ is isomorphic to the centre of T, where an element d in the centre of T corresponds to the scalar matrix $dI_k \in \mathcal{M}_k(T)$. Since F lies in the centre of S, we find that F is isomorphic to a subfield of the centre of T.

8.2 Clifford's theorem

This section aims to give a brief reminder of Clifford's theorem, a theorem which describes the FN-module structure of an irreducible FG-module, where N is a normal subgroup of a finite group G.

Theorem 8.4 (Clifford's theorem) *Let N be a normal subgroup of a finite group G and let V be an irreducible FG-module for some field F. Then*

$$V = Y_1 \oplus \cdots \oplus Y_k$$

as an FN-module, where the subspaces Y_i are permuted transitively under the action of G; there exist a multiplicity m and non-isomorphic irreducible FN-modules $X_1, \ldots, X_k$ such that

$$Y_i \cong mX_i$$

(where $mX_i = X_i \oplus \cdots \oplus X_i$ (m copies)).

Clifford's theorem is proved in Section 49 of Curtis and Reiner [20]. We provide a sketch proof of the theorem.

Sketch proof: Choose a minimal FN-submodule X of V. Then gX is an FN-submodule of V for any g in G by normality of N. (Note that $h(gX) = g(g^{-1}hg)X$ for h in N.) Now $\sum_{g \in G} gX$ is a G-invariant submodule of V. So by the irreducibility of V we have that

$$V = \sum_{g \in G} gX.$$

Since gX is an irreducible FN-submodule, V may be expressed as a direct sum of some of the gX. Let $X_1, \ldots, X_k$ be the non-isomorphic modules occurring among the gX and for each i let Y_i be the homogeneous FN-submodule of V which is the sum of the gX that are isomorphic to X_i. Then

$$V = Y_1 \oplus \cdots \oplus Y_k.$$

It is not difficult to see that the subspaces $Y_1, \ldots, Y_k$ are permuted transitively under the action of G. Therefore the Y_i have the same dimension and so we get a multiplicity m such that $Y_i \cong mX_i$ for $i = 1, \ldots, k$.

8.3 The Skolem–Noether theorem

This section aims to prove some results concerning simple F-algebras that we will need in the chapters to come. In particular, we aim to prove the Skolem–Noether theorem which asserts that isomorphic simple F-subalgebras of certain F-algebras are conjugate. We also prove a theorem to the effect that every automorphism of a simple F-algebra that fixes its centre is inner. The proofs in this section make use of the notion of a tensor product $R \otimes S$ of F-algebras R and S. We begin with a brief reminder of this notion.

Informally, the elements of $R \otimes S$ are F-linear combinations of elements of the form $r \otimes s$ where $r \in R$ and $s \in S$ with the rules that

$$r \otimes (s_1 + s_2) = r \otimes s_1 + r \otimes s_2 \text{ for all } r \in R \text{ and all } s_1, s_2 \in S$$

$$(r_1 + r_2) \otimes s = r_1 \otimes s + r_2 \otimes s \text{ for all } r_1, r_2 \in R \text{ and all } s \in S \text{ and}$$

$$(\alpha r) \otimes s = r \otimes (\alpha s) = \alpha(r \otimes s) \text{ for all } r \in R, s \in S \text{ and } \alpha \in F.$$

Multiplication in $R \otimes S$ is the bilinear operation such that

$$(r_1 \otimes s_1)(r_2 \otimes s_2) = (r_1 r_2 \otimes s_1 s_2)$$

for all $r_1, r_2 \in R$ and $s_1, s_2 \in S$. For a proof that this multiplication is well-defined, see Curtis and Reiner [20] Section 12, especially page 72. We now summarise some of the basic properties of the tensor product of F-algebras that we will use in this section.

If $x_1, x_2, \ldots, x_k$ and $y_1, y_2, \ldots, y_\ell$ are F-bases of R and S respectively, then an F-basis of $R \otimes S$ is given by $x_i \otimes y_j$ where $1 \leqslant i \leqslant k$ and $1 \leqslant j \leqslant \ell$. Thus $\dim_F R \otimes S = (\dim_F R)(\dim_F S)$. Using the basis $x_i \otimes y_j$, it is easy to show that any element $z \in R \otimes S$ may be written uniquely in the form

$$z = \sum_{i=1}^{k} x_i \otimes s_i$$

for some $s_1, s_2, \ldots, s_k \in S$.

The F-subalgebras $R \otimes 1$ and $1 \otimes S$ of $R \otimes S$ are isomorphic to R and S respectively; these subalgebras together generate $R \otimes S$ as an F-algebra. Let M be both an R-module and an S-module, and suppose the actions commute, so

$$r(sm) = s(rm) \text{ for all } r \in R, s \in S \text{ and } m \in M.$$

Then M is naturally an $R \otimes S$-module by defining

$$(r \otimes s)m = r(sm)$$

for all $r \in R, s \in S$ and $m \in M$. One important special case of this construction is as follows. We have that R is naturally a (left) R-module. Moreover, R is also a right R-module, and hence a left R^o-module. (Recall that R^o, the opposite algebra of R, has the same underlying set and addition as R but with multiplication $*$ defined by $r_1 * r_2 = r_2 r_1$ for all $r_1, r_2 \in R$.) The actions of R and R^o on R commute since multiplication is associative, so R is an $R \otimes R^o$-module. The algebra $R \otimes R^o$ is known as the enveloping algebra R^e of R. Our first proposition characterises $R \otimes R^o$ and its action on R, in the case when R is a simple F-algebra with centre F.

Proposition 8.5 *Let R be a simple F-algebra, and suppose the centre of R is F. Let $k = \dim_F R$. Then the homomorphism from the enveloping algebra R^e to $\mathrm{End}_F(R)$ induced by the natural action of R^e on R is an isomorphism. In particular, $R^e \cong \mathcal{M}_k(F)$.*

Proof: It is enough to show that the homomorphism from R^e to $\mathrm{End}_F(R)$ is surjective, for the fact that $\dim_F R^e = \dim_F \mathcal{M}_k(F) = k^2$ then implies we have an isomorphism.

Let $x_1, x_2, \ldots, x_k$ be an F-basis for R. Let $\ell \in \{1, 2, \ldots, k\}$ and let $y_1, y_2, \ldots,$ $y_\ell \in R$. We claim that there exists $\phi \in R^e$ mapping x_i to y_i for all $i \in \{1, 2, \ldots, \ell\}$. The case $\ell = k$ of our claim implies that the homomorphism from R^e to $\mathrm{End}_F(R)$ is onto, as required.

Consider the case when $\ell = 1$. Now, Rx_1R is a non-zero two-sided ideal of R, and so $Rx_1R = R$ as R is simple. So $y_1 \in R = Rx_1R$ thus $y_1 = \sum_{j=1}^n r_j x_1 r'_j = \sum_{j=1}^n (r_j \otimes r'_j)x_1$, and so we may take $\phi = \sum_{j=1}^n (r_j \otimes r'_j)$ in this case. So our claim holds when $\ell = 1$.

Suppose that $\ell > 1$ and our claim holds for all smaller values of ℓ. So there exist $\psi_1, \psi_2, \ldots, \psi_{\ell-1} \in R^e$ such that for all $i, j \in \{1, 2, \ldots, \ell-1\}$

$$\psi_i x_j = \begin{cases} 1 & \text{if } i = j, \\ 0 & \text{otherwise.} \end{cases}$$

Define $z_1, z_2, \ldots, z_{\ell-1} \in R$ by $z_i = \psi_i x_\ell$.

We now show that there exists $\psi \in R^e$ such that $\psi x_1 = \psi x_2 = \cdots = \psi x_{\ell-1} = 0$ and such that $\psi x_\ell \neq 0$. We consider two cases. Firstly, suppose that $z_u \notin F$ for some $u \in \{1, 2, \ldots, \ell-1\}$. Then (since F is the centre of R) there exists $r \in R$ that does not commute with z_u and we may take $\psi = (r \otimes 1)\psi_u - (1 \otimes r)\psi_u$. Secondly, suppose that $z_i \in F$ for all $i \in \{1, 2, \ldots, \ell-1\}$. Then we may define $\psi = \sum_{i=1}^{\ell-1} (x_i \otimes 1)\psi_i - (1 \otimes 1)$. Here x_ℓ is mapped to a non-zero element by ψ because the x_i are linearly independent over F.

Let $\psi' \in R^e$ map ψx_ℓ to 1. Such an element exists, by the argument used for the case when $\ell = 1$. Let $\phi_\ell = \psi'\psi$ and define $\phi_1, \phi_2, \ldots, \phi_{\ell-1}$ by $\phi_i = \psi_i - (z_i \otimes 1)\phi_\ell$. It is easy to check that for all $i, j \in \{1, 2, \ldots, \ell\}$,

$$\phi_i x_j = \begin{cases} 1 & \text{if } i = j, \\ 0 & \text{if } i \neq j. \end{cases}$$

But then setting $\phi = \sum_{i=1}^\ell (y_i \otimes 1)\phi_i$ we find that $\phi x_i = y_i$ for all $i \in \{1, 2, \ldots, \ell\}$, as required. Our claim, and hence the proposition, now follows by induction on ℓ.

Proposition 8.6 *Let R and S be simple F-algebras. Suppose that the centre of R is F. Then $R \otimes S$ is a simple F-algebra.*

Proof: Let I be a two-sided ideal in $R \otimes S$, and suppose that there exists $z \in I$ that is non-zero. We must show that $I = R \otimes S$. Let $x_1, x_2, \ldots, x_k$ be a basis for R. We may write $z = \sum_{i=1}^{k} x_i \otimes s_i$ for some elements $s_i \in S$. Since $z \neq 0$, there exists $u \in \{1, 2, \ldots, k\}$ such that $s_u \neq 0$. Let ϕ be the linear transformation of R mapping x_u to 1 and mapping x_i to 0 if $i \neq u$. By Proposition 8.5, there is an element of R^e that induces this linear transformation. Hence there exist $r_1, r'_1, r_2, r'_2, \ldots, r_k, r'_k \in R$ such that

$$\sum_{i=1}^{k} r_i x_j r'_i = \begin{cases} 1 & \text{if } j = u, \\ 0 & \text{if } j \neq u. \end{cases} \qquad (8.2)$$

Let $y \in R \otimes S$ be defined by

$$y = \sum_{i=1}^{k} (r_i \otimes 1) z (r'_i \otimes 1).$$

Since I is a two-sided ideal, $y \in I$. But (8.2) implies that $y = 1 \otimes s_u$ and so y is a non-zero element of $(1 \otimes S) \cap I$. So $I \cap (1 \otimes S)$ is a non-zero two-sided ideal of $1 \otimes S$. Since $1 \otimes S$ is isomorphic to the simple F-algebra S, $1 \otimes S \subseteq I$. But, since I is a left ideal,

$$R \otimes S = (R \otimes 1)(1 \otimes S) \subseteq (R \otimes 1)I = I.$$

Hence $I = R \otimes S$ and so $R \otimes S$ is simple, as required.

Theorem 8.7 *Let R and S be simple F-algebras, and suppose that the centre of R is F. Let $\theta_1, \theta_2 : S \to R$ be F-algebra homomorphisms. Then there exists a unit $u \in R$ such that*

$$\theta_2(s) = u^{-1}(\theta_1(s))u \text{ for all } s \in S. \qquad (8.3)$$

We recall that the definition of an F-algebra homomorphism means that θ_i must map $1 \in S$ to $1 \in R$.

Proof: We may regard R as an $R \otimes S^o$-module V_1 by defining

$$(r \otimes s)m = rm(\theta_1(s)) \text{ for all } r, m \in R \text{ and all } s \in S$$

and extending the action by linearity to the whole of $R \otimes S^o$. Similarly, we may regard R as an $R \otimes S^o$-module V_2 by defining

$$(r \otimes s)m = rm(\theta_2(s)) \text{ for all } r, m \in R \text{ and all } s \in S.$$

We claim that $V_1 \cong V_2$ as $R \otimes S^o$-modules. Now, S^o is a simple F-algebra since S is a simple F-algebra. By Proposition 8.6, $R \otimes S^o$ is a simple F-algebra. By Theorem 8.1, there is a unique irreducible $R \otimes S^o$-module V. Since $R \otimes S^o$ is semisimple, every $R \otimes S^o$-module is completely reducible. So

$$V_1 \cong k_1 V \text{ and } V_2 \cong k_2 V$$

for some positive integers k_1 and k_2. But the F-dimensions of V_1 and V_2 are equal, and so $k_1 = k_2$ and therefore $V_1 \cong V_2$. This proves our claim.

Let $\phi : V_1 \to V_2$ be an $R \otimes S^o$-module isomorphism from V_1 to V_2. So for all $r, m \in R$ and all $s \in S$,

$$\phi(rm(\theta_1(s))) = r(\phi(m))(\theta_2(s)). \tag{8.4}$$

Define $u = \phi(1)$. Then (8.4) with $m = 1$ and $s = 1$ implies that $\phi(r) = ru$ for all $r \in R$. When $r = \phi^{-1}(1)$, this equality shows that u is a unit. Setting $r = 1$ and $m = 1$ in (8.4) we find that

$$\theta_1(s)u = u(\theta_2(s))$$

for all $s \in S$. Thus Equation (8.3) holds and the theorem is established.

We finish this section by proving two corollaries. Corollary 8.8 will be used in the next section, and Corollary 8.9 will be used in Chapter 13.

Corollary 8.8 *Let R be a simple F-algebra with centre F. Let S_1 and S_2 be isomorphic simple F-subalgebras of R. Then there exists a unit $u \in R$ such that $S_2 = u^{-1}S_1 u$.*

Proof: Let $\alpha : S_1 \to S_2$ be an isomorphism of F-algebras. The corollary follows from Theorem 8.7, where we take $\theta_1 : S_1 \to R$ to be the inclusion mapping for S_1, and $\theta_2 : S_1 \to R$ to be the composition of α with the inclusion mapping for S_2.

Corollary 8.9 (Skolem–Noether) *Let R be a simple F-algebra with centre F. Let α be an F-algebra automorphism of R. Then α is an inner automorphism.*

Proof: The corollary follows by applying Theorem 8.7 in the case when $S = R$, the homomorphism θ_1 is the identity mapping and $\theta_2 = \alpha$.

8.4 Every finite skew field is a field

This section aims to prove Wedderburn's theorem that a finite skew field is a field. The theorem is proved at the end of Section 68 of Curtis and Reiner [20], using properties of cyclotomic polynomials. We give a different proof, based on Corollary 8.8 above. We first prove a result that gives structural information about skew fields that are not necessarily finite (but are finite-dimensional over their centre).

Proposition 8.10 *Let T be a skew field with centre F. Let K be a maximal subfield of T. Then $\dim_F T = (\dim_F K)^2$.*

Proof: Let $k = \dim_F K$. We must show that $\dim_F T = k^2$.

Consider the algebra $\mathcal{M}_k(T)$ of $k \times k$ matrices over T. Note that $\mathcal{M}_k(T)$ is a simple F-algebra with centre F. We begin by exhibiting two subalgebras S_1 and S_2 of $\mathcal{M}_k(T)$ that are isomorphic to K, and computing their centralisers in $\mathcal{M}_k(T)$.

Let $I_k \in \mathcal{M}_k(T)$ be the $k \times k$ identity matrix. Let $S_1 = \{aI_k : a \in K\}$. Then S_1 is an F-subalgebra of $\mathcal{M}_k(T)$ that is isomorphic to K. Since K is a maximal subfield of T, we have that K is its own centraliser in T. Using this fact, it is not difficult to check that the centraliser of S_1 in $\mathcal{M}_k(T)$ is $\mathcal{M}_k(K)$.

We may regard K as a k-dimensional vector space over F. We choose an F-basis $a_1, a_2, \ldots, a_k$ of K such that $a_1 = 1$. Every element of K gives rise to an F-linear transformation on K by associating $a \in K$ with the map $x \mapsto ax$. We may represent this map by a $k \times k$ matrix $\phi(a)$ with respect to the basis $a_1, a_2, \ldots, a_k$. If we write e_i for the ith standard basis vector, we have that $\phi(a_i)e_1 = e_i$. The map $\phi : K \to \mathcal{M}_k(F) \subseteq \mathcal{M}_k(T)$ is an injective homomorphism of F-algebras. Define $S_2 = \phi(K)$. Then S_2 is an F-subalgebra of $\mathcal{M}_k(T)$ that is isomorphic to K.

We claim that the centraliser of S_2 in $\mathcal{M}_k(T)$ is the set $\langle S_2 \rangle_T$ of all T-linear combinations of elements in S_2. Note that any two matrices in S_2 commute. Since all the elements of S_2 are matrices over F, and F is contained in the centre of T, it is clear that $\langle S_2 \rangle_T$ is contained in the centraliser.

Let c be an element of the centraliser of S_2 in $\mathcal{M}_k(T)$. We must show that $c \in \langle S_2 \rangle_T$. We remarked above that, writing e_i for the ith standard basis vector, $\phi(a_i) \in S_2$ maps e_1 to e_i. So for any vector $v = (t_1, t_2, \ldots, t_k)^{\text{tr}} \in T^k$ there exists an element of $\langle S_2 \rangle_T$ that maps e_1 to v, namely the matrix $\sum_{i=1}^k t_i \phi(a_i)$. Hence there exists $d \in \langle S_2 \rangle_T$ such that $ce_1 = de_1$. But now, for all $i \in \{1, 2, \ldots, k\}$,

$$ce_i = c\phi(a_i)e_1 = \phi(a_i)ce_1 \qquad \text{(since c centralises S_2)}$$

$$= \phi(a_i)de_1 = d\phi(a_i)e_1 \qquad \text{(since d centralises S_2)}$$

$$= de_i,$$

and so $c = d \in \langle S_2 \rangle_T$. Thus the centraliser of S_2 in $\mathcal{M}_k(T)$ is $\langle S_2 \rangle_T$, and so our claim follows.

Since $S_1 \cong K \cong S_2$, and since both S_1 and S_2 are simple F-subalgebras of $\mathcal{M}_k(T)$, Corollary 8.8 implies that S_1 and S_2 are conjugate, and hence their centralisers are conjugate. In particular, the centralisers of S_1 and S_2 have the same F-dimension. Now, $\dim_F \mathcal{M}_k(K) = k^2 \dim_F K = k^3$ and $\dim_F \langle S_2 \rangle_T = k \dim_F T$ (since $\phi(a_1), \phi(a_2), \ldots, \phi(a_k)$ form a T-basis for $\langle S_2 \rangle_T$). Hence $k^3 = k \dim_F T$, and so $\dim_F T = k^2$, as required.

Theorem 8.11 (Wedderburn) *Let T be a finite skew field. Then T is a field.*

Proof: Suppose, for a contradiction, that T is not a field. Let F be the centre of T. Then F is a finite field of order q, where q is a power of some prime. By Proposition 8.10 the F-dimension $\dim_F T$ of T is a square: k^2 say, where $k > 1$. Moreover, every maximal subfield of T has F-dimension k. Hence any maximal subfield of T is isomorphic to $\mathbb{F}_{q^k}$ and so Corollary 8.8 implies that the maximal subfields of T are all conjugate. Let K be a maximal subfield of T. The number of conjugates of K is at most the index $(q^{k^2} - 1)/(q^k - 1)$ of the centraliser of $K \setminus \{0\}$ in the multiplicative group $T \setminus \{0\}$ of T. But then, considering the union of the maximal subfields L of T, we find that

$$|\textstyle\bigcup_L L| = q + |\textstyle\bigcup_L L \setminus F| \leqslant q + (q^k - q)(q^{k^2} - 1)/(q^k - 1) < q^{k^2},$$

since $k > 1$. So there exists an element $x \in T$ that is not contained in a subfield of T. But x is contained in the subfield $F(x)$ of T. This contradiction establishes the theorem.

9

Primitive soluble linear groups

This chapter proves some basic facts about soluble subgroups of a general linear group over a finite field. Our approach parallels that of D. A. Suprunenko [88]. Throughout this chapter $\mathbb{F}_q$ is a finite field with q elements. We take G to be a finite soluble group and V to be a primitive $\mathbb{F}_q G$-module which is faithful as a G-module.

In fact, all the results of this chapter are valid for any field if the final assertion of Proposition 9.1 is dropped and if the word 'cyclic' in Proposition 9.4 is replaced by 'abelian'.

We fix the following notation:

- V – an $\mathbb{F}_q G$-module on which G acts faithfully and primitively. We identify G with a subset of $\mathrm{End}_{\mathbb{F}_q} V$, the algebra of all $\mathbb{F}_q$-linear transformations on V;
- d – the dimension of V over $\mathbb{F}_q$;
- A – a subgroup of G that is maximal subject to being abelian and normal in G;
- K – the $\mathbb{F}_q$-subalgebra $\langle A \rangle_{\mathbb{F}_q}$ of $\mathrm{End}_{\mathbb{F}_q} V$ generated by A;
- X – an irreducible $\mathbb{F}_q A$-submodule of V (which is unique up to isomorphism, by Clifford's theorem);
- d_1 – the $\mathbb{F}_q$-dimension of X;
- d_2 – the multiplicity of X in the $\mathbb{F}_q A$-module V (by Clifford's theorem we have that $V \cong d_2 X$, and so $d_2 = d/d_1$);
- C – the centraliser $C_G(A)$ of A in G;
- B – a subgroup of C that is maximal with respect to B/A being an abelian normal subgroup of G/A.

9.1 Some basic structure theory

Proposition 9.1 *The algebra K is isomorphic (as an $\mathbb{F}_q$-algebra) to the field $\mathbb{F}_{q^{d_1}}$. The subgroup A is cyclic.*

Proof: The $\mathbb{F}_q A$-module X is naturally a K-module. Indeed, X is an irreducible K-module, since $A \subseteq K$. Since V is faithful as a K-module, and since $V \cong d_2 X$, we find that X is faithful as a K-module. So K is an $\mathbb{F}_q$-algebra with a faithful irreducible representation, which means that K is simple by Proposition 8.2. Theorem 8.3 now implies that K is isomorphic as an $\mathbb{F}_q$-algebra to $\mathcal{M}_k(T)$ for some integer k and some skew field T. Since A is abelian, K is commutative. Thus T must be a field, and $k = 1$. But then $K \cong T$ and so K is a field.

Now, X is an irreducible K-module, and so is a vector space over K of dimension 1. Hence $\dim_{\mathbb{F}_q} K = \dim_{\mathbb{F}_q} X = d_1$ and K has order q^{d_1}. Finally, A is a multiplicative subgroup of the group of units of a finite field, and so A is cyclic.

Proposition 9.2 *The subgroup C of G may be realised as a subgroup of* $GL(d_2, K)$, *by regarding V as a K-module. From this perspective, the field K is identified with the set of scalar transformations on V.*

Proof: We observed above that V is a K-module – a vector space over K. Since $\dim_{\mathbb{F}_q} V = d$ we have that $\dim_K V = (\dim_{\mathbb{F}_q} V)/(\dim_{\mathbb{F}_q} K) = d_2$.

We remarked in the proof of Proposition 9.1 that $V \cong d_2 X$ as an A-module, where X is irreducible. So A, and therefore K, acts in the same way on each factor X in this decomposition. Hence K acts as the set of scalar transformations when V is regarded as a K-module.

Since C centralises A, and A is a spanning set for K, we have that C centralises K. Hence for all $g \in C$ and $z \in K$,

$$g(zu) = z(gu) \text{ for all } u \in V.$$

Moreover, $g(u + v) = gu + gv$ for all $u, v \in V$ since C acts $\mathbb{F}_q$-linearly on V. Thus C may be regarded as a group of K-linear transformations of V, and so C is isomorphic to a subgroup of $GL(d_2, q^{d_1})$.

Proposition 9.3 *The quotient G/C is isomorphic to a subgroup of the Galois group* $\mathrm{Gal}(K : \mathbb{F}_q)$.

Proof: We define a homomorphism $\phi : G \to \mathrm{Gal}(K : \mathbb{F}_q)$ by mapping $g \in G$ to the automorphism $\phi(g)$ such that $x \mapsto gxg^{-1}$ for all $x \in K$. We see that this map is well-defined as follows. Firstly, since K is spanned by A and A is normal in G, conjugation by G permutes the elements of K and so $\phi(g) : K \to K$ for all $g \in G$. Secondly, conjugation respects addition and multiplication, and so $\phi(g)$ is a field automorphism for all $g \in G$. Finally, $\mathbb{F}_q \subseteq$

K is represented by the scalar transformations in $\mathrm{End}_{\mathbb{F}_q} V$, and G centralises the scalar transformations associated with $\mathbb{F}_q$, so the elements of $\mathbb{F}_q$ are fixed by $\phi(g)$. Thus $\phi(g) \in \mathrm{Gal}(K : \mathbb{F}_q)$ for all $g \in G$.

Since K is spanned by A, the automorphism $\phi(g)$ is the identity if and only if $g \in C_G(A) = C$. Hence $\ker \phi = C$ and the proposition follows.

9.2 The subgroup B

Recall that the subgroup B is defined to be a subgroup of C that is maximal subject to B/A being abelian and normal in G/A. The aim of this section is to provide some structural information about B.

Proposition 9.4 *The subgroup B is nilpotent of class at most 2, and $Z(B) = A$. Let p be a prime dividing $|B|$ and let P be the Sylow p-subgroup of B. Then either P is cyclic or $P/Z(P)$ has exponent p and $|P'| = p$.*

We comment that if the word 'cyclic' is replaced by the word 'abelian' in the statement of Proposition 9.4, the resulting statement is true over any field.

Proof: Since $B \leqslant C$, the subgroup A is centralised by B and so $A \leqslant Z(B)$. But $Z(B)$, being a characteristic subgroup of a normal subgroup of G, is normal in G. Since $Z(B)$ is abelian, we must have $Z(B) = A$ by the maximality of A.

The quotient B/A is abelian by our choice of B, and so $B' \subseteq A = Z(B)$. Hence B is nilpotent of class at most 2.

Let p be a prime dividing $|B|$ and let P be the Sylow p-subgroup of B (we know that P is unique since B is nilpotent). Note that P is a characteristic subgroup of the normal subgroup B, and so P is normal in G. Moreover, $Z(P) = P \cap A$. If P is abelian, then $P = Z(P) \leqslant A$ and so P is cyclic. Suppose that P is not abelian; so P is nilpotent of class 2. We claim that the exponent of P' divides the exponent of $P/Z(P)$. Let the exponent of $P/Z(P)$ be p^ℓ. Since P is nilpotent of class 2, Equation (3.8) in Lemma 3.3 implies that

$$[x^i, y] = [x, y]^i = [x, y^i]$$

for all integers i and all $x, y \in P$. In particular, for all $x, y \in P$,

$$[x, y]^{p^\ell} = [x^{p^\ell}, y] = 1,$$

since $x^{p^\ell} \in Z(P)$. Thus P' is an abelian group generated by elements of order dividing p^ℓ, and our claim follows.

Assume, for a contradiction, that $\ell > 1$. We claim that $P^{p^{\ell-1}}$, the subgroup generated by the set $\{x^{p^{\ell-1}} : x \in P\}$, is abelian. For let $x^{p^{\ell-1}}, y^{p^{\ell-1}} \in P^{p^{\ell-1}}$ where $x, y \in P$. Then

$$[x^{p^{\ell-1}}, y^{p^{\ell-1}}] = [x, y^{p^{\ell-1}}]^{p^{\ell-1}} = [x, y]^{p^{2\ell-2}} = 1,$$

since P' has exponent dividing p^ℓ and $2\ell - 2 \geqslant \ell$. Thus $P^{p^{\ell-1}}$ is abelian, as required.

Now, $P^{p^{\ell-1}}$ is a characteristic subgroup of the normal subgroup P, and so $P^{p^{\ell-1}}$ is normal in G. Thus $P^{p^{\ell-1}}A$ is normal in G, and is abelian since $P^{p^{\ell-1}} \leqslant B \leqslant C$. By the maximality of A, we have that $P^{p^{\ell-1}} \leqslant A$. Hence $P^{p^{\ell-1}} \leqslant Z(P)$. So the exponent of $P/Z(P)$ divides $p^{\ell-1}$. This contradiction shows that $P/Z(P)$ has exponent p.

Since P has nilpotency class 2, we have that P' is non-trivial and $P' \leqslant Z(P) \leqslant A$, and so P' is cyclic. Moreover, P' has exponent p since we have shown that the exponent of P' divides the exponent of $P/Z(P)$. Since P' is non-trivial, $|P'| = p$ as required.

Proposition 9.5 *Let $M = \langle B \rangle_K$, the subset of $\mathrm{End}_{\mathbb{F}_q}(V)$ consisting of all K-linear combinations of B. Then $|B/A| = \dim_K M$. In particular, we have that $|B/A| \leqslant d_2^2$.*

Proof: Let $k = |B/A|$ and let $x_1, x_2, \ldots, x_k$ be a transversal of A in B. We claim that this set forms a K-basis for M. Clearly $B \subseteq \langle x_1, x_2, \ldots, x_k \rangle_K$, as every element of B may be written in the form zx_i for some $z \in A \subseteq K$ and some $i \in \{1, 2, \ldots, k\}$. Hence $x_1, x_2, \ldots, x_k$ form a K-spanning set for M. To prove our claim, it suffices to show that $x_1, x_2, \ldots, x_k$ are linearly independent.

Suppose, for a contradiction, that there exist $z_1, z_2, \ldots, z_k \in K$ such that

$$z_1 x_1 + z_2 x_2 + \cdots + z_k x_k = 0, \tag{9.1}$$

where not all of the z_i are 0. Moreover, suppose that the number of non-zero coefficients z_i is as small as possible.

Note that we cannot have $z_i x_i = 0$ where $z_i \neq 0$, as both z_i and x_i are invertible linear transformations. Hence at least two of the z_i are non-zero: let $u, v \in \{1, 2, \ldots, k\}$ be such that $z_u \neq 0$, $z_v \neq 0$ and $u \neq v$.

Let $y \in B$ be such that $[x_u, y] \neq [x_v, y]$. We may choose such an element y, for if no such y exists then (since B has nilpotency class at most 2)

$$[x_u x_v^{-1}, y] = [x_u, y][x_v, y]^{-1} = 1$$

for all $y \in B$. But this would imply that $x_u x_v^{-1}$ lies in $Z(B) = A$. This cannot happen since $x_1, x_2, \ldots, x_k$ form a transversal for A in B.

Since B/A is abelian, there exist $r_1, r_2, \ldots, r_k \in A \subseteq K$ such that

$$y^{-1} x_i y = x_i r_i$$

for all $i \in \{1, 2, \ldots, k\}$. Since $[x_i, y] = r_i$, our choice of y implies that $r_u \neq r_v$. But now, using (9.1) we find that

$$0 = r_u(z_1 x_1 + z_2 x_2 + \cdots + z_k x_k) - y^{-1}(z_1 x_1 + z_2 x_2 + \cdots + z_k x_k)y$$

$$= (r_u - r_1)z_1 x_1 + (r_u - r_2)z_2 x_2 + \cdots + (r_u - r_k)z_k x_k.$$

Since $z_v \neq 0$ and $r_u \neq r_v$, the coefficient $(r_u - r_v)z_v$ of x_v in this sum is non-zero. However, the coefficient $(r_u - r_u)z_u$ of x_u is zero and so the number of non-zero coefficients has strictly decreased when compared to (9.1). This contradicts our choice of the coefficients z_i, and so $x_1, x_2, \ldots, x_k$ form a K-basis for M. Hence $|B/A| = k = \dim_K M$, as required.

To prove the final statement of the proposition, note that Proposition 9.2 implies that $B \subseteq C \subseteq \mathrm{GL}(d_2, K)$. This implies that M is a K-subspace of the set of all $d_2 \times d_2$ matrices over K and so $\dim_K M \leqslant d_2^2$.

Proposition 9.6 *Let $\alpha \in \mathrm{Aut}\,(B)$, and suppose that α fixes A pointwise and induces the identity automorphism on B/A. Then α is inner.*

Proof: By Proposition 9.4, B is a direct product of its (characteristic) Sylow p-subgroups. Hence it is sufficient to consider a prime p dividing $|B|$ and the Sylow p-subgroup P of B, and to prove that an automorphism $\alpha_1 \in \mathrm{Aut}\,(P)$ that fixes $Z(P)$ pointwise and induces the identity automorphism on $P/Z(P)$ is inner.

Let $\alpha_1 \in \mathrm{Aut}\,(P)$ be one such automorphism. If $P = Z(P)$ the result is trivial, and so we may suppose that P is non-abelian. Let the integer r be defined by $|P/Z(P)| = p^r$ and let $y_1, y_2, \ldots, y_r \in P$ be such that $y_1 Z(P), y_2 Z(P), \ldots, y_r Z(P)$ generate $P/Z(P)$. Since α_1 fixes the elements of $Z(P)$, the automorphism α_1 is determined by the images $\alpha_1(y_i)$ where $i \in \{1, 2, \ldots, r\}$. Since α_1 induces the identity automorphism on $P/Z(P)$, there exist $z_1, z_2, \ldots, z_r \in Z(P)$ such that

$$\alpha_1(y_i) = y_i z_i$$

for all $i \in \{1, 2, \ldots, r\}$. Proposition 9.4 implies that $P/Z(P)$ has exponent p, and so α_1 fixes the elements y_i^p. Thus

$$y_i^p = \alpha_1(y_i^p) = (\alpha_1(y_i))^p = (y_i z_i)^p = y_i^p z_i^p,$$

and hence $z_i^p = 1$. Since $Z(P)$ is cyclic, there are at most p choices for each element z_i. Thus the number of automorphisms α_1 that fix $Z(P)$ pointwise and induce the identity automorphism on $P/Z(P)$ is at most p^r. But there are $|P/Z(P)| = p^r$ inner automorphisms (and they all have the right form). So α_1 is always inner. This establishes the proposition.

Proposition 9.7 *The quotient C/B acts faithfully by conjugation on B/A.*

Proof: Let $\phi : C \to \mathrm{Aut}\,(B/A)$ be the map that takes $g \in C$ to the automorphism of B/A induced by conjugation by g. To prove the proposition, it suffices to show that $\ker \phi = B$. We will first show that $\ker \phi = BD$, where D is the centraliser of B in C.

Clearly $BD \subseteq \ker \phi$. We prove the reverse inclusion as follows. Let $g \in \ker \phi$. The automorphism α of B arising from conjugation by g fixes A pointwise since $g \in C$ and induces the identity automorphism on B/A since $g \in \ker \phi$. By Proposition 9.6, the automorphism α is inner – let conjugating by $h \in B$ give rise to α. But now $h^{-1}g \in D$, and so $g \in BD$, as required.

To show that $\ker \phi = B$, it suffices to show that $D \leqslant B$. Suppose, for a contradiction, that BD/B is non-trivial. Define $Z \leqslant C$ as the subgroup containing B such that Z/B is the last non-trivial term in the derived series of BD/B. Note that D is normal in G, as it is the intersection of the normal subgroups C and $C_G(B)$. Hence Z is normal in G. Since Z/B is abelian, $Z' \leqslant B$. Moreover, since $Z \leqslant BD$ and D centralises B, we have that $Z' \leqslant B'D' \leqslant AD' \leqslant D$. Thus $Z' \leqslant B \cap D = Z(B) = A$. Hence Z is a normal subgroup of G contained in C, it strictly contains B and Z/A is abelian. This contradicts the maximality of B. Hence $\ker \phi = B$ and so the proposition follows.

We comment that we are now able to deduce a significant amount of structural information about C/B. Proposition 9.4 shows that $B/A \cong P_1/Z(P_1) \times P_2/Z(P_2) \times \cdots \times P_k/Z(P_k)$, where P_i is a non-abelian Sylow p_i-subgroup of B for $1 \leqslant i \leqslant k$. Now, $P_i/Z(P_i)$ is of exponent p_i and so may be regarded as a vector space over $\mathbb{F}_{p_i}$; moreover, P_i' has order p_i and so may be identified with $\mathbb{F}_{p_i}$. The process of forming commutators in P_i induces a non-degenerate alternating form on $P_i/Z(P_i)$, and so the rank r_i of $P_i/Z(P_i)$ is even. Since conjugation behaves well with respect to forming commutators and since C centralises P_i', we find that C/B preserves the alternating form on $P_i/Z(P_i)$. Hence Proposition 9.7 implies that C/B is isomorphic to a subgroup of $\mathrm{Sp}(r_1, p_1) \times \mathrm{Sp}(r_2, p_2) \times \cdots \times \mathrm{Sp}(r_k, p_k)$.

10

The orders of groups

This chapter provides upper bounds on the order of a soluble subgroup of a symmetric group and of a general linear group. The first theorem we prove is due to John Dixon [23].

Theorem 10.1 *If $G \leqslant \mathrm{Sym}(n)$ and G is soluble then*

$$|G| \leqslant k^{n-1}$$

where $k = \sqrt[3]{24} = 2.88\ldots$

Proof: We use induction on n. Take $\Omega = \{1, \ldots, n\}$ and let $G \leqslant \mathrm{Sym}(\Omega)$. Suppose that G is not transitive. Then $\Omega = \Omega_1 \dot\cup \Omega_2$ where Ω_1 and Ω_2 are non-empty G-invariant subsets of Ω. Define $G_1 = G^{\Omega_1}$ and $G_2 = G^{\Omega_2}$. Then G is isomorphic to a subgroup of $G_1 \times G_2$. For $i \in \{1, 2\}$, we have that G_i is a soluble subgroup of $\mathrm{Sym}(\Omega_i)$ and $|\Omega_i| < n$. So, by the inductive hypothesis,

$$|G| \leqslant |G_1| \times |G_2| \leqslant k^{n_1-1} k^{n_2-1}$$

where $n_1 = |\Omega_1|$ and $n_2 = |\Omega_2|$. Hence $|G| \leqslant k^{n-1}$ and the result follows in this case.

Suppose that G is transitive but not primitive. (Recall the discussion in Section 6.3 on imprimitive permutation groups.) In this case, there is a G-invariant equivalence relation ρ on Ω which is non-trivial and non-universal. Let α be a fixed element of Ω, let Ω_1 be the congruence class of ρ containing α and let Ω_2 be the set of congruence classes of ρ. Let $G_2 = G^{\Omega_2}$ and take N to be the kernel of the map from G to G_2, so

$$N = \{g \in G : \omega \sim_\rho g\omega \text{ for all } \omega \in \Omega\}.$$

Now $\Omega \cong \Omega_1 \times \Omega_2$, so if $n_1 = |\Omega_1|$, $n_2 = |\Omega_2|$, then $n = n_1 n_2$. Let $\Delta_1, \Delta_2, \ldots, \Delta_{n_2}$ be the distinct congruence classes under ρ and let $N_i = N^{\Delta_i}$. Then N is

94

isomorphic to a subgroup of $N_1 \times N_2 \times \cdots \times N_{n_2}$. Note that for all i, we have that N_i is a soluble subgroup of $\text{Sym}(\Delta_i)$ and $|\Delta_i| = n_1 < n$. Hence, by the inductive hypothesis,

$$|G| = |N| |G_2| \leqslant \left(k^{n_1-1}\right)^{n_2} k^{n_2-1} = k^{n-1},$$

as required.

Suppose finally that G is primitive. Let M be a minimal normal subgroup of G, let $\alpha \in \Omega$ and define $H = G_\alpha$. Since G is soluble and primitive, Proposition 6.13 implies that G is the semidirect product of M by H and $n = |\Omega| = |M| = p^d$ for some prime number p and positive integer d. Moreover, H is isomorphic to a soluble subgroup of $\text{GL}(d, p)$. Hence

$$|G| = |M||H|$$

$$\leqslant n \, |\text{GL}(d, p)|$$

$$< n p^{d^2}$$

$$\leqslant n^{1+\log n}$$

$$< k^{n-1}$$

provided that $n \geqslant 32$. For $n < 32$ we check by hand that $|G| \leqslant k^{n-1}$ with equality only for $\text{Sym}(4)$ of degree 4. The result is almost trivial if n is prime since then $|G| \leqslant n(n-1)$. This leaves only the cases $n = 4, 8, 9, 16, 25, 27$, which are easy to deal with.

We remark that the bound of Theorem 10.1 is achieved whenever n is a power of 4, the group being the iterated wreath product of $\text{Sym}(4)$.

The following theorem was proved independently by P. P. Pálfy and T. R. Wolf [80, 97].

Theorem 10.2 *If $G \leqslant \text{GL}(d, q)$ and G is both soluble and completely reducible, then*

$$|G| \leqslant q^{3d-2}.$$

Proof: We use induction and assume that the result holds for groups of smaller degree (perhaps over larger fields). Let $V = (\mathbb{F}_q)^d$, and let G act faithfully on V.

Suppose first that G is not irreducible, so $V = V_1 \oplus V_2$, where V_1 and V_2 are non-trivial $\mathbb{F}_q G$-modules. Define d_1 and d_2 by $d_i = \dim V_i$, so $d_1 + d_2 = d$. Then G is isomorphic to a subgroup of the direct product of a soluble subgroup of $\mathrm{GL}(d_1, q)$ and a soluble subgroup of $\mathrm{GL}(d_2, q)$. Hence our inductive hypothesis implies that

$$|G| \leqslant q^{3d_1-2} q^{3d_2-2} < q^{3(d_1+d_2)-2} = q^{3d-2}.$$

We may therefore assume that G is irreducible.

Suppose that G is imprimitive, so G preserves a decomposition

$$V = V_1 \oplus \cdots \oplus V_{m_2}$$

where $\dim V_i = m_1$ for all i. Note that in this case we have that $m_2 > 1$ and $d = m_1 m_2$. By Proposition 6.15 we have that $G \leqslant G_1 \mathrm{wr}\, G_2$, where G_1 is a soluble subgroup of $\mathrm{GL}(m_1, q)$ and G_2 is a (transitive) soluble subgroup of $\mathrm{Sym}(m_2)$. So

$$|G| = |G_1|^{m_2} |G_2| \leqslant \left(q^{3m_1-2} \right)^{m_2} k^{m_2-1} \leqslant q^{3d-2},$$

since $m_2 \geqslant 2$ (where k is as in Theorem 10.1).

Finally, suppose that G is primitive. Let d be the degree of G. Let A be a maximal abelian normal subgroup of G, let $C = C_G(A)$ and let K be the subalgebra of $\mathrm{End}_{\mathbb{F}_q} V$ generated by A. By Proposition 9.1, $K \cong \mathbb{F}_{q^{d_1}}$ for some positive divisor d_1 of d. Define $d_2 = d/d_1$. Propositions 9.2 and 9.3 imply that C may be thought of as a (soluble) subgroup of $\mathrm{GL}(d_2, q^{d_1})$ and G/C may be identified with a subgroup of the Galois group $\mathrm{Gal}(\mathbb{F}_{q^{d_1}} : \mathbb{F}_q)$ of $\mathbb{F}_{q^{d_1}}$ over $\mathbb{F}_q$. Thus G/C is a cyclic group of order at most d_1. When $d_1 > 1$, our inductive hypothesis implies that

$$|G| = |G/C| \times |C| \leqslant d_1 \left(q^{d_1} \right)^{3d_2-2} < q^{3d_1 d_2-2} = q^{3d-2},$$

and so we may assume that $d_1 = 1$. In particular, this implies that $G = C$ and $|A| \leqslant q-1$.

Let the subgroup B of G have the property that B/A is a maximal abelian normal subgroup of G/A. Propositions 9.5 and 9.7 imply that

$$|B/A| \leqslant d^2$$

and that G/B acts faithfully on B/A. So

$$|G/B| \leqslant |\mathrm{Aut}(B/A)| \leqslant d^{4\log d}.$$

Hence

$$
\begin{aligned}
|G| &= |A| \times |B/A| \times |G/B| \\
&\leqslant (q-1) d^2 d^{4\log d} \\
&\leqslant q^{1+2\log d+4(\log d)^2} \\
&\leqslant q^{3d-2}
\end{aligned}
$$

if $d \geqslant 45$. Now if $d < 45$ then $d \in \{4, 9, 16, 25, 36\}$ and G/B is isomorphic to a subgroup of $\mathrm{Sp}(2, 2)$, $\mathrm{Sp}(2, 3)$, $\mathrm{Sp}(4, 2)$, $\mathrm{Sp}(2, 5)$ and $\mathrm{Sp}(2, 2) \times \mathrm{Sp}(2, 3)$ accordingly; see the comment after the proof of Theorem 9.7. In each of these cases it may be checked that the theorem holds.

Theorem 10.3 *If $G \leqslant \mathrm{Sym}(n)$ is soluble and primitive then $|G| \leqslant n^4$.*

Proof: Let G be a soluble primitive subgroup of $\mathrm{Sym}(n)$, let M be a minimal normal subgroup of G and let H be the stabiliser of a point α. Proposition 6.13 shows that $n = p^d = |M|$ for some prime p, and that $|G| = |M||H|$, where H may be identified with an irreducible soluble subgroup of $\mathrm{GL}(d, p)$. In particular, H is completely reducible and so Theorem 10.2 implies that

$$|H| \leqslant p^{3d-2} < (p^d)^3 = n^3.$$

Thus $|G| = |M||H| \leqslant nn^3 = n^4$, as required.

In fact, by arguing more carefully, the upper bound of Theorem 10.3 may be reduced to $24^{-1/3} n^c$, where c is approximately $3.24399\ldots$; see Pálfy [80, Theorem 1].

11

Conjugacy classes of maximal soluble subgroups of symmetric groups

In this chapter we shall give bounds for the number of conjugacy classes of maximal soluble subgroups in symmetric groups. We begin by proving a simple combinatorial lemma.

Let n be a positive integer. An *ordered multiplicative partition* of n is a sequence $(n_1, n_2, \ldots, n_k)$ of integers such that $n_i > 1$ for all i and such that $\prod_{i=1}^{k} n_i = n$.

Lemma 11.1 *Let $m(n)$ be the number of ordered multiplicative partitions of n. Then $m(n) \leqslant n^2$.*

Proof: We shall prove the lemma by induction. The only multiplicative partition of 1 arises from the empty sequence. So $m(1) = 1$. Assume that the result holds for all numbers less than n. Then

$$m(n) = \sum_{n_1 \mid n \text{ and } n_1 \geqslant 2} m(n/n_1)$$

$$\leqslant \sum_{n_1 \mid n \text{ and } n_1 \geqslant 2} \frac{n^2}{n_1^2} \qquad \text{(by our inductive hypothesis)}$$

$$\leqslant n^2 \sum_{t \geqslant 2} \frac{1}{t^2} < n^2, \text{ since } \pi^2/6 - 1 < 1.$$

We now state and prove a theorem due to L. Pyber [82]. We introduce the following notation:

$m_{\text{ps}}(n) = $ the number of conjugacy classes of primitive maximal soluble
  subgroups in $\text{Sym}(n)$;

$m_{\text{ts}}(n) = $ the number of conjugacy classes of transitive maximal soluble
  subgroups in $\text{Sym}(n)$;

$m_{\text{ss}}(n) = $ the number of conjugacy classes of maximal soluble subgroups
  in $\text{Sym}(n)$.

Theorem 11.2 *Using the above notation,*

$$m_{ps}(n) \leqslant 2^{(\log n)^2(3\log n - 2)},$$

$$m_{ts}(n) \leqslant 2^{3(\log n)^3},$$

$$m_{ss}(n) \leqslant 2^{16n}.$$

Proof: We note that $m_{ps}(n) = 0$ unless $n = p^d$ for some prime p.

Let G be a primitive soluble subgroup of $\mathrm{Sym}(n)$, let M be a minimal normal subgroup of G and let H be the stabiliser of a point. Proposition 6.13 implies that M is transitive and $n = p^d = |M|$ for some prime p. Moreover, $G = HM$ and H may be identified with a soluble subgroup of $\mathrm{GL}(d, p)$.

The conjugacy class of G in $\mathrm{Sym}(n)$ is determined by the conjugacy class of H in $\mathrm{GL}(d, p)$. We know from Theorem 10.2 that

$$|H| \leqslant p^{3d-2} \leqslant \frac{n^3}{4}.$$

Let $d(H)$ be the minimal number of generators of H. Then we have

$$d(H) \leqslant 3\log(n) - 2.$$

Therefore

$m_{ps}(n) \leqslant$ the number of $(3\log(n) - 2)$ generator subgroups of $\mathrm{GL}(d, p)$

$$\leqslant p^{d^2(3\log(n)-2)}$$

$$\leqslant 2^{(\log(n))^2(3\log(n)-2)}.$$

Now let $\Omega = \{1, 2, \ldots, n\}$ and let G be a transitive maximal soluble subgroup of $\mathrm{Sym}(\Omega)$. Further, let $\rho_0 < \rho_1 < \cdots < \rho_k$ be a maximal chain of congruences on Ω. So ρ_0 must be the trivial congruence and ρ_k is the universal congruence. If we define sets Ω_i and primitive permutation groups $G_i \subseteq \mathrm{Sym}(\Omega_i)$ as in the statement of Proposition 6.14, then Proposition 6.14 implies that we may identify Ω with $\Omega_1 \times \Omega_2 \times \cdots \times \Omega_k$ in such a way that $G \leqslant G_1 \mathrm{wr}\, G_2 \mathrm{wr} \cdots \mathrm{wr}\, G_k$. Since G is a maximal soluble subgroup of $\mathrm{Sym}(\Omega)$, each G_i is a maximal primitive soluble subgroup of $\mathrm{Sym}(\Omega_i)$ and $G = G_1 \mathrm{wr}\, G_2 \mathrm{wr} \cdots \mathrm{wr}\, G_k$. In particular, the conjugacy class of G in $\mathrm{Sym}(\Omega)$ is determined by the conjugacy classes of the subgroups G_i in $\mathrm{Sym}(\Omega_i)$.

Let $n_i = |\Omega_i|$. Then $(n_1, \ldots, n_k)$ is an ordered multiplicative partition of n.

So

$$m_{ts}(n) = \sum_{(n_1,\dots,n_k)} \prod_{i=1}^{k} m_{ps}(n_i)$$

$$\leqslant \sum_{(n_1,\dots,n_k)} \prod_{i=1}^{k} \left(2^{3(\log(n_i))^3 - 2(\log(n_i))^2} \right).$$

Now,

$$\prod_{i=1}^{k} 2^{3(\log(n_i))^3} \leqslant 2^{3(\sum_{i=1}^{k} \log(n_i))^3} = 2^{3(\log(n))^3}. \tag{11.1}$$

Let c be a constant and let $x_1, x_2, \dots, x_k$ be real variables. If $x_i \geqslant 1$ for all i and $\sum_{i=1}^{k} x_i = c$ then $\sum_{i=1}^{k} x_i^2 \geqslant c$. Using this fact with $c = \log n$ and $x_i = \log n_i$, we find that

$$\prod_{i=1}^{k} 2^{-2\log(n_i)^2} \leqslant 2^{-2\log n} = n^{-2}. \tag{11.2}$$

The inequalities (11.1) and (11.2) together imply that

$$m_{ts}(n) \leqslant \sum_{(n_1,\dots,n_k)} 2^{3(\log(n))^3} n^{-2}$$

$$= m(n) 2^{3(\log(n))^3} n^{-2}$$

$$\leqslant 2^{3(\log(n))^3}$$

by Lemma 11.1.

Now let G be an arbitrary maximal soluble subgroup of $\text{Sym}(\Omega)$. Then G has orbits $\Omega_1, \dots, \Omega_t$ where if $n_i = |\Omega_i|$ then the n_i form a partition of n and if $G_i = G^{\Omega_i}$ then $G = G_1 \times \cdots \times G_t$. As before, the G_i are maximal soluble subgroups of $\text{Sym}(\Omega_i)$ and are also transitive. Further, the conjugacy class of G is determined by the (unordered) set of conjugacy classes of the G_i. Thus

$$m_{ss}(n) = \sum_{\{n_1,\dots,n_t\}} \prod_{i=1}^{t} m_{ts}(n_i)$$

$$\leqslant \sum_{\{n_1,\dots,n_t\}} \prod_{i=1}^{t} 2^{3(\log(n_i))^3}$$

$$\leqslant \sum_{\{n_1,\dots,n_t\}} \prod_{i=1}^{t} 2^{15n_i} \qquad \text{(since } (\log(n))^3 \leqslant 5n\text{)}$$

$$= p(n) 2^{15n},$$

where $p(n)$ is the number of (unordered additive) partitions of n. But $p(n) \leqslant 2^{n-1}$ by Corollary 5.10, and so $m_{ss}(n) \leqslant 2^{16n}$, as required.

Comments

1. The constant 3 appearing in the bounds for $m_{ps}(n)$ and $m_{ts}(n)$ may be reduced considerably, by improving the upper bound on $d(H)$ in the proof of Theorem 11.2. Indeed, this upper bound may be reduced by first bounding $\mu(|H|)$ (using a little of the structure theory of irreducible soluble subgroups of $GL(d, p)$) and then using Kovács' theorem [59] which states that $d(X) \leqslant \mu(|X|) + 1)$ for a finite soluble group X.
2. One may conjecture that $m_{ss}(n) \leqslant 2^n$ and that in fact $m_{ss}(n) = 2^{O(\sqrt{n})}$. One may also be able to show that there exists a constant k such that

$$\frac{m_{ss}(n)}{p(n)} \longrightarrow k \ \text{ as } \ n \longrightarrow \infty.$$

3. Since the number of maximal soluble subgroups of $\mathrm{Sym}(n)$ is at most $n! m_{ss}(n)$, Theorem 11.2 shows that the number of maximal soluble subgroups of $\mathrm{Sym}(n)$ is polynomial in $n!$. This contrasts with the situation when we count all soluble subgroups of $\mathrm{Sym}(n)$. The number of subgroups (all soluble) of $C_2 \times \cdots \times C_2$ (d times) is greater than $2^{\frac{1}{4}d^2}$. So there are at least $2^{\frac{1}{4}d^2}$ soluble subgroups of $\mathrm{Sym}(2d)$, a number which is not bounded by a polynomial in $(2d)!$.

12
Enumeration of finite groups with abelian Sylow subgroups

In this chapter we will concentrate on enumeration of finite soluble A-groups. There are two reasons for doing this. The first is that A-groups form an important subclass of groups, especially from the enumeration point of view. The behaviour of the enumeration function on this subclass is very different from its behaviour on the class of p-groups or the class of all groups. The second is that some of the techniques used in the enumeration of finite soluble A-groups will serve as an introduction to the techniques used in enumerating finite groups in the class of all groups.

As in Section 7.4, let $f_A(n)$ denote the number of (isomorphism classes of) A-groups of order n and let $f_{A, sol}(n)$ denote the number of soluble A-groups of order n up to isomorphism.

We shall show that $f_{A, sol}(n) \leqslant n^{11\mu+13}$. Since $\mu(n) \leqslant \lambda(n)$, in most cases this bound will be better than the bound for the isomorphism classes of soluble A-groups given in Chapter 7, Theorem 7.18. A similar bound will arise from Corollary 15.6 of Theorem 15.5 in Chapter 15. This bound in fact has a better leading term though it is worse in the error term. On the other hand, the tools required to prove Pyber's theorem are much more complicated than those used in this chapter.

We shall prove our bound for $f_{A, sol}(n)$ in Section 12.4. The methods used here will be an adaptation to soluble A-groups of the techniques used by Pyber [82]. His approach in turn draws on an earlier work of G. A. Dickenson [21]. We shall also use some of the results proved in previous chapters and so will end up with a weaker bound than that given by Venkataraman [94]. But the main aim of the section will be to understand the essential aspects of Pyber's proof by working within the easier framework of soluble A-groups.

12.1 Counting soluble A-groups: an overview

Let n be a positive integer and let $n = \prod_{i=1}^{k} p_i^{\alpha_i}$ be the prime factorisation of n. Consider a soluble A-group G of order n. We can embed G into a soluble A-group $\hat{G}$ which is a direct product of k 'nice' soluble A-groups $\hat{G}_i$, where $i = 1, \ldots, k$.

By 'nice' we mean that for each i, the group $\hat{G}_i$ has as its Sylow p_i-subgroup a group P_i which is isomorphic to the Sylow p_i-subgroup of G. Further, $\hat{G}_i$ can be regarded as a semidirect product of P_i by M_i, where M_i is a maximal soluble p_i'-A-subgroup of $A_i = \text{Aut}(P_i)$.

Let $\Phi(P_i)$ denote the Frattini subgroup of P_i. There is a natural homomorphism from A_i to $\text{Aut}(P_i/\Phi(P_i))$, and by Lemma 3.13 the kernel of this homomorphism is a p_i-subgroup. Also, as a consequence of Lemma 3.12, we have that $\text{Aut}(P_i/\Phi(P_i))$ is isomorphic to $\text{GL}(d_i, p_i)$ where d_i is the minimal number of generators required to generate P_i.

We estimate the choices for G as a subgroup of $\hat{G}$ using a method of 'Sylow systems' which was introduced by Pyber in [82]. The choices for $\hat{G}$ depend on those for the groups $\hat{G}_i$. Once P_i is fixed we need to enumerate the choices for M_i as a maximal soluble p_i'-A-subgroup of A_i. The natural homomorphism described above and some simple results in Section 12.3 help us to 'linearise' the environment for counting the choices for M_i. Finally, putting together the various estimates gives us an upper bound for $f_{A, \text{sol}}(n)$.

In the next section we count the number of conjugacy classes of subgroups that are maximal amongst soluble p'-A-subgroups of $\text{GL}(d, q)$, where p is a prime and q is a power of p.

12.2 Soluble A-subgroups of the general linear group and the symmetric group

Lemma 12.1 *The number of transitive soluble subgroups of* $\text{Sym}(n)$ *is at most* $3^{(n^2-1)/2} \, 2^{n \log(n) + 3(\log(n))^3}$.

Proof: Let H be a transitive soluble subgroup of $\text{Sym}(n)$. Then H is contained in a transitive maximal soluble subgroup M of $\text{Sym}(n)$. By Theorem 11.2, the number $m_{\text{ts}}(n)$ of conjugacy classes of transitive maximal soluble subgroups of $\text{Sym}(n)$ is at most $2^{3(\log(n))^3}$. Any conjugacy class of transitive maximal soluble subgroups of $\text{Sym}(n)$ can contain at most $n!$ subgroups. Thus we have less than or equal to $2^{n \log(n) + 3(\log(n))^3}$ choices for M. Also by Theorem 10.1 we know that $|M| \leqslant 3^{n-1}$. By Theorem 6.11, any subgroup of $\mathrm{Sym}(n)$ can be generated by $(n+1)/2$ elements. Thus once M is chosen, we have at most $3^{(n^2-1)/2}$ choices for H. Putting together all these estimates we get the required upper bound.

In the next proposition we will bound the number of conjugacy classes of subgroups of $\mathrm{GL}(d, q)$ that are maximal amongst irreducible soluble p'-A-subgroups. The structure of such A-groups is relatively simple as compared with that of maximal soluble subgroups of $\mathrm{GL}(d, q)$ that are irreducible. So there is much less theory involved in the proof of this proposition.

Proposition 12.2 *Let $T = \mathrm{GL}(d, q)$, where q is a power of some prime p. Then the number of conjugacy classes of irreducible soluble p'-A-subgroups of T is at most $q^{6d^2} 2^{13d}$.*

Proof: Let G be a subgroup of T that is maximal among irreducible soluble p'-A-subgroups of T. Let $V = (\mathbb{F}_q)^d$ and let $F = F(G)$, where $F(G)$ is the Fitting subgroup of G. (Section 6.2 of Chapter 6 contains information on the Fitting subgroup.) By Proposition 6.3, the Fitting subgroup of G is the unique maximal normal nilpotent subgroup of G and since G is an A-group, F is abelian. In the first part of this proof we show that there are relatively few possibilities for F up to conjugacy.

By Clifford's theorem (Theorem 8.4), regarding V as an F-module we have

$$V = Y_1 \oplus Y_2 \oplus \cdots \oplus Y_r$$

where the subspaces Y_i are permuted transitively by the action of G. Further, there is a multiplicity ℓ and there are pairwise non-isomorphic irreducible $\mathbb{F}_q F$-submodules $X_1, \ldots, X_r$ of V such that $Y_i = \ell X_i$ for all i.

Now let K_i be the kernel of the action of F on Y_i. Since $Y_i = \ell X_i$, when we restrict the action of F to X_i, we find that the kernel of this action is still K_i. Thus F/K_i acts faithfully on Y_i and when its action is restricted to X_i, it acts faithfully and irreducibly on X_i. Let E_i denote the subalgebra generated by F/K_i in $\mathrm{End}_{\mathbb{F}_q}(Y_i)$. As $Y_i = \ell X_i$ we find that X_i is a faithful E_i-module. Thus E_i is an $\mathbb{F}_q$-algebra with a faithful irreducible representation and so E_i is simple by Proposition 8.2. Since F/K_i is abelian, E_i is commutative. Thus (see Theorem 8.3) E_i is a field.

Now, X_i is an irreducible E_i-module, and so is a vector space over E_i of dimension 1. Hence $\dim_{\mathbb{F}_q} E_i = \dim_{\mathbb{F}_q} X_i$. Thus if $\dim_{\mathbb{F}_q} X_i = s$, then $E_i \cong \mathbb{F}_{q^s}$. Note that $d = \ell rs$. We will bound the number of possibilities for F (up to conjugacy), once ℓ, r and s are fixed.

Let E_i^* denote the multiplicative group of the field E_i. Now $F/K_i \leqslant E_i^* \leqslant$ $\mathrm{GL}(Y_i)$. Therefore

$$F \leqslant F/K_1 \times F/K_2 \times \cdots \times F/K_r$$

$$\leqslant E_1^* \times E_2^* \times \cdots \times E_r^*$$

$$\leqslant \mathrm{GL}(Y_1) \times \mathrm{GL}(Y_2) \times \cdots \times \mathrm{GL}(Y_r)$$

$$\leqslant \mathrm{GL}(V).$$

Let $E = E_1^* \times \cdots \times E_r^*$. Then $|E| = (q^s - 1)^r$. Note that as an $\mathbb{F}_q E$-module

$$V = \ell X_1 \oplus \cdots \oplus \ell X_r$$

where for all i, E_i^* acts faithfully and irreducibly on X_i and $\dim_{E_i} X_i = 1$. Further, for $i \neq j$, E_i^* acts trivially on X_j. It is not difficult to show that there is only one conjugacy class of subgroups like E in $\mathrm{GL}(V)$.

So once ℓ, r and s are chosen such that $d = \ell r s$, up to conjugacy there is only one choice for E. Since E is a direct product of r isomorphic cyclic groups, any subgroup of E can be generated by r elements. In particular, F can be generated by r elements and so the number of choices for F as a subgroup of E is at most $|E|^r$, that is $(q^s - 1)^{r^2}$.

Now G acts transitively on $\{Y_1, \ldots, Y_r\}$. So we have a homomorphism from G into $\mathrm{Sym}(r)$. Let N be the kernel of this map. Then G/N is a transitive soluble A-subgroup of $\mathrm{Sym}(r)$. Clearly $F \leqslant N$.

For any g in G, if $gY_i = Y_j$ then $gK_ig^{-1} = K_j$. Further, since E_i is the subalgebra generated by F/K_i in $\mathrm{End}_{\mathbb{F}_q}(Y_i)$, we will have $gE_ig^{-1} = E_j$. Thus for all $g \in N$, since $gY_i = Y_i$, we get that $gK_ig^{-1} = K_i$ and $gE_ig^{-1} = E_i$ for all i. Thus N acts by conjugation on each E_i and this action clearly leaves each element of the ground field $\mathbb{F}_q$ fixed. So for each i, we have a homomorphism from N into $\mathrm{Gal}(E_i : \mathbb{F}_q)$ with kernel $N_i = C_N(E_i)$. We claim that $\bigcap_{i=1}^r N_i = F$.

Since F is abelian and E_i is the subalgebra generated by F/K_i, we have that $F \leqslant N_i$ for all i. On the other hand, let $t \in \bigcap_{i=1}^r N_i$ and let $h \in F$. Then for all i,

$$hK_i = t(hK_i)t^{-1} = tht^{-1}tK_it^{-1} = tht^{-1}K_i.$$

Thus $h^{-1}tht^{-1} \in \bigcap_{i=1}^r K_i = \{1\}$. So for all h in F we have $th = ht$. Thus $t \in C_G(F)$. But since G is soluble we have that $C_G(F) \leqslant F$, by Corollary 6.5. Hence we have proved our claim.

So N/F is isomorphic to a subgroup of $\mathrm{Gal}(E_1 : \mathbb{F}_q) \times \cdots \times \mathrm{Gal}(E_r : \mathbb{F}_q)$. Since for each i, $\mathrm{Gal}(E_i : \mathbb{F}_q)$ is isomorphic to a cyclic group of order s, we can regard N/F as a subgroup of a direct product of r isomorphic cyclic groups. Consequently, N/F can be generated by r elements.

We now proceed to bound the choices for N given a choice of F. Let X be the subgroup of $T = \mathrm{GL}(d, q)$ which normalises F and preserves the direct sum decomposition

$$V = Y_1 \oplus \cdots \oplus Y_r.$$

So $X = \{x \in N_T(F) \mid xY_i = Y_i, \text{ for all } i\}$. Note that since F is fixed, X is known and $F \leqslant N \leqslant X$. We will now determine the choices for N as a subgroup of X.

It is easy to see that for each i, we have that X acts by conjugation on E_i and leaves each element of the ground field $\mathbb{F}_q$ fixed. So we have a homomorphism from X into $\mathrm{Gal}(E_i : \mathbb{F}_q)$ with kernel $C_X(E_i)$. Let $C = \bigcap_{i=1}^r C_X(E_i)$. Note that $N \cap C = \bigcap_{i=1}^r N_i = F$. Also we can regard X/C as a subgroup of $\mathrm{Gal}(E_1 : \mathbb{F}_q) \times \cdots \times \mathrm{Gal}(E_r : \mathbb{F}_q)$. So $|X/C| \leqslant s^r$. Further, for each i we have that C centralises E_i and so we have a homomorphism from C into $\mathrm{GL}_{E_i}(Y_i)$. Thus we get that C is isomorphic to a subgroup of $\mathrm{GL}_{E_1}(Y_1) \times \cdots \times \mathrm{GL}_{E_r}(Y_r)$. Since $\dim_{E_i} Y_i = \ell$ and $E_i \cong \mathbb{F}_{q^s}$, we get that $|C| \leqslant q^{s\ell^2 r}$ and $|X| \leqslant s^r q^{s\ell^2 r}$.

Now $NC/C \cong N/(N \cap C) = N/F$. So NC/C can be generated by r elements and we have $|X/C|^r$ choices for NC/C as a subgroup of X/C and this is clearly less than or equal to s^{r^2}.

Once a choice of NC/C is made as a subgroup of X/C, we choose a set of r (or fewer) generators for NC/C. Since $N \cap C = F$, as a subgroup of X, N is determined by F and r other elements that map to the generating set of NC/C that we have chosen. There are $|C|$ choices for an element of X that maps to any fixed member of X/C. So there are at most $|C|^r$ possibilities in all for N as a subgroup of X once NC/C has been chosen. Putting all the estimates together, once F is fixed we have at most $q^{s\ell^2 r^2} s^{r^2}$ choices for N as a subgroup of X.

We are now left with determining the choices for G given that F and N are fixed. Let Y be the subgroup of T that permutes the Y_i and normalises F. So $N \leqslant G \leqslant Y$. Further, there is a homomorphism from Y into $\mathrm{Sym}(r)$ with kernel X. Thus Y/X may be regarded as a subgroup of $\mathrm{Sym}(r)$.

Clearly $G \cap X = N$ and so $GX/X \cong G/N$. Thus GX/X is a transitive soluble A-subgroup of $\mathrm{Sym}(r)$ and by Lemma 12.1 we have at most $3^{(r^2-1)/2}$ $2^{r\log(r)+3(\log(r))^3}$ choices for GX/X as a subgroup of Y/X. Theorem 6.11 shows that GX/X can be generated by $(r+1)/2$ elements. Once the choice for GX/X as a subgroup of Y/X is fixed and $(r+1)/2$ generators for GX/X have been chosen, G is determined as a subgroup of Y by N and $(r+1)/2$ elements of Y that map to the generating set of GX/X. Thus we have at most $|X|^{(r+1)/2}$ choices for G as a subgroup of Y. This implies that once F and

N are fixed we have at most $3^{(r^2-1)/2} \, 2^{r\log(r)+3(\log(r))^3} \, s^{r(r+1)/2} q^{s\ell^2 r(r+1)/2}$ choices for G as a subgroup of Y.

Putting together all the estimates, we get the following. The number of conjugacy classes of subgroups that are maximal amongst irreducible soluble p'-A-subgroups of T is at most

$$\sum_{(\ell,r,s)} (q^s - 1)^{r^2} \, q^{s\ell^2 r^2} \, s^{r^2} \, 3^{(r^2-1)/2} \, 2^{r\log(r)+3(\log(r))^3} \, s^{r(r+1)/2} \, q^{s\ell^2 r(r+1)/2}$$

where (ℓ, r, s) ranges over the ordered triples of positive integers satisfying $d = \ell r s$.

This part is devoted to simplifying the above expression. We note that

- $(q^s - 1)^{r^2} \, s^{r^2} \leqslant q^{2sr^2 - r^2}$ (since $s \leqslant q^{s-1}$);
- $3^{(r^2-1)/2} \leqslant 2^{(r^2-1)}$.

It is easy to see that we have at most $6d^{5/6}$ choices for (ℓ, r, s). Also $\log r \leqslant r - 1$. Therefore the number of conjugacy classes of subgroups that are maximal amongst irreducible soluble p'-A-subgroups of T is at most

$$6d^{5/6} \, q^{6d^2} \, 2^{3(\log d)^3 - 1}.$$

Using the fact that $3(\log(d))^3 + (5/6)\log(d) + \log(6) - 1$ is at most $13d$, we get the required result.

Corollary 12.3 *Let q be a power of a prime p and let $T = \mathrm{GL}(d, q)$. Then the number of conjugacy classes of subgroups that are maximal amongst soluble p'-A-subgroups of T is at most $q^{6d^2} \, 2^{14d-1}$.*

Proof: Let G be maximal amongst soluble p'-A-subgroups of T. Now G is completely reducible, by Maschke's theorem. Thus $G = G_1 \times \cdots \times G_k$ where for each i, we have that G_i is maximal amongst irreducible soluble p'-A-subgroups of $\mathrm{GL}(d_i, q)$ and $d = d_1 + \cdots + d_k$. Further, the conjugacy classes of G_i in $\mathrm{GL}(d_i, q)$ respectively determine the conjugacy class of G in T. Therefore the number of conjugacy classes of subgroups of T that are maximal among soluble p'-A-subgroups of T is at most

$$\sum \prod_{i=1}^{k} q^{6d_i^2} \, 2^{13d_i} \leqslant \sum q^{6d^2} \, 2^{13d},$$

where the sum is over all (unordered) partitions $d_1, d_2, \ldots, d_k$ of d. Since the number of partitions of d is at most 2^{d-1} by Corollary 5.10, the result follows.

12.3 Maximal soluble p'-A-subgroups

We require three more small results to establish our upper bound for $f_{A,\,\mathrm{sol}}(n)$. These are presented in this section.

For any group G and a prime p, let $M_{s,p',A}(G)$ be the set of subgroups of G that are maximal amongst soluble p'-A-subgroups of G and let $[M_{s,p',A}(G)]$ denote the set of conjugacy classes of subgroups of G that are maximal amongst soluble p'-A-subgroups of G.

Lemma 12.4 *Let G be a finite group and let H be a subgroup of G. Then* $|M_{s,p',A}(H)| \leqslant |M_{s,p',A}(G)|.$

Proof: Let M belong to $M_{s,p',A}(H)$. Then there exists $\hat{M} \in M_{s,p',A}(G)$ such that $M \leqslant \hat{M}$. If $M_1 \in M_{s,p',A}(H)$ and $M_1 \leqslant \hat{M}$ then

$$T = \langle M, M_1 \rangle \leqslant \hat{M} \cap H.$$

Consequently, T is a soluble p'-A-subgroup of H and hence $T = M = M_1$. Therefore $|M_{s,p',A}(H)| \leqslant |M_{s,p',A}(G)|.$

Lemma 12.5 *Let G be a finite group and N a normal p-subgroup of G. Then* $|[M_{s,p',A}(G)]| \leqslant |[M_{s,p',A}(G/N)]|.$

Proof: Let M belong to $M_{s,p',A}(G)$. Since $MN/N \cong M/(M \cap N) \cong M$ we have that MN/N is a soluble p'-A-subgroup of G/N. We prove that MN/N is maximal with respect to this property.

Now let T/N be any soluble p'-A-subgroup of G/N such that $MN/N \leqslant T/N$. Clearly T is soluble and N is the Sylow p-subgroup of T. So by the Schur–Zassenhaus theorem (see Corollary 7.17 and the remark following it), there exists a p'-subgroup H of T such that $T = HN$. Clearly H is a Hall p'-subgroup of T. So by Theorem 6.1 we have that $M \leqslant tHt^{-1}$ for some t in T. But H is a soluble p'-A-subgroup of G and $M \in M_{s,p',A}(G)$, so $M = tHt^{-1}$. Therefore $T/N = HN/N = tHt^{-1}N/N = MN/N$ and thus $MN/N \in M_{s,p',A}(G/N)$.

We can now define a map from $[M_{s,p',A}(G)]$ to $[M_{s,p',A}(G/N)]$ which takes $[M]$ in $[M_{s,p',A}(G)]$ to $[MN/N]$ in $[M_{s,p',A}(G/N)]$. We shall show that this map is injective.

Suppose $[M_1N/N] = [M_2N/N]$ for some M_1 and M_2 in $M_{s,p',A}(G)$. Then there exists g in G such that $gM_1g^{-1}N = M_2N$. It is easy to see that M_2N is a soluble subgroup of G and by the Schur–Zassenhaus theorem there exists b in M_2N such that $gM_1g^{-1} = bM_2b^{-1}$. Consequently $[M_1] = [M_2]$ and the lemma follows.

For a group G and a prime p, let $O_{p'}(G)$ denote the largest normal p'-subgroup of G and let $O_p(G)$ denote the largest normal p-subgroup of G. Recall from Section 6.2 of Chapter 6 that for a finite group G of order n, the Fitting subgroup of G, denoted by $F(G)$, is the direct product of the subgroups $O_p(G)$ for all primes p dividing n.

Lemma 12.6 *Let G be a soluble A-group of order $\prod_{i=1}^k p_i^{\alpha_i}$ where the p_i are distinct primes and let $G_i = G/O_{p'_i}(G)$. Then G_i is a soluble A-group and*

(i) $G \leqslant G_1 \times \cdots \times G_k$ *as a subdirect product;*

(ii) $O_{p'_i}(G_i) = \{1\}$ *and G_i is a semidirect product of its Sylow p_i-subgroup by a p'_i-subgroup and any Sylow p_i-subgroup of G_i is isomorphic to a Sylow p_i-subgroup of G.*

Proof: Part (i) follows from the fact that $\bigcap_{i=1}^k O_{p'_i}(G) = \{1\}$. It is obvious that $O_{p'_i}(G_i) = \{1\}$ and this implies that $F(G_i) = O_{p_i}(G_i)$. Let P_i be a Sylow p_i-subgroup of G_i. Since G_i is a soluble A-group, by Corollary 6.5 we have

$$P_i \leqslant C_{G_i}(O_{p_i}(G_i)) = C_{G_i}(F(G_i)) = F(G_i).$$

Thus $P_i = O_{p_i}(G_i)$ and so G_i is an extension of a p_i-subgroup by a p'_i-subgroup. The result follows by the Schur–Zassenhaus theorem.

Note 12.7 Let G_i and P_i be as in Lemma 12.6. Let H_i be a Hall p'_i-subgroup of G_i. Then $G_i = P_i H_i$ and H_i acts on P_i by conjugation. By Lemma 12.6 (ii) this action is faithful. Thus H_i is isomorphic to a subgroup of Aut (P_i).

Note 12.8 Let $\Phi(P_i)$ denote the Frattini subgroup of P_i and let $A_i = \text{Aut}(P_i)$. Then there exists a natural homomorphism from A_i into $\text{Aut}(P_i/\Phi(P_i))$. By Lemma 3.13, the kernel B_i of this homomorphism is a p_i-group. If $d_i = d(P_i) = d(P_i/\Phi(P_i))$ then $\text{Aut}(P_i/\Phi(P_i)) \cong \text{GL}(d_i, p_i)$ and so we can regard A_i/B_i as a subgroup of $\text{GL}(d_i, p_i)$.

12.4 Enumeration of soluble A-groups

We shall prove an upper bound for $f_{A,\,\text{sol}}(n)$ here. Pyber's paper [82] has been the main source of inspiration for the proof of Theorem 12.9.

Theorem 12.9 *Let $n = \prod_{i=1}^k p_i^{\alpha_i}$ be the prime factorisation of n. Then*

$$f_{A,\,\text{sol}}(n) \leqslant n^{3\mu+13} \prod_{i=1}^k p_i^{8\alpha_i^2}.$$

Proof: Let G be a soluble A-group of order n and let $G_i = G/O_{p_i'}(G)$. Then by Lemma 12.6, we have that $G \leqslant G_1 \times \cdots \times G_k$. Further by Note 12.7, for each i, we have $G_i = P_i H_i$ where P_i is the Sylow p_i-subgroup of G_i and H_i is a p_i'-A-subgroup of $A_i = \mathrm{Aut}(P_i)$. So there exists $M_i \leqslant A_i$ such that M_i is maximal among soluble p_i'-A-subgroups of A_i and $H_i \leqslant M_i$. Let $\hat{G}_i = P_i M_i$ and let $\hat{G} = \hat{G}_1 \times \cdots \times \hat{G}_k$. Then $G \leqslant \hat{G}$. By our construction $\hat{G}$ is a soluble A-group.

We shall first enumerate the possibilities for $\hat{G}$ up to isomorphism and then estimate the number of subgroups of $\hat{G}$ of order n up to isomorphism. For the former we need to count the number of $\hat{G}_i$ up to isomorphism and this depends only on the isomorphism class of the P_i and the conjugacy class of M_i in A_i. So the number of choices for $\hat{G}_i$ up to isomorphism is at most

$$\sum_{\substack{|P_i| = p_i^{\alpha_i} \\ P_i \text{ abelian}}} |[M_{s,p_i',\mathrm{A}}(A_i)]| . \tag{12.1}$$

Fix P_i. Let d_i and B_i be as in Note 12.8 and let $T_i = \mathrm{GL}(d_i, p_i)$. By Lemma 12.5 we have $|[M_{s,p_i',\mathrm{A}}(A_i)]| \leqslant |[M_{s,p_i',\mathrm{A}}(A_i/B_i)]|$. But

$$\begin{aligned}
|[M_{s,p_i',\mathrm{A}}(A_i/B_i)]| &\leqslant |M_{s,p_i',\mathrm{A}}(A_i/B_i)| \\
&\leqslant |M_{s,p_i',\mathrm{A}}(T_i)| &&\text{(by Lemma 12.4)} \\
&\leqslant p_i^{d_i^2} |[M_{s,p_i',\mathrm{A}}(T_i)]| \\
&\leqslant p_i^{7d_i^2} 2^{14d_i - 1} &&\text{(by Corollary 12.3)} \\
&\leqslant p_i^{7\alpha_i^2} 2^{14\alpha_i - 1} .
\end{aligned}$$

Thus

$$|[M_{s,p_i',\mathrm{A}}(A_i)]| \leqslant p_i^{7\alpha_i^2} 2^{14\alpha_i - 1} . \tag{12.2}$$

The number of choices for P_i is equal to the number of unordered partitions of α_i and by Corollary 5.10 this is at most $2^{\alpha_i - 1}$. Thus by (12.1) and (12.2) the number of choices for $\hat{G}_i$ up to isomorphism is less than or equal to

$$p_i^{7\alpha_i^2} 2^{15\alpha_i - 2}$$

and the number of choices for $\hat{G}$ is at most

$$\prod_{i=1}^{k} p_i^{7\alpha_i^2} 2^{15\alpha_i - 2} .$$

Now let $\hat{G}$ be fixed. Let $\{S_1, \ldots, S_k\}$ be a Sylow system for G. So S_i is a Sylow p_i-subgroup of G and for all i, j we have $S_i S_j = S_j S_i$. Thus $G = S_1 S_2 \cdots S_k$. Further, by Theorem 6.2 there exist $Q_1, \ldots, Q_k$, part of a Sylow system for $\hat{G}$, such that $S_i \leqslant Q_i$. Any two Sylow systems for $\hat{G}$ are conjugate. Consequently the number of choices for G as a subgroup of $\hat{G}$ (and up to conjugacy) is at most

$$\left| \{S_1, \ldots, S_k \mid S_i \leqslant Q_i \text{ and } |S_i| = p_i^{\alpha_i}\} \right| \leqslant \prod_{i=1}^{k} |Q_i|^{\alpha_i}$$

where Q_i is a Sylow p_i-subgroup of $\hat{G}$ and the Q_i form part of a Sylow system for $\hat{G}$.

For each i, we have $Q_i = R_{i1} \times \cdots \times R_{ik}$ for some Sylow p_i-subgroups R_{ij} of $\hat{G}_j$. Note that $|R_{ii}| = p_i^{\alpha_i}$. Let

$$X_j = \prod_{\substack{i=1 \\ i \neq j}}^{k} R_{ij}.$$

Since the Q_i form part of a Sylow system for $\hat{G}$, we have that $Q_i Q_j = Q_j Q_i$. So X_j is well-defined as (a soluble p'_j-A-) subgroup of $\hat{G}_j = P_j M_j$. So X_j is isomorphic to a subgroup of M_j. Further,

$$M_j \cong M_j B_j / B_j \leqslant A_j / B_j \leqslant \mathrm{GL}(d_j, p_j).$$

Note that by Maschke's theorem M_j may be regarded as a completely reducible soluble subgroup of $\mathrm{GL}(d_j, p_j)$. Thus by Theorem 10.2

$$|X_j| \leqslant |M_j| \leqslant p_j^{3d_j - 2} \leqslant p_j^{3\alpha_j - 2}.$$

Let $\mu = \mu(n)$. Then

$$\prod_{i=1}^{k} |Q_i|^{\alpha_i} = \prod_{i=1}^{k} \prod_{j=1}^{k} |R_{ij}|^{\alpha_i} \leqslant \prod_{t=1}^{k} (|R_{tt}|^{\alpha_t}) \left(\prod_{i \neq j} |R_{ij}| \right)^{\mu}$$

$$= \prod_{i=1}^{k} p_i^{\alpha_i^2} \prod_{j=1}^{k} |X_j|^{\mu},$$

since $|X_j| = \prod_{\substack{i=1 \\ i \neq j}}^{k} |R_{ij}|$. Hence

$$\prod_{i=1}^{k} |Q_i|^{\alpha_i} \leqslant \prod_{i=1}^{k} p_i^{\alpha_i^2 + 3\alpha_i \mu - 2\mu} = n^{3\mu} \prod_{i=1}^{k} p_i^{\alpha_i^2 - 2\mu} \leqslant n^{3\mu - 2} \prod_{i=1}^{k} p_i^{\alpha_i^2}.$$

Putting together all the estimates, we get

$$f_{A, \text{sol}}(n) \leqslant (\text{number of choices for } \hat{G}) \, n^{3\mu-2} \prod_{i=1}^{k} p_i^{\alpha_i^2}$$

$$\leqslant n^{3\mu-2} \prod_{i=1}^{k} p_i^{8\alpha_i^2} \, 2^{15\alpha_i-2}$$

$$\leqslant n^{3\mu-2} \prod_{i=1}^{k} p_i^{8\alpha_i^2+15\alpha_i} \, .$$

Therefore

$$f_{A, \text{sol}}(n) \leqslant n^{3\mu+13} \prod_{i=1}^{k} p_i^{8\alpha_i^2} \, .$$

We have the following obvious corollary to the above theorem:

Corollary 12.10

$$f_{A, \text{sol}}(n) \leqslant n^{11\mu+13} \, .$$

We need to deliberate on whether a bound of the form $n^{a\mu+b}$, as in Corollary 12.10, is a 'good' bound for the number of soluble A-groups up to isomorphism.

For any prime p and any $\alpha \in \mathbb{N}$, $f_A(p^\alpha)$ is just the number of isomorphism classes of abelian groups of order p^α. Thus $f_A(p^\alpha) = f_{A, \text{sol}}(n)(p^\alpha) \leqslant p^\alpha$. Therefore there exist infinitely many n with $\{\mu(n)\}$ unbounded such that $f_{A, \text{sol}}(n) \leqslant n$. Thus it is impossible to find a constant $c > 0$ such that $n^{c\mu} \leqslant f_{A, \text{sol}}(n)$ for all n.

Let

$$\ell = \overline{\lim}_{n \to \infty} \log f_{A, \text{sol}}(n)/(\mu(n) \log n).$$

Then given any $\epsilon > 0$, for all sufficiently large n we have $f_{A, \text{sol}}(n) < n^{(\ell+\epsilon)\mu(n)}$ and given any positive integer m, there exists $n \geqslant m$ such that $f_{A, \text{sol}}(n) > n^{(\ell-\epsilon)\mu(n)}$. It would be interesting to determine the exact value of ℓ. An upper bound for ℓ is a consequence of Corollary 12.10, and a lower bound follows from the results contained in Chapter 18; see the discussion at the end of Section 18.1.

A correct leading term in the case of soluble A-groups should lead us to the correct leading term for A-groups in general. As suggested in Pyber's paper [82], this in turn should lead to better estimates for the error term in the bound for $f(n)$.

13

Maximal soluble linear groups

In Chapter 9, we established some of the structure of a soluble primitive subgroup G of $\mathrm{GL}(d, q)$. In this chapter, we will assume in addition that G is maximal subject to being soluble. This allows us to prove more comprehensive structural results about G. We will use the notation of Chapter 9. In particular, we define q, d, V, A, B, C, K, d_1 and d_2 in the same way. In addition, we define the following notation:

- $\langle H \rangle$ – for any subset $H \subseteq \mathrm{GL}(d, q)$, we write $\langle H \rangle$ for the $\mathbb{F}_q$-subalgebra of $\mathrm{End}_{\mathbb{F}_q}(V)$ generated by H;
- L – the centre of $\langle G \rangle$;
- ℓ – the dimension of L over $\mathbb{F}_q$.

13.1 The field K and a subfield of K

Proposition 13.1 *The subgroup A is cyclic of order $q^{d_1} - 1$. Moreover, $K = A \cup \{0\}$.*

Proof: Proposition 9.1 shows that A is cyclic, and that $|K| = q^{d_1}$. So it is sufficient to show that $K = A \cup \{0\}$.

Let $K^* = K \setminus \{0\}$. Since K is a field, K^* is a cyclic subgroup of $\mathrm{GL}(d, q)$. Since G normalises A, the definition of K implies that K is preserved under conjugation by elements of G. Hence K^* is normalised by G and so $GK^* \leqslant \mathrm{GL}(d, q)$. Moreover, GK^* is soluble, since K^* is abelian and since GK^*/K^* is isomorphic to the quotient $G/G \cap K^*$ of the soluble group G. By the maximality of G we have that $K^* \leqslant G$. But K^* is abelian, normal in G and contains A. So the definition of A implies that $K^* = A$. Thus $K = A \cup \{0\}$, as required.

113

Proposition 13.2 *The algebra L is a subfield of K, and $\langle G \rangle$ is isomorphic to the algebra $\mathcal{M}_{d/\ell}(L)$ of $d/\ell \times d/\ell$ matrices over L.*

Proof: Since G acts irreducibly, V is a faithful irreducible $\langle G \rangle$-module and so $\langle G \rangle$ is simple by Proposition 8.2. Hence by Theorem 8.3 we have that $\langle G \rangle$ is isomorphic to the set $\mathcal{M}_r(F)$ of all $r \times r$ matrices over a skew field F, for some integer r. Since F is finite, F is a field by Theorem 8.11. The centre of a full matrix ring over a field consists of the set of scalar matrices, and so the centre L of $\langle G \rangle$ is isomorphic to F. Thus $\langle G \rangle \cong \mathcal{M}_r(L)$. The natural $\mathcal{M}_r(L)$-module (where $r \times r$ matrices over L act on the vector space L^r) is irreducible. Since $\mathcal{M}_r(L)$ is simple, Theorem 8.1 implies that all irreducible $\mathcal{M}_r(L)$-modules are isomorphic and so, in particular, their $\mathbb{F}_q$-dimensions are equal. But V is an irreducible $\mathcal{M}_r(L)$ module, so $d = r\ell$ and therefore $r = d/\ell$. This establishes the second statement of the proposition.

Now, L^* is an abelian subgroup of $\mathrm{GL}(d, q)$. Since L commutes with every element of $\langle G \rangle$, the subgroup L^* is centralised by G. As in the proof of Proposition 13.1, the group GL^* is soluble, and so $L^* \leqslant G$. Hence

$$L^* \leqslant Z(G) \leqslant A \leqslant K^*,$$

and so L is a subfield of K. This establishes the proposition.

13.2 The quotient G/C and the algebra $\langle C \rangle$

Lemma 13.3 *Let $\langle C \rangle$ be the $\mathbb{F}_q$-subalgebra of $\mathrm{End}_{\mathbb{F}_q}(V)$ generated by C. Then*

$$\dim_{\mathbb{F}_q} \langle G \rangle = |G/C| \cdot \dim_{\mathbb{F}_q} \langle C \rangle.$$

Proof: Let $g_1, g_2, \ldots, g_r$ be a transversal for C in G. Clearly

$$\langle G \rangle = \langle C \rangle g_1 + \langle C \rangle g_2 + \cdots + \langle C \rangle g_r.$$

To establish the lemma, it suffices to show that this sum is direct. So suppose, for a contradiction, that

$$c_1 g_1 + c_2 g_2 + \cdots + c_r g_r = 0 \tag{13.1}$$

for some elements $c_1, c_2, \ldots, c_r \in \langle C \rangle$ that are not all zero. Assume that the number of elements c_i that are non-zero is as small as possible.

There must exist distinct $u, v \in \{1, 2, \ldots, r\}$ such that $c_u \neq 0$ and $c_v \neq 0$ (for $c_i g_i = 0$ implies that $c_i = 0 g_i^{-1} = 0$). Since g_u and g_v lie in distinct cosets of C, conjugation by g_u^{-1} and conjugation by g_v^{-1} yield distinct automorphisms of A. So there exists $z \in A$ such that $g_u z g_u^{-1} \neq g_v z g_v^{-1}$. For all $i \in \{1, 2, \ldots, r\}$, define $z_i \in A$ by $z_i = g_i z g_i^{-1}$. Then

$$0 = (c_1 g_1 + c_2 g_2 + \cdots + c_r g_r) z - z_u (c_1 g_1 + c_2 g_2 + \cdots + c_r g_r)$$

$$= \sum_{i=1}^{r} (c_i g_i z - z_u c_i g_i)$$

$$= \sum_{i=1}^{r} (c_i z_i g_i - z_u c_i g_i)$$

$$= \sum_{i=1}^{r} (z_i - z_u) c_i g_i,$$

the last equality following since c_i is a linear combination of elements in C and so commutes with $z_i \in A$. But the equality

$$\sum_{i=1}^{r} (z_i - z_u) c_i g_i = 0$$

contradicts the minimality of the number of non-zero coefficients in (13.1), since the coefficient corresponding to g_u is zero but the coefficient corresponding to g_v is non-zero. Thus the lemma follows.

Proposition 13.4 *The quotient G/C is isomorphic to $\mathrm{Gal}(K : L)$. The $\mathbb{F}_q$-algebra $\langle C \rangle$ is isomorphic to $\mathcal{M}_{d_2}(K)$.*

Proof: The fact that $K^* \subseteq C$ implies that $\langle C \rangle$ is equal to the set $\langle C \rangle_K$ of K-linear combinations of elements of C. Since C may be regarded as a subgroup of $\mathrm{GL}(d_2, K)$, we must have that $\langle C \rangle_K \subseteq \mathcal{M}_{d_2}(K)$. Hence

$$\dim_{\mathbb{F}_q} \langle C \rangle \leqslant d_2^2 \dim_{\mathbb{F}_q} K = d_2 d,$$

with equality if and only if $\langle C \rangle = \mathcal{M}_{d_2}(K)$.

By Proposition 9.3, the quotient G/C may be realised as a subgroup of $\mathrm{Gal}(K : \mathbb{F}_q)$ by associating $gC \in G/C$ with the field automorphism $x \mapsto gxg^{-1}$ of K. Since L is centralised by G, we have in fact embedded G/C into $\mathrm{Gal}(K : L)$. Now, $|\mathrm{Gal}(K : L)| = d_1/\ell$. Thus $|G/C| \leqslant d_1/\ell$, and $G/C \cong \mathrm{Gal}(K : L)$ if and only if $|G/C| = d_1/\ell$.

By Proposition 13.2, we have that $\langle G \rangle$ is isomorphic to $\mathcal{M}_{d/\ell}(L)$. Hence

$$\dim_{\mathbb{F}_q} \langle G \rangle = (d^2/\ell^2) \dim_{\mathbb{F}_q} L = d^2/\ell.$$

But by Lemma 13.3,

$$\dim_{\mathbb{F}_q} \langle G \rangle = |G/C| \dim_{\mathbb{F}_q} \langle C \rangle$$

$$\leqslant (d_1/\ell)\, d_2 d$$

$$= d^2/\ell,$$

and since equality holds we must have that $|G/C| = d_1/\ell$ and $\dim_{\mathbb{F}_q} \langle C \rangle = d_2 d$. By the last sentences of the previous two paragraphs, this establishes the proposition.

13.3 The quotient B/A

Lemma 13.5 *The $\mathbb{F}_q$-subalgebra $\langle B \rangle$ is simple with centre K.*

Proof: The subgroup B is a normal subgroup of the primitive group G. Hence we may argue, just as we did for A in the proof of Proposition 9.1, that V may be decomposed as a direct sum of one or more copies of a $\langle B \rangle$-module X, and $\langle B \rangle$ acts faithfully and irreducibly on X. So, since X is an irreducible faithful $\langle B \rangle$-module, we find that $\langle B \rangle$ is simple.

Since $K^* = A \leqslant B \subseteq \langle B \rangle$, and since B centralises A, we have that K is contained in the centre of $\langle B \rangle$. Conversely, suppose that c is an element in the centre of $\langle B \rangle$. Let $x_1, x_2, \ldots, x_k$ be a transversal of A in B where $x_1 = 1$. We may write

$$c = z_1 x_1 + z_2 x_2 + \cdots + z_k x_k$$

for some $z_1, z_2, \ldots, z_k \in K$. To show that $c \in K$, it is sufficient to show that $z_u = 0$ whenever $u \neq 1$. Let $u \in \{2, 3, \ldots, k\}$ be fixed. Since $x_u \notin A = Z(B)$, there exists an element $y \in B$ that does not commute with x_u. For all $i \in \{1, 2, \ldots, k\}$, define $r_i \in A$ by $r_i = [y, x_i^{-1}]$. Note that $r_i \in A$ for all i, since B/A is abelian. Moreover, note that $r_u \neq 1$, by our choice of y. Since c is in the centre of $\langle B \rangle$, we have that

$$0 = c - y^{-1}cy$$

$$= \sum_{i=1}^{k} z_i (1 - r_i) x_i.$$

The proof of Proposition 9.5 showed that the elements x_i are linearly independent over K. So $z_i(1 - r_i) = 0$ for all $i \in \{1, 2, \ldots, k\}$. Since $r_u \neq 1$, we find that $z_u = 0$, as required.

Lemma 13.6 *Let $\langle B \rangle$ and $\langle C \rangle$ be the $\mathbb{F}_q$-subalgebras of $\text{End}_{\mathbb{F}_q}(V)$ generated by B and C respectively. Then $\langle B \rangle = \langle C \rangle$.*

Proof: Clearly $\langle B \rangle \subseteq \langle C \rangle$. To show the reverse inclusion, it suffices to show that $C \subseteq \langle B \rangle$.

We define a function $\phi : C/A \to C_{\text{GL}(d,q)}(B)/A$ as follows. Let $u \in C$. The map $x \mapsto uxu^{-1}$ is an automorphism of $\langle B \rangle$ that fixes K. Lemma 13.5 states that K is the centre of $\langle B \rangle$. Further, since $K = Z(\langle B \rangle)$, the subalgebra $\langle B \rangle$ is also a K-subalgebra and by Lemma 13.5 it is simple. The Skolem–Noether theorem (Corollary 8.9) now implies that $x \mapsto uxu^{-1}$ is an inner automorphism: let $v \in \langle B \rangle$ be such that $vxv^{-1} = uxu^{-1}$ for all $x \in \langle B \rangle$. Note that v is determined up to multiplication by an invertible element in the centre of $\langle B \rangle$; in other words, v is determined up to multiplication by an element of A. We define $\phi(uA) = v^{-1}uA$. Note that ϕ is well-defined, since conjugation by u gives rise to the same automorphism as conjugation by uz for any $z \in A$.

Now, ϕ is a homomorphism. For let $u_1, u_2 \in C$ and let $v_1, v_2 \in \langle B \rangle$ be such that $u_i x u_i^{-1} = v_i x v_i^{-1}$ for all $x \in \langle B \rangle$. So $\phi(u_i A) = v_i^{-1} u_i A$. For all $x \in \langle B \rangle$ we find that

$$(u_1 u_2) x (u_1 u_2)^{-1} = u_1 (u_2 x u_2^{-1}) u_1^{-1}$$
$$= u_1 (v_2 x v_2^{-1}) u_1^{-1}$$
$$= v_1 (v_2 x v_2^{-1}) v_1^{-1},$$

the last equality following since $v_2 x v_2^{-1} \in \langle B \rangle$. This shows that $\phi(u_1 u_2 A) = (v_1 v_2)^{-1} u_1 u_2 A$. Since $v_2^{-1} \in \langle B \rangle$ and $v_1^{-1} u_1 \in C_{\text{GL}(d,q)}(B)$, we find that v_2^{-1} and $v_1^{-1} u_1$ commute and so

$$\phi(u_1 u_2 A) = v_2^{-1} v_1^{-1} u_1 u_2 A$$
$$= v_1^{-1} u_1 v_2^{-1} u_2 A$$
$$= \phi(u_1 A) \phi(u_2 A).$$

Thus ϕ is indeed a homomorphism.

We claim that $gA\phi(uA)(gA)^{-1} = \phi(gug^{-1}A)$ for all $u \in C$ and all $g \in G$. To see this, let $v \in \langle B \rangle$ be such that $uxu^{-1} = vxv^{-1}$ for all $x \in \langle B \rangle$; so $\phi(uA) = v^{-1}uA$. Note that $g^{-1}xg \in g^{-1}\langle B \rangle g = \langle B \rangle$ since g^{-1} normalises B. It is now easy to check, as in the previous paragraph, that $(gug^{-1})x(gug^{-1})^{-1} = (gvg^{-1})x(gvg^{-1})^{-1}$ for all $x \in \langle B \rangle$ and so

$$\phi(gug^{-1}A) = (gvg^{-1})^{-1}gug^{-1}A = gv^{-1}ug^{-1}A = gA\phi(u)(gA)^{-1},$$

and our claim follows.

Let $D \leqslant C_{\mathrm{GL}(d,q)}(B)$ be such that $D/A = \operatorname{im} \phi$. Since ϕ is a homomorphism, $\operatorname{im} \phi$ is soluble and so D is soluble. Moreover, since

$$\phi(gug^{-1}A) = gA\phi(uA)(gA)^{-1} \quad \text{for all } u \in C \text{ and } g \in G,$$

we find that G/A normalises $\operatorname{im} \phi$ and so G normalises D. Hence GD is soluble and so by the maximality of G we know that $D \leqslant G$. We now use, three times, the fact that D centralises B. Firstly,

$$D \leqslant C_G(B) \leqslant C_G(A) = C.$$

Secondly, since we now know that D is contained in the centraliser of B in C, Proposition 9.7 implies that $D \leqslant B$. Finally, we find that $D \leqslant Z(B) = A$. By the definition of ϕ, we have that for all $u \in C$ there exists $v \in \langle B \rangle$ such that $v^{-1}uA \in D/A = A$. Hence $u \in vA \subseteq \langle B \rangle A = \langle B \rangle$. Thus $C \subseteq \langle B \rangle$ and the proposition follows.

Proposition 13.7 *The quotient B/A has order d_2^2 and $\langle B \rangle$ is isomorphic (as an $\mathbb{F}_q$-algebra) to $\mathcal{M}_{d_2}(K)$.*

Proof: By Proposition 13.4, we have that $\langle C \rangle$ is isomorphic to $\mathcal{M}_{d_2}(K)$. By Lemma 13.6, $\langle B \rangle = \langle C \rangle$ and so the second statement of the proposition follows. Note that since $K^* = A \leqslant B$, we have that $\langle B \rangle = \langle B \rangle_K$, the subset of $\mathrm{End}_{\mathbb{F}_q}(V)$ consisting of K-linear combinations of B. Proposition 9.5 and Lemma 13.6 now imply that

$$|B/A| = \dim_K \langle B \rangle = \dim_K \langle C \rangle = d_2^2,$$

and the proposition follows.

Proposition 13.8 *The quotient B/A is its own centraliser in $C_{\mathrm{GL}(d,q)}(A)/A$.*

Proof: Define $\widehat{C} = C_{\mathrm{GL}(d,q)}(A)$. Let $y \in \widehat{C}$ and suppose that yA centralises B/A. We must show that $yA \in B/A$.

Since $y \in \widehat{C}$ we have that y centralises K and so y may be regarded as an element of $\mathrm{GL}(d_2, K)$. Hence

$$y \in \mathrm{GL}(d_2, K) \subseteq \langle B \rangle,$$

by Proposition 13.7.

Now the automorphism $\alpha : B \to B$ defined by $\alpha(x) = y^{-1}xy$ fixes A since $y \in \widehat{C}$ and induces the identity automorphism on B/A since yA centralises B/A. By Proposition 9.6, this automorphism is inner: let $h \in B$ be such that $\alpha(x) = h^{-1}xh$ for all $x \in B$. But now yh^{-1} centralises B. Moreover, $yh^{-1} \in \langle B \rangle$, and so yh^{-1} is contained in the centre of $\langle B \rangle$. But the centre of

$\langle B \rangle$ is K, and so yh^{-1} is an invertible element in K. Hence $yh^{-1} \in A$ and so $yA = hA \in B/A$, as required.

13.4 The subgroup B

In this section, we determine the structure of the subgroup B. We begin by considering B as an abstract group, and then consider its representation as a subgroup of $\mathrm{GL}(d, q)$.

We define the following groups (which will occur as possible Sylow subgroups of B). Let p be a prime (not, for the purposes of this section, the characteristic of $\mathbb{F}_q$) and let r and s be positive integers. We define $M(p, r, s)$ to be the group with generating set $\{x_1, x_2, \ldots, x_r, y_1, y_2, \ldots, y_r, z\}$ subject to the following relations:

$$x_i^p = 1 \text{ for all } i \in \{1, 2, \ldots, r\}, \tag{13.2}$$

$$y_i^p = 1 \text{ for all } i \in \{1, 2, \ldots, r\}, \tag{13.3}$$

$$z^{p^s} = 1, \tag{13.4}$$

$$[x_i, z] = 1 \text{ for all } i \in \{1, 2, \ldots, r\}, \tag{13.5}$$

$$[y_i, z] = 1 \text{ for all } i \in \{1, 2, \ldots, r\}, \tag{13.6}$$

$$[x_i, x_j] = 1 \text{ for all } i, j \in \{1, 2, \ldots, r\}, \tag{13.7}$$

$$[y_i, y_j] = 1 \text{ for all } i, j \in \{1, 2, \ldots, r\}, \tag{13.8}$$

$$[x_i, y_j] = 1 \text{ for all } i, j \in \{1, 2, \ldots, r\} \text{ such that } i \neq j \text{ and} \tag{13.9}$$

$$[x_i, y_i] = z^{p^{s-1}} \text{ for all } i \in \{1, 2, \ldots, r\}. \tag{13.10}$$

It is not difficult to see that $|M(p, r, s)| \leqslant p^{2r+s}$, since every word in the generators and their inverses can be brought into the form

$$x_1^{\alpha_1} x_2^{\alpha_2} \cdots x_r^{\alpha_r} y_1^{\beta_1} y_2^{\beta_2} \cdots y_r^{\beta_r} z^{\gamma}$$

where $\alpha_i, \beta_i \in \{0, 1, \ldots, p-1\}$ and where $\gamma \in \{0, 1, \ldots, p^s - 1\}$. In fact, $|M(p, r, s)| = p^{2r+s}$ and the elements of $M(p, r, s)$ may be written in the form above in a unique manner. This may be established by using similar techniques to those in Section 4.1. It is not difficult to show that $M(p, r, s)$ is an amalgamated product of $M(p, r, 1)$ with a cyclic group of order p^s, amalgamating the central subgroup of order p.

When $s = 1$ and $p = 2$, we define $N(r)$ to be the group generated by the set $\{x_1, x_2, \ldots, x_r, y_1, y_2, \ldots, y_r, z\}$ subject to the relations (13.4) to (13.10) together with the relations

$$x_1^2 = z, \tag{13.11}$$

$$x_i^2 = 1 \text{ for all } i \in \{2, 3, \dots, r\}, \tag{13.12}$$

$$y_1^2 = z \text{ and} \tag{13.13}$$

$$y_i^2 = 1 \text{ for all } i \in \{2, 3, \dots, r\}. \tag{13.14}$$

It is easy to see that every word in the generators and their inverses may be brought into the form

$$x_1^{\alpha_1} x_2^{\alpha_2} \cdots x_r^{\alpha_r} y_1^{\beta_1} y_2^{\beta_2} \cdots y_r^{\beta_r} z^{\gamma}$$

where $\alpha_i, \beta_i, \gamma \in \{0, 1\}$ and so $N(r)$ has order at most 2^{2r+1}. In fact, as before, $|N(r)| = 2^{2r+1}$. Moreover, $N(r)$ and $M(2, r, 1)$ are not isomorphic: it is possible to show that they contain a different number of elements of order 2.

Proposition 13.9 *The subgroup B has order $d_2^2(q^{d_1} - 1)$ and is the direct product of its Sylow subgroups. Let p be a prime dividing $|B|$. Let r be the largest power of p dividing d_2 and let s be the largest power of p dividing $q^{d_1} - 1$. Let P be the Sylow p-subgroup of B; so $|P| = p^{2r+s}$. Let z generate $Z(P)$. Then one of the following cases occurs.*

(1) *P is cyclic and $r = 0$.*
(2) *$r > 0$, $s > 0$ and there exist $x_1, \dots, x_r, y_1, y_2, \dots, y_r \in P$ that generate P modulo $Z(P)$ and satisfy the relations (13.2) to (13.10). In particular, $P \cong M(p, r, s)$.*
(3) *$p = 2$, $r > 0$, $s = 1$ and there exist $x_1, \dots, x_r, y_1, y_2, \dots, y_r \in P$ that generate P modulo $Z(P)$ and satisfy the relations (13.11) to (13.14) and (13.4) to (13.10). In particular, $P \cong N(r)$.*

In particular, Proposition 13.9 shows that once q, d_1 and d_2 are fixed the isomorphism class of the Sylow p-subgroup of B is determined when $p > 2$ and there are at most two possibilities for the Sylow 2-subgroup of B. So there are at most two choices for the isomorphism class of B once q, d_1 and d_2 are fixed.

Proof of Proposition 13.9: Proposition 13.7 implies that $|B/A| = d_2^2$, and $|A| = q^{d_1} - 1$ by Proposition 13.1. So $|B| = d_2^2(q^{d_1} - 1)$. Proposition 9.4 implies that B is nilpotent and so B is the direct product of its Sylow subgroups. It remains to show that for a fixed prime p and Sylow p-subgroup P of B, one of the cases listed at the end of the statement of the proposition holds.

Proposition 9.4 implies that $Z(B) = A$, and that $Z(P) = P \cap A$. Now, $|P \cap A| = p^s$ and A is cyclic. Hence $Z(P)$ is cyclic of order p^s. Since $|P| = p^{2r+s}$

we have that P is abelian (and cyclic) if and only if $r = 0$. Thus when $r = 0$, Case 1 holds. From now on, we assume that $r > 0$ and so P is non-abelian. We must show that either Case 2 or Case 3 holds.

Since P is non-abelian, Proposition 9.4 implies that $|P'| = p$ and $P/Z(P)$ is elementary abelian of order p^{2r}. Note that $P' \leqslant Z(P)$, and so $s > 0$.

We will now use some of the theory of alternating forms that we developed in Section 3.4. Since z generates $Z(P)$, the subgroup P' is generated by $z^{p^{s-1}}$. By identifying $i \in \mathbb{F}_p$ with $z^{ip^{s-1}}$, we may regard the elements of P' as elements of $\mathbb{F}_p$. Since $P/Z(P)$ is elementary abelian, $P/Z(P)$ may be thought of as a vector space over $\mathbb{F}_p$. The process of forming commutators in P induces a map $[\,,\,] : P/Z(P) \times P/Z(P) \to P'$; regarding the elements of P' as elements of $\mathbb{F}_p$, it is easy to check that this map is an alternating form – this follows from the fact that $[ab, c] = [a, c][b, c]$ and $[a, bc] = [a, b][a, c]$ for all $a, b, c \in P$ since P is nilpotent of class 2. Moreover, the representative of any coset of $Z(P)$ corresponding to an element in the radical of $[\,,\,]$ commutes with every element of P and so lies in $Z(P)$. Hence the radical of $[\,,\,]$ is trivial, and so $[\,,\,]$ is non-degenerate.

Let $x_1, x_2, \ldots, x_r, y_1, y_2, \ldots, y_r \in P$ be chosen so that their images $x_i Z(P)$ and $y_i Z(P)$ form a symplectic basis for $P/Z(P)$. We say that the elements x_i and y_i have the *symplectic basis property*. The definition of the alternating form in terms of commutators, together with the fact that z is central, imply that the relations (13.5) to (13.10) hold. Moreover, (13.4) holds since $z \in Z(P)$ and $Z(P)$ has order p^s. Since the cosets $x_i Z(P)$ and $y_i Z(P)$ form a basis of $P/Z(P)$, the elements x_i and y_i generate P modulo $Z(P)$. Since z generates $Z(P)$, we find that $z, x_1, x_2, \ldots, x_r, y_1, y_2, \ldots, y_r$ form a generating set for P.

Assume we are not in the situation when $s = 1$ and $p = 2$. We must show that Case 2 always occurs. We claim that for all $x \in P$, there exists $z' \in Z(P)$ such that $x^p = (z')^p$. Establishing this claim suffices to prove the proposition in the situation we are considering; we may see this as follows. Our claim implies that for all $i \in \{1, 2, \ldots, r\}$ there exists $z_i \in Z(P)$ such that $z_i^p = x_i^p$. Replacing x_i by $x_i z_i^{-1}$ does not affect the symplectic basis property, and so we may assume that $x_i^p = 1$. Similarly, we may assume that $y_i^p = 1$ for all $i \in \{1, 2, \ldots, r\}$. But now the elements $z, x_1, x_2, \ldots, x_r, y_1, y_2, \ldots, y_r$ generate P and satisfy the relations (13.2) and (13.3) in addition to the relations (13.4) to (13.10). Hence P is a quotient of $M(p, r, s)$. But $|M(p, r, s)| \leqslant p^{2r+s} = |P|$, so $|P| = |M(p, r, s)|$ and therefore $P \cong M(p, r, s)$.

We now establish our claim. Let $\pi : P/Z(P) \to Z(P)/Z(P)^p$ be defined by $\pi(xZ(P)) = x^p Z(P)^p$ for all $x \in P$. Note that π is well-defined since $P/Z(P)$ has exponent p and since for all $x \in P$ and $c \in Z(P)$ we have that $(xc)^p = x^p \mod Z(P)^p$. Our claim may be rephrased as the statement that π is trivial.

We show that the map π is a homomorphism. For since P has nilpotency class 2, Lemma 3.3 implies that

$$(xy)^p = x^p y^p [y, x]^{\frac{1}{2}p(p-1)}.$$

Now, $[y, x] \in P'$ and $|P'| = p$, so $[y, x]$ has order dividing p. When $p > 2$, we have that p divides $\frac{1}{2}p(p-1)$ and so $(xy)^p = x^p y^p$. So π is a homomorphism when $p > 2$. When $p = 2$, we are assuming that $s > 1$ and so $P' \leqslant Z(P)^2$ since $Z(P)$ is cyclic. In this case $(xy)^2 Z(P)^2 = x^2 y^2 Z(P)^2$ and so π is a homomorphism when $p = 2$.

Suppose, for a contradiction, that π is not trivial. Since $|Z(P)/Z(P)^p| = p$, the kernel of π has index p in $P/Z(P)$. So there exists a normal subgroup N of P such that $\ker \pi = N/Z(P)$, and clearly N has index p in P. Indeed, the definition of π shows that N is a characteristic subgroup of P, and so $Z(N)$ is a characteristic subgroup of P. Moreover, P is a characteristic subgroup of B and B is normal in G, so $Z(N)$ is normal in G. Hence $Z(N)A$ is an abelian normal subgroup of G and so $Z(N) \leqslant A$. Now, $Z(N)/Z(P)$ is the radical of the restriction of the alternating form $[\,,\,]$ to $N/Z(P)$. Since $Z(N) \leqslant A$, we have shown that the radical of the restriction of the form is trivial and so $[\,,\,]$ is non-degenerate on $N/Z(P)$. But since N has index p in P, the subspace $N/Z(P)$ has codimension 1 in $P/Z(P)$. In particular, $N/Z(P)$ has odd dimension and so (see Proposition 3.18) there can be no non-degenerate alternating form defined on it. This contradiction establishes our claim, and so the proposition follows whenever we do not have both $p = 2$ and $s = 1$.

It remains to consider the situation when $s = 1$ and $p = 2$. Note that $Z(P) = P' = \{1, z\}$ in this case, so that $x^2 \in \{1, z\}$ for all $x \in P$. If we are allowed to use the theory of extra-special groups (see [36, Section 5.5] for example), we may deduce that Case 2 or Case 3 must occur since P is extra-special and since $M(2, r, 1)$ and $N(r)$ are the two non-isomorphic extra-special groups of order 2^{2r+1}. However, for those readers unfamiliar with this theory, we give a direct proof as follows.

We claim that we may choose our elements x_i, y_i so that $x_i^2 = y_i^2$ for all $i \in \{1, 2, \ldots, r\}$. For suppose that $x_i^2 = z$ and $y_i^2 = 1$. Then replacing x_i by $x_i y_i$ preserves the symplectic basis property and

$$(x_i y_i)^2 = x_i^2 y_i^2 [y_i, x_i] = z 1 z = 1.$$

Similarly, if $x_i^2 = 1$ and $y_i^2 = z$ we may replace y_i by $y_i x_i$. Thus our claim follows and we may assume that $x_i^2 = y_i^2$ for all $i \in \{1, 2, \ldots, r\}$.

We claim that we may choose our elements x_i and y_i to have the property that $x_u^2 = y_u^2 = z$ for at most one value of $u \in \{1, 2, \ldots, r\}$. For suppose that there exists $u, v \in \{1, 2, \ldots, r\}$ such that $u \neq v$ and such that $x_u^2 = y_u^2 =$

$x_v^2 = y_v^2 = z$. Replacing x_u, y_u, x_v and y_v by $x_u x_v, x_u x_v y_u, x_v y_u y_v$ and $y_u y_v$ respectively, it is not difficult to check that we still have the symplectic basis property but now $x_u^2 = y_u^2 = x_v^2 = y_v^2 = 1$. Each time we modify the elements x_i and y_i in this way, the number of values $u \in \{1, 2, \ldots, r\}$ such that $x_u^2 = y_u^2 = z$ is reduced by two, and so our claim follows. By relabelling our indices if necessary, we may assume that $x_i^2 = y_i^2 = 1$ whenever $i > 1$.

Suppose that $x_i^2 = y_i^2 = 1$ for $i \in \{1, 2, \ldots, r\}$. Then the relations (13.2) and (13.3) hold. Hence P is a quotient of $M(p, r, 1)$. Since $M(p, r, 1)$ has order at most p^{2r+1} and P has order p^{2r+1}, we find that $P \cong M(p, r, 1)$ and Case 2 holds.

Suppose that $x_1^2 = y_1^2 = z$ and $x_i^2 = y_i^2 = 1$ for all $i \in \{2, 3, \ldots, r\}$. Then the relations (13.11) to (13.14) hold. Hence P is a quotient of $N(r)$. Since $N(r)$ has order at most p^{2r+1} and P has order p^{2r+1}, we find that $P \cong N(r)$ and Case 3 holds.

We have shown that when $s = 1$ and $p = 2$, either Case 2 or Case 3 holds. This establishes the proposition.

Corollary 13.10 *Let $B_1, B_2 \leqslant \mathrm{GL}(d_2, K)$ be such that $Z(B_1) = Z(B_2)$ and $B_1 \cong B_2 \cong B$. Define Z by $Z = Z(B_1) = Z(B_2)$. Then there exist elements $h_{1,1}, h_{2,1}, \ldots, h_{k,1} \in B_1$, elements $h_{1,2}, h_{2,2}, \ldots, h_{k,2} \in B_2$ and prime numbers $p_1, p_2, \ldots, p_k$ with the following properties.*

1. *For $a \in \{1, 2\}$, every element of B_a / Z may be written uniquely in the form*

$$h_{1,a}^{\alpha_1} h_{2,a}^{\alpha_2} \cdots h_{k,a}^{\alpha_k} Z$$

   where $\alpha_i \in \{0, 1, \ldots, p_i - 1\}$.
2. *For all $i \in \{1, 2, \ldots, k\}$,*

$$h_{i,1}^{p_i} = h_{i,2}^{p_i}. \tag{13.15}$$

   Moreover, $h_{i,1}^{p_i} \in Z$.
3. *For all $i, j \in \{1, 2, \ldots, k\}$,*

$$[h_{i,1}, h_{j,1}] = [h_{i,2}, h_{j,2}]. \tag{13.16}$$

   Moreover, all these commutators lie in Z.

Proof: For each prime p dividing d_2, let z_p be a generator of the Sylow p-subgroup of Z. Let P_1 and P_2 be the Sylow p-subgroups of B_1 and B_2, respectively.

Suppose that $P_1 \cong P_2 \cong M(p, r, s)$. Let $a \in \{1, 2\}$. Then Proposition 13.9 applied to P_a implies that there exist elements $x_{i,a}, y_{i,a} \in P_a$ of order p that generate P_a modulo Z and satisfy the commutator relations

$$[x_{i,a}, x_{j,a}] = 1 \text{ for all } i, j \in \{1, 2, \ldots, r\}, \tag{13.17}$$

$$[y_{i,a}, y_{j,a}] = 1 \text{ for all } i, j \in \{1, 2, \ldots, r\}, \tag{13.18}$$

$$[x_{i,a}, y_{j,a}] = 1 \text{ for all } i, j \in \{1, 2, \ldots, r\} \text{ such that } i \neq j, \tag{13.19}$$

$$[x_{i,a}, y_{i,a}] = (z_p)^{p^{s-1}} \text{ for all } i \in \{1, 2, \ldots, r\}. \tag{13.20}$$

Note that every element of $P_a/Z(P_a)$ may be expressed uniquely in the form

$$x_{1,a}^{\alpha_1} x_{2,a}^{\alpha_2} \ldots x_{r,a}^{\alpha_r} y_{1,a}^{\beta_1} y_{2,a}^{\beta_2} \ldots y_{r,a}^{\beta_r} Z(P_a)$$

where $\alpha_i, \beta_i \in \{0, 1, \ldots, p-1\}$. The relations (13.17) to (13.20) imply that the commutator of any pair of elements in the set $\{x_{i,1}, y_{i,1}\}$ is equal to the commutator of the corresponding pair of elements in the set $\{x_{i,2}, y_{i,2}\}$. Moreover, the pth power of an element in the set $\{x_{i,1}, y_{i,1}\}$ is equal to the pth power of the corresponding element in the set $\{x_{i,2}, y_{i,2}\}$, as all pth powers are trivial.

In the same way, by Proposition 13.9, when $P_a \cong N(r)$ for $a \in \{1, 2\}$ there exist elements $x_{i,1}, y_{i,1} \in P_1$ and elements $x_{i,2}, y_{i,2} \in P_2$ whose corresponding commutators and squares are equal. (The proposition implies that we may choose $x_{1,a}^2 = y_{1,a}^2 = z$, and all the other elements to be of order 2, and so the squares of corresponding elements are still equal even though not all of the elements have order 2.)

The corollary now follows, by taking the set of elements $h_{i,1}$ to be equal to the union of the sets $\{x_{i,1}, y_{i,1}\} \subseteq B_1$ over each prime p dividing d_2 and the elements $h_{i,2}$ to be the corresponding elements of B_2.

We have already proved (in Proposition 13.9) that there are at most two possibilities for the subgroup B as an abstract group. We now show that there are at most two possibilities for B as a linear group.

Proposition 13.11 *Let the isomorphism class of B be fixed. Then there is a unique choice for B as a subgroup of* $\mathrm{GL}(d_2, K)$, *up to conjugation in* $\mathrm{GL}(d_2, K)$.

Proof: By Propositions 9.2 and 13.1, the subgroup A is equal to the set of non-zero scalar matrices in $\mathrm{GL}(d_2, K)$. Clearly, any representation of B that we are considering must contain A.

Let B_1 and B_2 be subgroups of $\mathrm{GL}(d_2, K)$ such that $A \leqslant B_1$, $A \leqslant B_2$ and such that $B_1 \cong B_2 \cong B$. We must show that there exists $g \in \mathrm{GL}(d_2, K)$ such that $g^{-1} B_1 g = B_2$.

Let the elements $h_{1,1}, \ldots, h_{k,1} \in B_1$, the elements $h_{1,2}, \ldots, h_{k,2} \in B_2$ and integers $p_1, \ldots, p_k$ satisfy the conditions in the statement of Corollary 13.10.

Define, for $a \in \{1, 2\}$, the set S_a by

$$S_a = \{h_{1,a}^{\alpha_1} h_{2,a}^{\alpha_2} \cdots h_{k,a}^{\alpha_k} : \alpha_i \in \{0, 1, \ldots, p_i - 1\}\}.$$

Since S_a is a transversal for A in B_a / A, Propositions 9.5 and 13.7 imply that S_a is a basis for the set M of $d_2 \times d_2$ matrices over K. Define $\theta : M \to M$ to be the K-linear transformation such that

$$\theta(h_{1,1}^{\alpha_1} h_{2,1}^{\alpha_2} \cdots h_{k,1}^{\alpha_k}) = h_{1,2}^{\alpha_1} h_{2,2}^{\alpha_2} \cdots h_{k,2}^{\alpha_k}$$

for all $\alpha_i \in \{0, 1, \ldots, p_i - 1\}$. Since S_1 and S_2 are bases of M, we find that θ is a bijection. Given two elements of M that are expressed as linear combinations of the elements of the basis S_a, we may write their product in M with respect to the basis S_a using only our knowledge of $h_{i,a}^{p_i}$, and $[h_{i,a}, h_{j,a}]$. Thus the equalities (13.15) and (13.16) imply that θ is a K-algebra isomorphism of M. Clearly θ fixes K, and so the Skolem–Noether theorem (Corollary 8.9) implies that there exists $g \in \mathrm{GL}(d_2, K)$ such that $g^{-1} x g = \theta(x)$ for all $x \in M$. But $\theta(B_1) = B_2$, and so the proposition follows.

13.5 Structure of *G* determined by *B*

The following proposition shows that once the subgroup B has been chosen, the number of possibilities for G is rather restricted: $B \leqslant G \leqslant N$, where N is the normaliser of B in $\mathrm{GL}(d, q)$ and where $|N/B|$ is quite small.

Proposition 13.12 *Let G be a maximal soluble subgroup of $\mathrm{GL}(d, q)$, and suppose that G acts primitively. Let A be a maximal abelian normal subgroup of G. Let B be a maximal subgroup of $C_G(A)$ subject to being normal in G and to B/A being abelian. Define $N = N_{\mathrm{GL}(d,q)}(B)$. Then $|N/B| \leqslant d^{2\mu(d)+1}$.*

Proof: By Proposition 13.1, A is cyclic of order $q^{d_1} - 1$, where d_1 divides d. Moreover, defining $K = A \cup \{0\}$ we have that K is isomorphic (as an $\mathbb{F}_q$-algebra) to $\mathbb{F}_{q^{d_1}}$ by Propositions 9.1 and 13.1. By Proposition 9.4, we have that $A = Z(B)$ and so N normalises A. Since conjugation by an element of $\mathrm{GL}(d, q)$ respects matrix multiplication and $\mathbb{F}_q$-linear operations, we find that for all $h \in N$ the map $\phi(h) : K \to K$ defined by $z \mapsto hzh^{-1}$ for all $z \in K$ is contained in $\mathrm{Gal}(K : \mathbb{F}_q)$. It is easy to check that the map $\phi : N \to \mathrm{Gal}(K : \mathbb{F}_q)$ is a homomorphism. Define $\widehat{C} = C_N(A)$. Then clearly $\ker \phi = \widehat{C}$. Since $Z(B) = A$ we find that $B \leqslant \widehat{C}$. We have that

$$|N/\widehat{C}| \leqslant |\mathrm{im}\ \phi| \leqslant |\mathrm{Gal}(K : \mathbb{F}_q)| = d_1,$$

and so to prove the proposition it remains to show that $|\widehat{C}/B| \leqslant d_2 d^{2\mu(d)}$, where $d_2 = d/d_1$.

Let $S \leqslant \mathrm{Aut}\,(B/A)$ be the subgroup induced by those automorphisms of B that centralise A. Define the homomorphism $\psi : \widehat{C} \to S$ by setting $\psi(h)$ to be the automorphism of B/A induced by the map $x \mapsto hxh^{-1}$ in $\mathrm{Aut}\,(B)$. Proposition 13.8 implies that $\ker \psi = B$. Thus $\widehat{C}/B$ is isomorphic to a subgroup of S, and so the proposition will follow if we can show that $|S| \leqslant d_2 d^{2\mu(d)}$.

Let $d_2 = p_1^{\alpha_1} p_2^{\alpha_2} \cdots p_k^{\alpha_k}$ be the decomposition of d_2 into a product of distinct primes p_i. By Propositions 13.9 and 9.4, the quotient B/A has order d_2^2 and is the direct product of elementary abelian p_i-groups $V_1, V_2, \ldots, V_k$ where $|V_i| = p_i^{2\alpha_i}$.

Now, the process of forming commutators in B equips V_i with a non-degenerate alternating form that is preserved by S. Hence S is contained in the direct product

$$\mathrm{Sp}(2\alpha_1, p_1) \times \mathrm{Sp}(2\alpha_2, p_2) \times \cdots \times \mathrm{Sp}(2\alpha_k, p_k).$$

By Proposition 3.19,

$$\begin{aligned}
|\mathrm{Sp}(2\alpha_i, p_i)| &= (p_i^{2\alpha_i} - 1) p_i^{2\alpha_i - 1} (p_i^{2\alpha_i - 2} - 1) p_i^{2\alpha_i - 3} \cdots (p_i^2 - 1) p_i \\
&\leqslant p_i^{2\alpha_i} p_i^{2\alpha_i - 1} p_i^{2\alpha_i - 2} \cdots p_i^2 p_i \\
&= p_i^{\alpha_i(2\alpha_i + 1)}.
\end{aligned}$$

Hence

$$|S| \leqslant \prod_{i=1}^{k} p_i^{\alpha_i(2\alpha_i + 1)} = d_2 \prod_{i=1}^{k} \left(p_i^{\alpha_i}\right)^{2\alpha_i} \leqslant d_2 \prod_{i=1}^{k} \left(p_i^{\alpha_i}\right)^{2\mu(d_2)} \leqslant d_2 d_2^{2\mu(d_2)} \leqslant d_2 d^{2\log d},$$

and so the proposition follows.

14

Conjugacy classes of maximal soluble subgroups of the general linear group

We define the following notation:

$m_{\mathrm{lps}}(d, q) = $ the number of conjugacy classes of maximal soluble subgroups of $\mathrm{GL}(d, q)$ which are primitive;

$m_{\mathrm{lis}}(d, q) = $ the number of conjugacy classes of maximal soluble subgroups of $\mathrm{GL}(d, q)$ which are irreducible;

$m_{\mathrm{lss}}(d, q) = $ the number of conjugacy classes of maximal soluble subgroups of $\mathrm{GL}(d, q)$.

This chapter aims to prove the following result, due to P. P. Pálfy, which provides upper bounds on these values.

Theorem 14.1 *With the notation defined above, there exists $c > 0$ such that for all q and d*

$$m_{\mathrm{lps}}(d, q) \leqslant d^{4(\log d)^3 + 4(\log d)^2 + \log d + 2},$$

$$m_{\mathrm{lis}}(d, q) \leqslant d^{4(\log d)^3 + 4(\log d)^2 + \log d + 3},$$

$$m_{\mathrm{lss}}(d, q) \leqslant 2^{cd}.$$

Proof: We begin by proving the bound on $m_{\mathrm{lps}}(d, q)$. The theorem is trivial when $d = 1$, since $\mathrm{GL}(d, q)$ is soluble in this case. Hence we may assume that $d > 1$.

Let G be a maximal soluble subgroup of $\mathrm{GL}(d, q)$ which is also primitive. Let A be a maximal abelian normal subgroup of G, let B be a subgroup of $C_G(A)$ that is maximal subject to being normal in G and having the property that B/A is abelian.

By Proposition 13.1, A has order $q^{d_1} - 1$ where d_1 divides d. There are at most d possibilities for d_1. Once d_1 is fixed, Proposition 13.9 implies that there are at most two possibilities for the isomorphism class of B; Proposition 13.11 implies that each possibility may be embedded in $\mathrm{GL}(d/d_1, q^{d_1})$ in precisely one way up to conjugacy. Since we may regard $\mathrm{GL}(d/d_1, q^{d_1})$ as a subgroup of $\mathrm{GL}(d, q)$, once d_1 is fixed there are at most two choices for B as a subgroup of $\mathrm{GL}(d, q)$, up to conjugacy.

Suppose that B is fixed, and let $N = N_{\mathrm{GL}(d,q)}(B)$. Clearly, $B \leqslant G \leqslant N$, and so G is determined by the corresponding subgroup G/B of N/B. Now, N/B has at most $|N/B|^{\log|N/B|}$ subgroups (as every subgroup of N/B can be generated by $\log|N/B|$ elements). Proposition 13.12 implies that $|N/B| \leqslant d^{2\mu(d)+1}$, and so the number of choices for G/B once B is fixed is at most

$$\left(d^{2\mu(d)+1}\right)^{\log(d^{2\mu(d)+1})} = \left(d^{2\mu(d)+1}\right)^{(2\mu(d)+1)\log d} \leqslant d^{4(\log d)^3 + 4(\log d)^2 + \log d},$$

the inequality following from the fact that $\mu(d) \leqslant \log d$. Since the number of choices for B is at most $2d$, and $2d \leqslant d^2$, we find that the number of choices $m_{\mathrm{lps}}(d, q)$ for G is at most $d^{4(\log d)^3 + 4(\log d)^2 + \log d + 2}$, as required.

We prove the remaining two bounds of the theorem by induction on d. As we observed above, the theorem is trivial when $d = 1$. Assume, as an inductive hypothesis, that the theorem is true for all smaller dimensions d. An irreducible but imprimitive linear group G preserves a direct sum decomposition

$$V = V_1 \oplus \cdots \oplus V_k.$$

Choose a decomposition where k is as large as possible, and let $m = \dim V_1$. By Proposition 6.15, such a group is conjugate to a subgroup of $H \mathrm{wr} Q$ where $Q \leqslant \mathrm{Sym}(k)$ is the transitive permutation group induced by the action of G on the direct summands V_i, and where $H \leqslant \mathrm{GL}(m, q)$ is the irreducible group induced by the stabiliser of V_1 in G. Indeed H is primitive, by the remark after the proof of Proposition 6.15. Since G is soluble, Q and H are soluble. Since G is a maximal soluble subgroup, G is conjugate to $H \mathrm{wr} Q$, rather than to a general subgroup of $H \mathrm{wr} Q$. Moreover, the maximality of G implies that Q is a maximal soluble subgroup of $\mathrm{Sym}(k)$ and H is a maximal soluble subgroup of $\mathrm{GL}(V_1)$.

So for any maximal soluble irreducible subgroup G of $\mathrm{GL}(d, q)$ we have that G is conjugate to $H \mathrm{wr} Q$ where H is a maximal soluble subgroup of $\mathrm{GL}(m, q)$ for some divisor m of d, where H is primitive and where Q is a transitive subgroup of $\mathrm{Sym}(d/m)$. Hence

$$m_{\mathrm{lis}}(d, q) \leqslant \sum_{m|d} m_{\mathrm{lps}}(m, q) m_{\mathrm{ts}}(d/m)$$

$$\leqslant \sum_{m|d} m^{4(\log m)^3 + 4(\log m)^2 + \log m + 2} \, 2^{3(\log d - \log m)^3},$$

the second inequality following from Theorem 11.2. Each term in the sum above is at most $d^{4(\log d)^3+4(\log d)^2+\log d+2}$. (This is trivial when $m = d$. When $m < d$, this follows from the fact that $m \leqslant d/2$ and $3(\log d - \log m)^3 \leqslant 4(\log d)^3 + 4(\log d)^2 + \log d + 2$.) Since there are at most d terms in the sum,

$$m_{\text{lis}}(d, q) \leqslant d^{4(\log d)^3+4(\log d)^2+\log d+3},$$

as required.

Finally, let G be a (possibly reducible) maximal soluble subgroup of $\mathrm{GL}(d, q)$. Let

$$\{0\} = V_0 < V_1 < \cdots < V_k = \mathbb{F}_q^d$$

be an $\mathbb{F}_q G$ composition series. Let G_i be the group induced by G on V_i/V_{i-1}. Note that G_i is isomorphic to a quotient of G, and so is soluble. Moreover, G_i acts irreducibly on V_i/V_{i-1} since the V_i form a composition series. Let $d_i = \dim(V_i/V_{i-1})$. With respect to a suitable basis, G is a group of matrices of the form

$$\begin{pmatrix} A_1 & 0 & 0\dots & 0 \\ ? & A_2 & 0\dots & 0 \\ ? & \vdots & \ddots & 0 \\ ? & ? & ? & A_k \end{pmatrix}$$

where the $d_i \times d_i$ matrices A_i represent the groups G_i. Now, it is not difficult to check that the set L of all matrices of the form

$$\begin{pmatrix} I_{d_1} & 0 & 0\dots & 0 \\ ? & I_{d_2} & 0\dots & 0 \\ ? & \vdots & \ddots & 0 \\ ? & ? & ? & I_{d_k} \end{pmatrix},$$

where I_m is the $m \times m$ identity matrix, is a soluble group (indeed it is a p-group and so it is nilpotent). Moreover, since G preserves the subspaces V_i, G normalises L, and so $L \leqslant G$ by the maximality of G. Thus the conjugacy class of G is determined by the conjugacy classes of the groups G_i in $\mathrm{GL}(d_i, q)$. Moreover, the groups G_i must be maximal in order that G is maximal. So

$$m_{\mathrm{lss}}(d, q) \leqslant \sum_{(d_1, d_2, \ldots, d_k)} \prod m_{\mathrm{lis}}(d_i, q)$$

where $(d_1, d_2, \ldots, d_k)$ runs over the ordered partitions of d

$$\leqslant \sum_{(d_1, d_2, \ldots, d_k)} \prod d_i^{4(\log d_i)^3 + 4(\log d_i)^2 + \log d_i + 3}$$

$$= \sum_{(d_1, d_2, \ldots, d_k)} 2^{\sum_{i=1}^{k} \left(4(\log d_i)^4 + 4(\log d_i)^3 + (\log d_i)^2 + 3 \log d_i \right)}.$$

We wish to find an upper bound on the sum

$$\sum_{i=1}^{k} \left(4(\log d_i)^4 + 4(\log d_i)^3 + (\log d_i)^2 + 3 \log d_i \right). \tag{14.1}$$

If $(d_1, d_2, \ldots, d_k)$ is an ordered partition of d that maximises the sum (14.1), for all i we must have that

$$\left(4(\log d_i)^4 + 4(\log d_i)^3 + (\log d_i)^2 + 3 \log d_i \right)$$
$$\geqslant \lfloor d_i/2 \rfloor \left(4(\log 2)^4 + 4(\log 2)^3 + (\log 2)^2 + 3 \log 2 \right)$$
$$= 12 \lfloor d_i/2 \rfloor,$$

for otherwise we may replace d_i by $\lfloor d_i/2 \rfloor$ parts equal to 2 or 3 and the sum (14.1) increases. But this inequality implies that d_i is bounded. Indeed, we must have that $d_i \leqslant 2^{16}$. Because $(d_1, d_2, \ldots, d_k)$ has at most d parts, we therefore find that the sum (14.1) is at most $d(4(16)^4 + 4(16)^3 + (16)^2 + 3 \cdot 16) = 278\,832d$. Hence

$$m_{\mathrm{lss}}(d, q) < \sum_{(d_1, d_2, \ldots, d_k)} 2^{\sum_{i=1}^{k} (4(\log d_i)^4 + 4(\log d_i)^3 + (\log d_i)^2 + 3 \log d_i)}$$

$$< \sum_{(d_1, d_2, \ldots, d_k)} 2^{278\,832d}$$

$$= 2^{d-1} 2^{278\,832d} < 2^{278\,833d},$$

since (by Lemma 5.9) there are 2^{d-1} ordered partitions of d. So the last statement of the theorem follows (and we may take $c = 278\,833$ in our upper bound on $m_{\mathrm{lss}}(d, q)$).

Commentary:

1. Pyber [82, Lemma 3.4] gives this theorem (with attribution to Pálfy) for $q = p$ (prime) and with $20(\log d)^3$ as the leading term of the exponent for the bounds on $m_{\text{lps}}(d, q)$ and $m_{\text{lis}}(d, q)$ (and some other minor changes). We should remark that Pyber claims a slightly better bound for $m_{\text{lss}}(d, q)$ than that of Theorem 14.1. However, the proof he gives contains a small error, which when corrected leads to the bound we give here.

2. It is highly significant for our purposes that the bounds depend only on d and not on q. What lies behind this fact is that solubility of a subgroup of $\text{GL}(n, F)$ is, in general, inherited by its closure in the Zariski topology. Therefore maximal soluble subgroups are Zariski-closed, and so they are algebraic subgroups. Just as the maximal algebraic subgroups of an algebraic group can be classified into finitely many types (hence conjugacy classes), so the same is true of the maximal soluble subgroups.

3. It is possible to give a slightly better upper bound on the maximal number of generators required to generate a subgroup of N/B, by first deriving an upper bound on $\mu(|\text{Sp}(2\alpha, p)|)$ and then using facts about N/B established in the proof of Proposition 13.12 together with the result (see Corollary 16.7) that a group G of order n may be generated by $\mu(n) + 1$ elements. This leads to an improvement of the bounds of Theorem 14.1. However, the leading terms of the exponents of the bounds for $m_{\text{lps}}(d, q)$ and $m_{\text{lis}}(d, q)$ are unchanged, and the resulting explicit value of c is not significantly smaller than the value given above.

4. Our estimation of the sum (14.1) is very crude. In fact, we think the following may well be true:

$$m_{\text{lss}}(d, q) \leqslant O(2^d d); \quad \text{perhaps even} \quad m_{\text{lss}}(d, q) \leqslant O\left(2^d d^{1/2}\right).$$

15

Pyber's theorem: the soluble case

Let S be a family of groups, one for each prime power order. We want to enumerate soluble groups of order $n = p_1^{\alpha_1} \cdots p_k^{\alpha_k}$ whose Sylow subgroups are isomorphic to groups in S.

The technical machinery required to prove our main result is in place except for a few lemmas. We present these in Section 15.1. The next section will contain the proof of Pyber's theorem (for the soluble case). But first we set up notation that will be used throughout this chapter.

As before, we have $\mu = \max\{\alpha_1, \ldots, \alpha_k\}$. We focus on a soluble group G of order n whose Sylow subgroups are isomorphic to groups in S. The material presented in Section 6.2 of Chapter 6 can be referred to for information on $F(G)$, the Fitting subgroup of G.

For $i = 1, \ldots, k$ let

$$F_i = O_{p_i}(G),$$

$$F = F(G) = F_1 \times F_2 \times \cdots \times F_k,$$

$$A_i = \mathrm{Aut}(F_i),$$

$$A = \mathrm{Aut}(F) = A_1 \times A_2 \times \cdots \times A_k \text{ and}$$

$$B_i = \{\sigma \in A_i \mid \sigma g \equiv g \bmod \Phi(F_i) \text{ for all } g \in F_i\}.$$

For each i, there is a natural homomorphism from A_i into $\mathrm{Aut}\,(F_i/\Phi(F_i))$. By Lemma 3.13 the kernel B_i of this homomorphism is a p_i-group. Now let

$$d_i = d(F_i) = d(F_i/\Phi(F_i)).$$

Then by Lemma 3.12, $\mathrm{Aut}(F_i/\Phi(F_i))$ is isomorphic to $\mathrm{GL}(d_i, p_i)$ and so A_i/B_i may be regarded as a subgroup of $\mathrm{GL}(d_i, p_i)$.

Let $Z = Z(F)$ and $Z_i = Z \cap F_i$. Note that Z_i is the Sylow p_i-subgroup of Z. Further, let $H = G/Z$ and for $P_1, \ldots, P_k$, Sylow p_i-subgroups of G, let $Q_i = P_i/Z_i$. So the Q_i are isomorphic to Sylow p_i-subgroups of H.

15.1 Extensions and soluble subgroups

Lemma 15.1 *Suppose that Z, H are given (as is an action of H on Z). Then the number of groups G with Sylow subgroups $P_1, \ldots, P_k$ which are extensions of Z by H is at most $\prod_{i=1}^{k} p_i^{\alpha_i^2}$ and hence it is at most n^μ.*

Proof: Let $H_{\mathcal{S}}^2(H, Z)$ be the set of equivalence classes of extensions

$$E : 1 \longrightarrow Z \longrightarrow G \longrightarrow H \longrightarrow 1$$

with Sylow subgroups $P_1, \ldots, P_k$. Since $Z = Z_1 \times Z_2 \times \cdots \times Z_k$, by Proposition 7.12,

$$H^2(H, Z) \cong H^2(H, Z_1) \times H^2(H, Z_2) \times \cdots \times H^2(H, Z_k).$$

It is obvious that this isomorphism gives a one-to-one correspondence between $H_{\mathcal{S}}^2(H, Z)$ and $H_{\mathcal{S}_1}^2(H, Z_1) \times H_{\mathcal{S}_2}^2(H, Z_2) \times \cdots \times H_{\mathcal{S}_k}^2(H, Z_k)$, where $\mathcal{S}_i$ has p_i-group P_i and p_j-group Q_j for $j \neq i$.

Since Z_i is a p_i-subgroup, we get that $H^2(H, Z_i)$ is an abelian p_i-group. Thus for each i, by Corollary 7.16, the restriction map from $H^2(H, Z_i)$ into $H^2(Q_i, Z_i)$ is an injection. But we are only interested in extensions

$$1 \longrightarrow Z_i \longrightarrow P_i \longrightarrow Q_i \longrightarrow 1. \tag{15.1}$$

If $|Z_i| = p_i^{\beta_i}$ then the number of injective homomorphisms from Z_i into P_i (and whose images are normal subgroups of P_i) is less than $p_i^{\alpha_i \beta_i}$. Now P_i can be generated by $\alpha_i - \beta_i$ generators together with Z_i. So the number of homomorphisms from P_i onto Q_i with kernel Z_i is less than $|Q_i|^{\alpha_i - \beta_i}$. But $|Q_i| = p_i^{\alpha_i - \beta_i}$. Hence the number of exact sequences given by (15.1) is at most $p_i^{\alpha_i \beta_i} p_i^{(\alpha_i - \beta_i)^2} \leqslant p_i^{\alpha_i \beta_i} p_i^{\alpha_i(\alpha_i - \beta_i)} = p_i^{\alpha_i^2}$. Thus $|H_{\{P_i\}}^2(Q_i, Z_i)| \leqslant p_i^{\alpha_i^2}$ and so

$$H_{\mathcal{S}}^2(H, Z) \leqslant \prod_{i=1}^{k} p_i^{\alpha_i^2} \leqslant n^\mu.$$

For any group G, let $\mathrm{Mss}(G)$ denote the set of maximal soluble subgroups of G.

Lemma 15.2 *Let G be a group and H a subgroup of G. Then $|\mathrm{Mss}(H)| \leqslant |\mathrm{Mss}(G)|$.*

Proof: Let X_1 be a maximal soluble subgroup of H. Then there is a maximal soluble subgroup M of G such that $X_1 \leqslant M$. Now if X_2 is also a member of $\mathrm{Mss}(H)$ such that $X_2 \leqslant M$, then $\langle X_1, X_2 \rangle$ is a soluble subgroup of H containing X_1 and X_2. So by maximality we get that $X_1 = X_2$. This shows that $|\mathrm{Mss}(H)| \leqslant |\mathrm{Mss}(G)|$.

Lemma 15.3

(i) *If M is a soluble subgroup of A_i and $|M| = p_i^{m_i} x_i$ where p_i does not divide x_i, then $x_i \leqslant p_i^{3\alpha_i}$.*

(ii) $|\mathrm{Mss}(A_i)| \leqslant p_i^{\alpha_i^2 + 278\,833\alpha_i}$.

Proof: (i) By Theorem 6.1 we get that M has a Hall p_i'-subgroup, say X_i. So $|X_i| = x_i$. We have seen that A_i/B_i may be regarded as a subgroup of $\mathrm{GL}(d_i, p_i)$. Since B_i is a p_i-group, we get that X_i is isomorphic to a subgroup of A_i/B_i. Now p_i does not divide x_i, so by Maschke's theorem X_i is a completely reducible subgroup of $\mathrm{GL}(d_i, p_i)$ and is also soluble. Thus by Theorem 10.2 we have

$$|X_i| \leqslant p_i^{3d_i - 2} \leqslant p_i^{3\alpha_i}.$$

(ii) We claim that $|\mathrm{Mss}(A_i)| = |\mathrm{Mss}(A_i/B_i)|$. This can be seen using the following argument. Let M be a maximal soluble subgroup of A_i. Then MB_i is also a soluble subgroup of A_i and so $M = MB_i$. Thus B_i is contained in M and the map taking M to M/B_i gives a bijective map from $\mathrm{Mss}(A_i)$ to $\mathrm{Mss}(A_i/B_i)$.

It follows from Lemma 15.2 that

$$|\mathrm{Mss}(A_i)| = |\mathrm{Mss}(A_i/B_i)| \leqslant |\mathrm{Mss}(\mathrm{GL}(d_i, p_i))|.$$

We know from Theorem 14.1 that there exists an explicit constant $c > 0$ such that the number of conjugacy classes of maximal soluble subgroups of $\mathrm{GL}(d, q)$ is at most 2^{cd}. It was shown in the proof of the theorem that c could be taken to be equal to $278\,833$. Thus

$$|\mathrm{Mss}(A_i)| \leqslant |\mathrm{Mss}(\mathrm{GL}(d_i, p_i))|$$

$$\leqslant p_i^{d_i^2} 2^{278\,833 d_i}$$

$$\leqslant p_i^{d_i^2 + 278\,833 d_i}$$

$$\leqslant p_i^{\alpha_i^2 + 278\,833 \alpha_i}.$$

Corollary 15.4 $|\mathrm{Mss}(A)| \leqslant \prod_{i=1}^{k} p_i^{\alpha_i^2 + 278\,833\alpha_i} \leqslant n^{\mu + 278\,833}$.

Proof: Since a maximal soluble subgroup of A is a direct product of maximal soluble subgroups of each of the A_i, we get that

$$\mathrm{Mss}(A) = \mathrm{Mss}(A_1) \times \cdots \times \mathrm{Mss}(A_k).$$

Thus

$$\begin{aligned}
|\mathrm{Mss}(A)| &= \prod_{i=1}^{k} |\mathrm{Mss}(A_i)| \\
&\leqslant \prod_{i=1}^{k} p_i^{\alpha_i^2 + 278\,833\alpha_i} \\
&\leqslant \prod_{i=1}^{k} p_i^{\alpha_i(\mu + 278\,833)} \\
&= n^{\mu + 278\,833},
\end{aligned}$$

as required.

In their paper [10], Borovik, Pyber and Shalev have proved a conjecture of Laci Pyber that there is a constant c such that for all finite groups G we have

$$|\mathrm{Mss}(G)| \leqslant |G|^c.$$

Pyber now makes the conjecture that in fact $|\mathrm{Mss}(G)| \leqslant |G|$.

15.2 Pyber's theorem

Theorem 15.5 *The number of soluble groups of order n with Sylow subgroups $P_1, \ldots, P_k$ is at most $n^{8\mu + 278\,833}$.*

Proof: Before we embark on the details of a proof of Pyber's theorem, it is worthwhile to consider the main stages of the proof. We first present a summary of these.

Let $n = \prod_{i=1}^{k} p_i^{\alpha_i}$ be the decomposition of n into primes. Suppose that $P_1, \ldots, P_k$ are p_i-groups such that $|P_i| = p_i^{\alpha_i}$. Let G be a soluble group of order n with Sylow subgroups P_i.

Now G acts by conjugation on F, the Fitting subgroup of G, and so there is a homomorphism ϕ from G into A, the automorphism group of F. Clearly $\ker \phi = C_G(F)$. Further, since G is soluble, by Corollary 6.5 we know that $C_G(F) \leqslant F$. Thus $\ker \phi = Z$, the centre of F. Consequently, we may regard $H = G/Z$ as a subgroup of A.

Hence G is an extension of Z by H and so G is determined by H, Z, the action of H on Z and by E, an element of $H^2_s(H, Z)$. Note that the action of H on Z is determined by the action of H on F and once a choice for F is made, we know Z and A. Further, the choices for H and the action of H on F are taken into account by counting the choices for H as a subgroup of A.

Since the Sylow subgroups of G are isomorphic to the P_i, it is easy to determine the choices for F. Once this is done, we know Z and A. The main task lies in counting the choices for H as a subgroup of A. Since H is soluble, it will be contained in some maximal soluble subgroup M of A. So we first choose a maximal soluble subgroup M of A and then count the choices for H as a subgroup of M. This is done by considering Sylow systems (see Theorem 6.2) for H. It can be shown that the number of Sylow systems for M is fairly small and they contain the Sylow systems of H.

Let $Q_1, \ldots, Q_k$ be a Sylow system for H and let $R_1, \ldots, R_k$ be part of a Sylow system for M such that $Q_i \leqslant R_i$, for each i. Once M has been chosen and a Sylow system for M has been fixed, we could estimate the choices for H (up to conjugacy) as a subgroup of M by enumerating the possibilities for Q_i as a subgroup of R_i. Unfortunately, the R_i can be 'too large' for the purpose of enumerating the choices of Q_i and so instead of using Sylow systems for M, a small number of 'approximating Sylow systems' $S_1, \ldots, S_k$ are found such that S_i are p_i-subgroups of M of 'small enough order' and such that $Q_i \leqslant S_i$ for $i = 1, \ldots, k$.

In order to estimate the choices for H, as outlined above, we need to explore in depth the structural properties that H satisfies as a subgroup of A. Since H is soluble, there will be a maximal soluble subgroup M of A such that $H \leqslant M$. Further, $M = M_1 \times M_2 \times \cdots \times M_k$ where $M_j = M \cap A_j$ and M_j is a maximal soluble subgroup of A_j. By Theorem 6.2, we can choose a Sylow system $Q_1, \ldots, Q_k$ for H and (part of) a Sylow system $R_1, \ldots, R_k$ for M such that $Q_i = H \cap R_i$.

Since $M = M_1 \times M_2 \times \cdots \times M_k$, we get that for $i = 1, \ldots, k$,

$$R_i = R_{i1} \times R_{i2} \times \cdots \times R_{ik}$$

where R_{ij} is a Sylow p_i-subgroup of M_j.

Let $X_j = \prod_{\substack{i=1 \\ i \neq j}}^{k} R_{ij}$. Since the R_s form part of a Sylow system for M, we have that $R_s R_t = R_t R_s$. So X_j is well-defined as a soluble p'_j-subgroup of A_j. So by part (i) of Lemma 15.3, we get that $|X_j| \leqslant p_j^{3\alpha_j}$. Thus for $i \neq j$, we find that R_{ij} is 'small' but unfortunately the R_{ii} may be 'large'. So we replace the Sylow system R_i for M by an 'approximate Sylow system' S_i as follows.

Since $H = G/Z$, for each i we have that Q_i is a conjugate of P_iZ/Z. By choosing suitable conjugates of P_i in G, we can have Sylow p_i-subgroups for G, say $\hat{P}_1, \ldots, \hat{P}_k$, such that $Q_i = \hat{P}_iZ/Z$. So without loss of generality let us assume that $Q_i = P_iZ/Z$ for each i.

Define

$$S_{ij} = \begin{cases} R_{ij} & \text{if } i \neq j \\ \phi_i(P_i) & \text{if } i = j \end{cases}$$

where $\phi_i = \pi_i \circ \phi$ and π_i is the projection map from A to A_i. Note that $\phi_i(P_i) = \pi_i(P_iZ/Z) = \pi_i(Q_i)$. Define

$$S_i = S_{i1} \times S_{i2} \times \cdots \times S_{ik}.$$

Then S_i is a p_i-group and clearly $Q_i \leqslant S_i$.

For $i \neq j$, we have defined S_{ij} to be R_{ij} so

$$\prod_{i=1}^{k} |S_i| = \prod_{j=1}^{k} |X_j| \prod_{i=1}^{k} |S_{ii}|$$

$$\leqslant \prod_{j=1}^{k} p_j^{3\alpha_j} \prod_{i=1}^{k} p_i^{\alpha_i}$$

$$= \prod_{i=1}^{k} p_i^{4\alpha_i}$$

$$= n^4.$$

The group Q_i is a subgroup of S_i generated by at most α_i elements, so the number of possibilities for the sequence $Q_1, \ldots, Q_k$ is at most

$$\prod_{i=1}^{k} |S_i|^{\alpha_i}.$$

This is at most $\prod_{i=1}^{k} |S_i|^{\mu} \leqslant n^{4\mu}$.

Let

$$n_1 = \text{the number of possibilities for } F.$$

Given F, let

$$n_2 = \text{the number of possibilities for } M.$$

Given F and M, let

$$n_3 = \text{the number of possibilities for } S_1, \ldots, S_k.$$

Given $F, M, S_1, \ldots, S_k$, let

$$n_4 = \text{the number of possibilities for } Q_1, \ldots, Q_k.$$

We have shown above that $n_4 \leqslant n^{4\mu}$.

Since $F = F_1 \times F_2 \times \cdots \times F_k$, where $F_i = O_{p_i}(G) \leqslant P_i$, we get that

$$n_1 \leqslant \prod_{i=1}^{k} |P_i|^{\alpha_i} = \prod_{i=1}^{k} p_i^{\alpha_i^2} \leqslant n^{\mu}.$$

By Corollary 15.4 we have

$$n_2 = |\text{Mss} A| \leqslant n^{\mu + 278\,833}.$$

As all Sylow systems of M are conjugate, their number is at most $|M|$. Also

$$|M| \leqslant |A|$$

$$= \prod_{i=1}^{k} |A_i|$$

$$\leqslant \prod_{i=1}^{k} p_i^{\alpha_i^2}$$

$$\leqslant n^{\mu}.$$

So the number of choices for the R_i is at most n^{μ}. Further, when a choice for F was made, we fixed the embedding of F_i into P_i. The homomorphism ϕ_i is clearly just the map induced by the conjugation action of P_i on F_i and so this is fixed once F has been chosen. Thus the S_i are uniquely determined by the R_i. Consequently,

$$n_3 \leqslant |M| \leqslant n^{\mu}.$$

Thus the number of possibilities for H (as a subgroup of Aut F) is at most $n^{7\mu + 278\,833}$. Now, by Lemma 15.1, given $P_1, \ldots, P_k$, Z and H and the action of H on Z, the number of extensions of Z by H with Sylow subgroups isomorphic to the P_i is at most n^{μ}. Hence we have the required result.

Corollary 15.6

$$f_{A, \text{sol}}(n) \leqslant n^{8\mu + 278\,834}.$$

Proof: By Corollary 17.3 there are at most m choices for an abelian group of order m up to isomorphism. Hence there will be at most n choices that can be made for the abelian p_i-groups that will serve as the Sylow subgroups of a soluble A-group of order n. The result now follows from applying Theorem 15.5.

Comments

1. When $\mu(n)$ is large this corollary represents an improvement on Corollary 12.10. Moreover, it seems very likely that a combination of the methods of McIver and Neumann [67] with the methods of this chapter should yield the stronger theorem that $f_A(n) \leqslant n^{8\mu+278\,834}$, but we believe that this would still be quite far from best possible—see Question 22.22 on p. 266.

2. It seems likely that the exponent $8\mu + 278\,833$ in Theorem 15.5 should come down considerably, at least as far as $2\mu + O(1)$, and probably further.

3. Let $\mathcal{S}$ be a family of groups, one for each prime power order, and let $f_{\mathcal{S}}(n)$ denote the number of groups of order n whose Sylow subgroups are isomorphic to groups in $\mathcal{S}$. Let $n = p_1^{\alpha_1} \cdots p_k^{\alpha_k}$ be the prime decomposition of n. It had been hoped that in fact

$$f_{\mathcal{S}}(n) \leqslant p_1^{\alpha_1^2 + o(\alpha_1^2)} \, p_2^{\alpha_2^2 + o(\alpha_2^2)} \cdots p_k^{\alpha_1^2 + o(\alpha_k^2)} \leqslant n^{\mu + o(\mu)}$$

but Pyber in [82] has shown that this is false.

16

Pyber's theorem: the general case

In this chapter we aim to prove the general version of Pyber's theorem: the proof is contained in the final section. The three sections preceding the proof each deal with a different ingredient that is needed there. Section 16.1 contains theorems that bound the number of generators of a group in various contexts. Section 16.2 is concerned with central extensions (especially of perfect groups). Finally, in Section 16.3 we define and explore the notion of the generalised Fitting subgroup of a group.

16.1 Three theorems on group generation

This section contains proofs of three theorems, each of which makes statements about the existence of certain kinds of generating sets for finite groups. The first, due to Wolfgang Gaschütz [35], will be needed to prove the third theorem of this section. The second and third depend on the Classification of Finite Simple Groups; they will be used in the proof of the general case of Pyber's theorem in Section 16.4.

Theorem 16.1 Let G be a finite group, and let N be a normal subgroup of G. Suppose that G may be generated by r elements, and let $g_1, g_2, \ldots, g_r \in G$ be such that $g_1 N, g_2 N, \ldots, g_r N$ generate G/N. Then there exist generators $\{h_1, h_2, \ldots, h_r\}$ for G such that $h_i \in g_i N$ for $i \in \{1, 2, \ldots, r\}$.

Proof: We aim to show that the number of choices for $h_1, h_2, \ldots, h_r$ does not depend on $g_1, g_2, \ldots, g_r$ (but only on G, N and r). This will establish the result, since when $g_1, g_2, \ldots, g_r$ generate G there is clearly at least one choice for the h_i, namely $h_i = g_i$ for $i \in \{1, 2, \ldots, r\}$.

Define the set X by

$$X = \{(h_1, h_2, \ldots, h_r) \mid h_i \in g_i N \text{ for } i \in \{1, 2, \ldots, r\}\}.$$

For a subgroup U of G, define the subset X_U of X by

$$X_U = \{(h_1, h_2, \ldots, h_r) \in X \mid h_i \in U \text{ for } i \in \{1, 2, \ldots, r\}\}$$
$$= \{(h_1, h_2, \ldots, h_r) \mid h_i \in (g_i N) \cap U \text{ for } i \in \{1, 2, \ldots, r\}\}.$$

Note that $X_{U_1 \cap U_2} = X_{U_1} \cap X_{U_2}$ for any subgroups U_1 and U_2 of G. We claim that

$$|X_U| = \epsilon_U |N \cap U|^r, \text{ where } \epsilon_U = \begin{cases} 0 & \text{if } UN \neq G, \\ 1 & \text{if } UN = G. \end{cases}$$

To prove this claim, first note that if $UN \neq G$ then $(g_i N) \cap U = \emptyset$ for some $i \in \{1, 2, \ldots, r\}$, since the elements $g_i N$ generate G/N. Thus $X_U = \emptyset$ and the claim follows in this case. Moreover, if $UN = G$ then there exists $u_i \in (g_i N) \cap U$ for all $i \in \{1, 2, \ldots, r\}$. But then the map $x \mapsto u_i^{-1} x$ provides a bijection between the sets $(g_i N) \cap U$ and $N \cap U$. Therefore there are $|N \cap U|$ choices for each component of $(h_1, h_2, \ldots, h_r) \in X_U$, and so the claim follows.

Let $M_1, M_2, \ldots, M_k$ be the maximal subgroups of G. For $I \subseteq \{1, 2, \ldots, k\}$ define $M_I = \bigcap_{i \in I} M_i$. Now $\{h_1, h_2, \ldots, h_r\} \subseteq G$ is a generating set for G if and only if it is not contained in any maximal subgroup M_i of G. Hence, by the inclusion–exclusion principle (see, for example, Cameron [13, Section 5.1]), the number c of elements $(h_1, h_2, \ldots, h_r) \in X$ such that $h_1, h_2, \ldots, h_r$ generate G is given by

$$c = |X| + \sum_{I \subseteq \{1,2,\ldots,k\}} (-1)^{|I|} \left| \bigcap_{i \in I} X_{M_i} \right|$$

$$= |X| + \sum_{I \subseteq \{1,2,\ldots,k\}} (-1)^{|I|} \left| X_{M_I} \right|$$

$$= |N|^r + \sum_{I \subseteq \{1,2,\ldots,k\}} (-1)^{|I|} \epsilon_{M_I} |N \cap M_I|^r.$$

In particular, c does not depend on $g_1, g_2, \ldots, g_r$, and so the theorem follows.

We remark that Gaschütz [35] proves the stronger theorem, where G is no longer necessarily finite but where N is still assumed to be finite. The above argument establishes this more general result, once the observation is made that there are only finitely many maximal subgroups M of G such that $MN = G$ (as such a subgroup M is determined by the intersections $(g_i N) \cap M$ for $i \in \{1, 2, \ldots, r\}$ together with the intersection $N \cap M$) and that only maximal subgroups of this form are needed in the inclusion–exclusion argument.

The two remaining theorems we will prove in this section depend on the Classification of Finite Simple Groups. In both cases, the Classification is only needed to establish the truth of the following result:

Theorem 16.2 *Let L be a finite non-abelian simple group. Then there exists a Sylow 2-subgroup P of L and an element $x \in L$ such that L is generated by P and xPx^{-1}.*

Theorem 16.2 is due to Robert Guralnick. We do not give the proof here; the interested reader can consult Guralnick's paper [38]. We remark that the next theorem we consider, a theorem due to Michael Aschbacher and Robert Guralnick [2], follows from the weaker result where '2' is replaced by 'p'. Indeed, a proof of this weaker result makes up a significant proportion of the Aschbacher–Guralnick paper. In fact, we use the following corollary of Guralnick's theorem:

Corollary 16.3 *Let M be a direct product $L \times L \times \cdots \times L$ of several copies of a finite non-abelian simple group L. Then there exists a Sylow 2-subgroup P of M and an element $x \in M$ such that M is generated by P and xPx^{-1}.*

Proof: By Theorem 16.2, there exists a Sylow 2-subgroup Q of L and an element $x \in L$ such that Q and yQy^{-1} generate L. But then $P = Q \times Q \times \cdots \times Q$ is a Sylow 2-subgroup of M, and it is easy to check that P and xPx^{-1} generate M, where $x = (y, y, \ldots, y)$.

Theorem 16.4 *Let G be a finite group. Then there exists a soluble subgroup S of G and an element $x \in G$ such that G is generated by S and xSx^{-1}.*

Proof: We prove the result by induction on the order of G. Assume as an inductive hypothesis that the theorem holds for all groups of order smaller than $|G|$. Let M be a minimal normal subgroup of G. Suppose M is soluble. By our inductive hypothesis, there exists a subgroup S containing M and an element $x \in G$ such that G/M is generated by S/M and $(xM)(S/M)(xM)^{-1}$ and such that S/M is soluble. But then S is soluble and G is generated by S and xSx^{-1}. So the theorem holds for G in the case when M is soluble.

Suppose now that M is not soluble. We may identify M with a direct product $L \times L \times \cdots \times L$ of copies of a non-abelian simple group L. By Corollary 16.3, there exists a Sylow 2-subgroup P of M and an element $x \in M$ such that M is generated by P and xPx^{-1}.

Let N be the normaliser of P in G. By the Frattini argument (see Gorenstein [36, page 12, Theorem 3.7]) we find that $G = MN$. Now, P is not normal
in G (by the minimality of M) and so N is a proper subgroup of G. By our
inductive hypothesis, there exists a soluble subgroup S of N and an element
$z \in N$ such that N is generated by S and zSz^{-1}. Since PS is also a soluble
subgroup of N, we may assume that $P \leqslant S$. Let H be the subgroup of G
generated by S and $xzS(xz)^{-1}$. We claim that $H = G$. To see this, first note
that $xzS(xz)^{-1} \geqslant xzP(xz)^{-1} = xPx^{-1}$, the equality following since $z \in N$. So
H contains both P and xPx^{-1}, hence H contains M. In particular $x \in H$, so
H contains zSz^{-1} in addition to S. Therefore H contains N. Since $G = MN$,
we find that $H = G$ and our claim follows. So the theorem holds for G in
this case also. The theorem now follows by induction on the order of G.

We now turn to the final theorem of this section, a theorem proved independently by Robert Guralnick [39] and Andrea Lucchini [62], which provides a
bound on the number of generators of a finite group in terms of the maximum
number of elements needed to generate a Sylow subgroup. The theorem uses
some of the theory of the first cohomology group $H^1(G, M)$ from Chapter 7,
and we begin by making some easy observations about this group.

Let G be a group and let M be a finite G-module. Recall that the first
cohomology group $H^1(G, M)$ may be defined as the quotient of $Z^1(G, M)$ by
$B^1(G, M)$, where $Z^1(G, M)$ is the group of all derivations $f : G \to M$, and
where $B^1(G, M) \leqslant Z^1(G, M)$ is the subgroup of all inner derivations $\delta_m : G \to M$
M (where m runs through M). Now, a derivation $f : G \to M$ is determined by
the images of a generating set of G. So if G can be generated by d elements,
then $|Z^1(G, M)| \leqslant |M|^d$. The definition of an inner derivation makes clear that
$|B^1(G, M)| \leqslant |M|$. Moreover, the definition shows that $B^1(G, M)$ is trivial
if and only if M is a trivial G-module. So $|H^1(G, M)| \leqslant |Z^1(G, M)|$, with
equality if and only if M is a trivial G-module.

Lemma 16.5 *Let G be a finite group. Let $\rho(G)$ be the smallest integer k
such that every Sylow subgroup of G can be generated by k elements. Let M
be a normal p-subgroup of G and let r be an integer. Suppose that*

(i) *$p = 2$ and $r \geqslant \rho(G)$; or*
(ii) *$p \neq 2$ and $r \geqslant \rho(G) + 1$.*

*Then whenever G/M can be generated by r elements, G can be generated
by r elements.*

Proof: We prove the lemma in the case when M is a minimal normal
subgroup of G (and so is elementary abelian). It is clear that the lemma

follows from this special case by induction on the length of a G-invariant composition series for M.

Let $g_1, g_2, \ldots, g_r$ generate G modulo M. Let H be the subgroup generated by $g_1, g_2, \ldots, g_r$. So $G = HM$. The subgroup $H \cap M$ is clearly normalised by H, and is also normalised by M since M is abelian. But $G = HM$ and so $H \cap M$ is normal in G. Since M is a minimal normal subgroup of G we find that $H \cap M = M$ or $H \cap M = \{1\}$. If $H \cap M = M$ then $G = HM = H$, and so G is generated by the r elements $g_1, g_2, \ldots, g_r$. So we may assume that $H \cap M = \{1\}$. Thus H is a complement of M in G, and G is isomorphic to the semidirect product $M \rtimes H$ of M by H.

Let $q = p^\alpha = |M|$, where p is a prime. There are q^r sequences of the form $g_1 x_1, g_2 x_2, \ldots, g_r x_r$ with $x_1, x_2, \ldots, x_r \in M$. The argument of the paragraph above shows that the lemma follows when $g_1 x_1, g_2 x_2, \ldots, g_r x_r$ generates a non-complement to M in G. So we may assume that all such sequences generate such a complement. Distinct sequences generate distinct complements, and every complement arises from such a sequence. Hence M has q^r complements in G, so $|Z^1(H, M)| = q^r$ by Proposition 7.7. We aim to derive a contradiction from this equality.

We aim to give an upper bound on the order of $Z^1(H, M)$. Let P be a Sylow p-subgroup of H. Defining $Q = PM$, we see that Q is a Sylow p-subgroup of G. Corollary 7.16 states that $H^1(H, M)$ is isomorphic to a subgroup of $H^1(P, M)$, and so

$$|Z^1(H, M)| = |H^1(H, M)||B^1(H, M)| \leqslant |H^1(P, M)||B^1(H, M)|.$$

Let d be the minimum number of generators required to generate P. We remarked above the statement of the lemma that $|H^1(P, M)| \leqslant |Z^1(P, M)| \leqslant q^d$. Moreover, since $|B^1(H, M)| \leqslant q$, we find that

$$q^r = |Z^1(H, M)| \leqslant q^{d+1}.$$

Recall the Frattini subgroup $\Phi(P)$ of a group P introduced in Section 3.3. We have that $|P/\Phi(P)| = p^d$, by Lemma 3.12. Moreover, by Lemma 3.12 we have that $\Phi(Q) = Q^p Q'$. Using this it is easy to check that $\Phi(Q) = \Phi(P)[P, M]$. Writing p^a for the index of $[P, M]$ in M, we find that $|Q/\Phi(Q)| = p^{d+a}$, so any generating set for Q has at least $d + a$ generators. In particular, $d + a \leqslant \rho(G)$ and so

$$q^r = |Z^1(H, M)| \leqslant q^{\rho(G)-a+1}. \tag{16.1}$$

Note that $[P, M]$ is a proper subgroup of M since P and M are p-groups, and so $a \geqslant 1$. So (16.1) gives us our required contradiction whenever $r \geqslant \rho(G) + 1$. This proves the lemma when $p \neq 2$. Indeed, we have our required contradiction when $p = 2$, except in the case when $a = 1$ and $r = \rho(G)$. We now aim to derive a contradiction in this case.

Suppose that M is a non-trivial P-module. The remark above the statement of the lemma shows that $|H^1(P, M)| < |Z^1(P, M)| \leqslant q^d$ in this case, and the argument above then shows that $q^r = |Z^1(H, M)| < q^{\rho(G)-1+1} = q^{\rho(G)}$. This gives the contradiction we require. So we may assume that M is a trivial P-module. In this case $[P, M] = \{1\}$, and so $|M| = 2^a = 2$. Thus M is also a trivial H-module, so $|B^1(H, M)| = 1$. Therefore

$$q^r = |Z^1(H, M)| = |H^1(H, M)| \leqslant |H^1(P, M)| \leqslant q^d \leqslant q^{\rho(G)-a} = q^{\rho(G)-1}.$$

Again we have a contradiction, and so the lemma follows.

Theorem 16.6 *Let G be a finite group. Let $\rho(G)$ be the smallest integer k such that every Sylow subgroup of G can be generated by k elements. Then there exist generators $g_1, g_2, \ldots, g_{\rho(G)+1}$ for G such that the subgroup generated by $g_1, g_2, \ldots, g_{\rho(G)}$ contains a Sylow 2-subgroup of G. In particular, G can be generated by $\rho(G) + 1$ elements.*

Proof: We prove the result by induction on the order of G. Suppose the theorem holds for all groups of order smaller than $|G|$. Let M be a minimal normal subgroup of G. By the inductive hypothesis, there exist elements $h_1, h_2, \ldots, h_{\rho(G)+1} \in G$ that generate G modulo M and such that $h_1 M$, $h_2 M, \ldots, h_{\rho(G)} M$ generate a subgroup of G/M containing a Sylow 2-subgroup of G/M.

Firstly, suppose that M is a 2-group. Let H be the group generated by $h_1, h_2, \ldots, h_{\rho(G)}$ and M. Then H contains a Sylow 2-subgroup of G. Moreover, H/M can be generated by $\rho(G)$ elements and $\rho(H) \leqslant \rho(G)$, so there exist generators $g_1, g_2, \ldots, g_{\rho(G)}$ for H by Lemma 16.5. If we set $g_{\rho(G)+1} = h_{\rho(G)+1}$, we have found generators for G that satisfy the conditions of the theorem. So the inductive step follows in this case.

Secondly, suppose that M is a p-group, where $p \neq 2$. By Lemma 16.5, G can be generated by $\rho(G) + 1$ elements. By Gaschütz's theorem (Theorem 16.1), there exist generators $g_1, g_2, \ldots, g_{\rho(G)+1}$ for G such that $g_i \in h_i M$ for $i \in \{1, 2, \ldots, \rho(G) + 1\}$. So $g_1 M, g_2 M, \ldots, g_{\rho(G)} M$ generate a subgroup of G/M containing a Sylow 2-subgroup of G/M, by our choice of the generators h_i. Since M has odd order, this implies that $g_1, g_2, \ldots, g_{\rho(G)}$ generate a subgroup containing a Sylow 2-subgroup of G, and so the inductive step follows in this case also.

Finally, we assume that M is isomorphic to a direct product of copies of a non-abelian finite simple group. By Corollary 16.3, there exist a Sylow

2-subgroup P of M and an element $x \in M$ such that M is generated by P and xPx^{-1}. By the Frattini argument, $G = MN$, where N is the normaliser of P in G. So, without loss of generality, we may assume that $h_i \in N$ for $i \in \{1, 2, \ldots, \rho(G) + 1\}$. Let H be the subgroup generated by $h_1, h_2, \ldots, h_{\rho(G)}$ and P. Note that H contains a Sylow 2-subgroup of G. Since $H \leqslant N$, we have that P is normal in H and so H may be generated by $\rho(G)$ elements $g_1, g_2, \ldots, g_{\rho(G)}$ by Lemma 16.5. Set $g_{\rho(G)+1} = xh_{\rho(G)+1}$. We claim that $g_1, g_2, \ldots, g_{\rho(G)+1}$ generate G. The subgroup U generated by $g_1, g_2, \ldots, g_{\rho(G)+1}$ contains H, and so U contains both P and $g_{\rho(G)+1} P g_{\rho(G)+1}^{-1} = xPx^{-1}$ (the last equality following since $h_{\rho(G)+1} \in N$). So $M \leqslant U$. Moreover, G/M is generated by HM/M and $h_{\rho(G)+1}M$. Since $h_{\rho(G)+1}M = g_{\rho(G)+1}M$, we find that $U/M = G/M$. So $U = G$ and our claim follows. Since H contains a Sylow 2-subgroup of G, the generators $g_1, g_2, \ldots, g_{\rho(G)+1}$ satisfy the conditions of the theorem. The theorem now follows by induction on the order of G.

Corollary 16.7 *Let G be a finite group of order n. Then G may be generated by $\mu(n) + 1$ elements.*

16.2 Universal central extensions and covering groups

This section contains some results on the structure of central extensions by a perfect group G. (Recall: G is *perfect* if $G = G'$.) We will see that much of the structure of such extensions can be described by a universal object, the universal covering group $\tilde{G}$ of G. This material will be useful (in the case where G is a finite simple group) when we investigate the generalised Fitting subgroup in Section 16.3.

Recall the formal notion of a group extension, namely a short exact sequence

$$1 \longrightarrow M_1 \xrightarrow{i} E_1 \xrightarrow{\rho_1} G \longrightarrow 1 \tag{16.2}$$

of group homomorphisms. In this section, we will always assume that M_1 is contained in E_1, and that the map i is inclusion. We say that the extension (16.2) is a *central extension by* G if $M_1 \leqslant Z(E_1)$. Suppose that (16.2) is a central extension by G, and let

$$1 \longrightarrow M_2 \xrightarrow{i} E_2 \xrightarrow{\rho_2} G \longrightarrow 1 \tag{16.3}$$

be a second central extension by G. We say a homomorphism $\alpha : E_1 \to E_2$

is a *morphism of central extensions* if $\rho_1 = \rho_2 \alpha$. So α is a morphism if the following diagram commutes:

$$
\begin{array}{ccccccccc}
1 & \longrightarrow & M_1 & \longrightarrow & E_1 & \xrightarrow{\rho_1} & G & \longrightarrow & 1 \\
 & & \downarrow & & \alpha \downarrow & & \text{id} \downarrow & & \\
1 & \longrightarrow & M_2 & \longrightarrow & E_2 & \xrightarrow{\rho_2} & G & \longrightarrow & 1
\end{array}
\tag{16.4}
$$

where the left vertical arrow is the restriction of α to M_1.

A central extension (16.2) by G is *universal* if for all central extensions (16.3) by G there exists a unique morphism $\alpha : E_1 \to E_2$. Note that if a universal central extension exists, then it is unique up to isomorphism of central extensions. For if both (16.2) and (16.3) are universal, then there exist morphisms $\alpha : E_1 \to E_2$ and $\beta : E_2 \to E_1$. Since $\beta \alpha$ and the identity map on E_1 are both morphisms from E_1 to E_1, we find that $\beta \alpha$ is the identity map on E_1. Similarly $\alpha \beta$ is the identity map on E_2. Hence α is an isomorphism and so the diagram (16.4) is an isomorphism of central extensions.

Not all groups G possess a universal central extension. Indeed, G possesses such an extension if and only if G is perfect. This fact is a consequence of Lemma 16.8 and Theorem 16.10 below.

Lemma 16.8 *Let G be a group, and let*

$$
1 \longrightarrow M \longrightarrow \tilde{G} \xrightarrow{\sigma} G \longrightarrow 1
\tag{16.5}
$$

be a universal central extension by G. Then $\tilde{G}$ and G are perfect.

Proof: Let $H = \tilde{G} \times (\tilde{G}/\tilde{G}')$. The mapping $\rho : H \to G$ defined by $\rho((x, y)) = \sigma(x)$ is surjective. Defining $N = \ker \rho$, we have that $N \leqslant Z(H)$ and so

$$
1 \longrightarrow N \longrightarrow H \xrightarrow{\rho} G \longrightarrow 1
$$

is a central extension by G. Let $\alpha, \beta : \tilde{G} \to H$ be defined by $\alpha(x) = (x, 1\tilde{G}')$ and $\beta(x) = (x, x\tilde{G}')$ respectively. Then both α and β are morphisms. As (16.5) is universal, $\alpha = \beta$ and hence $\tilde{G}/\tilde{G}'$ is trivial. So $\tilde{G}$ is perfect and therefore G (being isomorphic to a quotient of a perfect group) is also perfect. This proves the lemma.

We now aim to construct a universal central extension whenever G is a perfect group. The following lemma will be useful.

Lemma 16.9 *Let G be a perfect group. Let E be a group (not necessarily finite) and let M be a subgroup of $Z(E)$ such that $E/M = G$. Then $E = E'M$, and E' is perfect.*

Proof: We have that $E'M/M = (E/M)' = G' = G = E/M$. Since $E'M$ clearly contains M, this implies that $E = E'M$.

To prove that E' is perfect, note that for $g_1, g_2 \in E'$ and $z_1, z_2 \in M$, we have that $[g_1z_1, g_2z_2] = [g_1, g_2]$ since $M \leqslant Z(E)$. Thus $(E')' = (E'M)' = E'$, so E' is perfect.

Theorem 16.10 *Let G be a perfect group. Let $G = F/R$ where F is a free group and R is a normal subgroup of F. Then there exists a universal central extension by G of the form (16.5) where $M = (F' \cap R)/[R, F]$ and where $\tilde{G} = F'/[R, F]$.*

Proof: Define $K = R/[R, F]$ and $E = F/[R, F]$. Note that $\tilde{G} = E'$, and $M = E' \cap K$. Also note that $K \leqslant Z(E)$. Now, $E/K \cong F/R = G$ and so there exists a natural central extension

$$1 \longrightarrow K \longrightarrow E \overset{\tau}{\longrightarrow} G \longrightarrow 1$$

by G. Here τ is the map induced from the natural homomorphism $\pi : F \to G$. This map is well-defined since $[R, F] \leqslant R = \ker \pi$.

Let σ be the restriction of τ to $\tilde{G}$. Now

$$\ker \sigma = \tilde{G} \cap \ker \tau = E' \cap K = M.$$

Moreover,

$$\operatorname{im} \sigma = \tau(\tilde{G}) = \tau(E') = G' = G,$$

since τ is surjective and since G is perfect. Moreover, M is central in $\tilde{G}$ since K is central in E. Thus (16.5) is a central extension. We note for later use that, since $\tilde{G} = E'$, Lemma 16.9 implies that $\tilde{G}$ is perfect.

We must now show that (16.5) is universal. To this end, suppose that

$$1 \longrightarrow N \longrightarrow H \overset{\rho}{\longrightarrow} G \longrightarrow 1$$

is a central extension by G. We construct a morphism $\alpha : \tilde{G} \to H$ as follows. Let X be a free generating set for F. For each $x \in X$, choose $\gamma(x) \in H$ so that $\rho(\gamma(x)) = \pi(x)$ (where as before $\pi : F \to G$ is the natural homomorphism). Extend γ to a homomorphism $\gamma : F \to H$. Clearly $\rho\gamma = \pi$, since $\rho\gamma$ and π agree on X. But $\ker \pi = R$ and so $\gamma(R) \leqslant \ker \rho = N \leqslant Z(H)$. Hence $\gamma([R, F]) \leqslant [Z(H), H] = \{1\}$. Since $[R, F] \leqslant \ker \gamma$, we find that γ induces a homomorphism $\hat{\alpha} : E \to H$. Note that $\rho\hat{\alpha} = \tau$. Defining α to be the restriction of $\hat{\alpha}$ to $\tilde{G}$ produces the morphism we require.

Suppose that $\beta : \tilde{G} \to H$ is another morphism. To finish the proof of the theorem, we need to show that $\alpha = \beta$. Define the function $\delta : \tilde{G} \to H$ by

$\delta(x) = \alpha(x)(\beta(x))^{-1}$ for all $x \in \tilde{G}$. To prove the theorem, it suffices to show that δ is trivial.

Since α and β are morphisms, $\rho\alpha = \sigma = \rho\beta$, and so $\rho\delta$ is trivial. So $\operatorname{im}\delta \leqslant \ker\rho = N \leqslant Z(H)$. Hence, for all $g_1, g_2 \in \tilde{G}$,

$$\delta(g_1 g_2) = \alpha(g_1)\alpha(g_2)(\beta(g_2))^{-1}(\beta(g_1))^{-1} = \alpha(g_1)\delta(g_2)(\beta(g_1))^{-1}$$
$$= \alpha(g_1)(\beta(g_1))^{-1}\delta(g_2) = \delta(g_1)\delta(g_2).$$

Thus δ is a homomorphism. Now $\operatorname{im}\delta \leqslant Z(H)$ and so $\tilde{G}/\ker\delta$ is isomorphic to an abelian group. But we observed above that $\tilde{G}$ is perfect, so cannot have any non-trivial abelian quotients. Thus $\ker\delta = \tilde{G}$ and hence δ is trivial. This proves the theorem.

The group $\tilde{G}$ in Theorem 16.10 is known as the *universal covering group* of G. Since a universal central extension is unique up to isomorphism of central extensions, the isomorphism class of $\tilde{G}$ is determined by G (so does not depend on the presentation $G = F/R$ we choose). The group M of Theorem 16.10 is known as the Schur multiplier of G, and is usually written as $M(G)$. Again, $M(G)$ is determined by the isomorphism class of G rather than by the presentation for G we have chosen. The Schur multiplier makes sense for any group, not just groups which are perfect. For a treatment of the Schur multiplier from a more general point of view and its definition using cohomology, see Suzuki [89, Chapter 2, Section 9].

The following theorem states that perfect central extensions by a perfect group G correspond to subgroups of the Schur multiplier $M(G)$.

Theorem 16.11 *Let G be a perfect group with universal covering group $\tilde{G}$ and Schur multiplier M. Let H be a perfect group, and let N be a subgroup of $Z(H)$ such that $H/N \cong G$. Then there exists a surjective homomorphism $\alpha : \tilde{G} \to H$ such that $\ker\alpha \leqslant M$. Moreover, if $Z(G) = \{1\}$ then $Z(\tilde{G}) = M$ and $\alpha(Z(\tilde{G})) = Z(H)$.*

Proof: By definition of a universal cover, there exists a universal central extension of the form (16.5). Moreover, our conditions on H imply that H may be realised as a central extension by G. Let $\rho : H \to G$ be an appropriate surjective homomorphism with kernel N. The definition of a universal central extension implies that there exists a homomorphism $\alpha : \tilde{G} \to H$ such that $\rho\alpha = \sigma$. Note that

$$\ker(\alpha) \leqslant \ker(\rho\alpha) = \ker\sigma = M.$$

We now show that α is surjective. Let $C = \operatorname{im} \alpha$. Now, $\rho(C) = \operatorname{im} \rho\alpha = \operatorname{im} \sigma = G$, and so $CN = H$. By Lemma 16.8, the group $\tilde{G}$ is perfect. Since C is a quotient of $\tilde{G}$ we find that C is perfect. But $C' = (CN)'$, since N is central (by the argument in the proof of Lemma 16.9). Hence $\operatorname{im} \alpha = C = C' = (CN)' = H' = H$. Thus α is surjective, and the first statement of the theorem is established.

Now suppose that $Z(G) = \{1\}$. Clearly $M \leqslant Z(\tilde{G})$. Moreover, $Z(\tilde{G})/M$ is a central subgroup of $\tilde{G}/M$, and so $Z(\tilde{G}) \leqslant M$. Thus $Z(\tilde{G}) = M$.

Since α is surjective, $\alpha(Z(\tilde{G})) \leqslant Z(H)$. So to prove the theorem, it suffices to show that $Z(H) \leqslant \alpha(Z(\tilde{G}))$. Suppose $h \in Z(H)$, and let $g \in \tilde{G}$ be such that $\alpha(g) = h$. Since $\rho(Z(H))$ is central in G, we find that $\rho(Z(H)) = \{1\}$ and so $\rho(h) = 1$. Now, $\sigma(g) = \rho\alpha(g) = 1$, and so $g \in M = Z(\tilde{G})$. Hence $Z(H) \leqslant \alpha(Z(\tilde{G}))$ and the theorem is proved.

The final result of this section depends on the Classification: it will be useful in Section 16.4.

Proposition 16.12 *Let L be a non-abelian finite simple group, and let $\tilde{L}$ be its universal cover. Then $|Z(\tilde{L})|$ divides $|L|$.*

Proof: Theorem 16.11 shows that $Z(\tilde{L})$ is equal to the Schur multiplier $M(L)$ of L. The proposition now follows from the Classification by examination of all cases (see the information on Schur multipliers in [19], for example).

Pyber [82] remarks that there is an alternative proof of this result that uses less information from the Classification. The Classification implies that the Schur multiplier of a non-abelian simple group is either cyclic or is the product of two cyclic groups. A result of Schur (see Schur's original paper [84] or Huppert [50, page 635]) states that the exponent e of $M(G)$ has the property that e^2 divides $|G|$. These two facts together imply the result we require.

16.3 The generalised Fitting subgroup

In this section we define the generalised Fitting subgroup of a finite group, and establish some of its properties. Our approach is based on Aschbacher [1, Section 31]. The results are independent of the Classification, but we do use the Feit–Thompson Theorem in the proof of Corollary 16.17.

Before defining the generalised Fitting subgroup, we prove some elementary facts concerning quasisimple groups. A group C is *quasisimple* if it is perfect and $C/Z(C)$ is isomorphic to a (non-abelian) simple group. Recall that a subgroup H is *subnormal* in G if there exists a chain $G = H_1 > H_2 > \cdots > H_k = H$ of subgroups, where H_{i+1} is normal in H_i for $i \in \{1, 2, \ldots, k-1\}$.

Lemma 16.13

(i) *Let G be a group, and suppose that $G/Z(G)$ is non-abelian and simple. Then $G = G'Z(G)$, and G' is quasisimple.*

(ii) *Let C be a quasisimple group, and suppose that H is a subnormal subgroup of C. Then either $H = C$ or $H \leqslant Z(C)$.*

Proof: Since $G/Z(G)$ is perfect, we find that $G = G'Z(G)$ and G' is perfect, by Lemma 16.9. Since $G/Z(G)$ has trivial centre, $Z(G') = Z(G) \cap G'$. So $G'/Z(G') = G'/(Z(G) \cap G') \cong G'Z(G)/Z(G) = G/Z(G)$. Hence $G'/Z(G')$ is simple and so G' is quasisimple. So part (i) of the lemma follows.

To prove part (ii), suppose that H is a subnormal subgroup of the quasi-simple group C, and that H is not contained in $Z(C)$. Then $HZ(C)/Z(C)$ is a non-trivial subnormal subgroup of the simple group $C/Z(C)$ and so $C = HZ(C)$. But then

$$C = C' = [HZ(C), HZ(C)] = [H, H] = H' \leqslant H,$$

and so $H = C$ and the lemma follows. ∎

A subgroup C of a group G is said to be a *component* of G if C is quasisimple and subnormal in G. The set of components of G is written as $\mathrm{Comp}(G)$. Note that if H is a subnormal subgroup of G then $\mathrm{Comp}(H) = \{C \in \mathrm{Comp}(G) \mid C \leqslant H\}$. The subgroup $E(G)$ of G is defined to be the subgroup generated by the components of G. We define the *generalised Fitting subgroup $F^*(G)$ of the finite group G* to be the group $F^*(G) = E(G)F(G)$. We begin by investigating the structure of $F^*(G)$, and conclude by proving a result that later allows us to use $F^*(G)$ in the same role for general groups as $F(G)$ took in the case of soluble groups in Chapter 15.

Lemma 16.14 *Let H be a subnormal subgroup of a finite group G. Let $C \in \mathrm{Comp}(G)$. Then either $C \in \mathrm{Comp}(H)$ or $[H, C] = \{1\}$.*

Proof: We prove the lemma by induction on the order of G. When $G = \{1\}$ the result is trivial. Assume, as an inductive hypothesis, that the lemma holds for all groups of order smaller than $|G|$.

When $G = C$ the assertion follows by Lemma 16.13 (ii), and when $G = H$ the assertion follows trivially. So we may assume that C and H are strictly contained in G. Let N_1 be the smallest normal subgroup of G containing C, and let N_2 be the smallest normal subgroup of G containing H. Since C and H are proper subnormal subgroups of G, we find that N_1 and N_2 are proper subgroups of G.

$N_1 \cap N_2$ is a normal subgroup of N_1, and so our inductive hypothesis (with G replaced by N_1, and with H replaced by $N_1 \cap N_2$) implies that either $C \in \text{Comp}(N_1 \cap N_2)$ or $[N_1 \cap N_2, C] = \{1\}$. In the first case, $C \in \text{Comp}(N_2)$ and so we may apply our inductive hypothesis (with G replaced by N_2) to deduce that either $C \in \text{Comp}(H)$ or $[H, C] = \{1\}$. So the inductive step follows in this case. In the second case, we have that $[N_2, C, C] \leqslant [N_1 \cap N_2, C] = \{1\}$. Also, $[C, N_2, C] = [N_2, C, C] = \{1\}$. By Lemma 3.4 (the Three Subgroup Lemma) we find that $[C, C, N_2] = \{1\}$. But C is perfect, and so $[C, N_2] = [C, C, N_2] = \{1\}$. So the inductive step follows in this case also. The lemma now follows by induction on $|G|$.

Proposition 16.15 *Let G be a finite group. Then $Z(F(G)) = Z(F^*(G))$.*

Proof: Now, $Z(F^*(G))$ is a soluble subgroup of $C_G(F(G))$ and is normal in G. Hence, as in the proof of Theorem 6.4, $Z(F^*(G)) \leqslant Z(F(G))$. To show the reverse containment, it is sufficient to show that $Z(F(G))$ centralises $E(G)$. But $Z(F(G))$ is a normal subgroup of G, and $\text{Comp}(Z(F(G))) = \emptyset$ since $Z(F(G))$ is soluble. So Lemma 16.14 implies that $Z(F(G))$ centralises each component C of G, whence $Z(F(G))$ centralises $E(G)$ as required.

We will now prove our main result concerning the structure of $F^*(G)$. Recall that a group G is a *central product* of subgroups $H_1, H_2, \ldots, H_k$ if G is generated by the subgroups H_i and $[H_i, H_j] = \{1\}$ for all distinct $i, j \in \{1, 2, \ldots, k\}$. Let X be the direct product $\prod_{i=1}^{k} H_i$. It is easy to see that G is a central product of the subgroups H_i if and only if G is isomorphic to a quotient of X by a subgroup K of $Z(X)$ such that K has a trivial intersection with each of the k factors of the direct product X.

Theorem 16.16 *Let G be a finite group. Let $\text{Comp}(G) = \{C_1, C_2, \ldots, C_t\}$. For $i \in \{1, 2, \ldots, t\}$, define $L_i = C_i / Z(C_i)$ and let $\tilde{L}_i$ be the universal covering group of L_i. Define*

$$X = \tilde{L}_1 \times \tilde{L}_2 \times \cdots \times \tilde{L}_t \times F(G).$$

Then $F^*(G)$ is isomorphic to a quotient X/K, for some subgroup $K \leqslant Z(X)$. Moreover, $F^*(G)$ is a central product of $F(G)$ and the components $C_i \in \text{Comp}(G)$.

Proof: Define $Y = C_1 \times C_2 \times \cdots \times C_t \times F(G)$. Since $F(G)$ is soluble $\mathrm{Comp}(F(G)) = \emptyset$, and so we find that $F(G)$ centralises all the components C_i by Lemma 16.14. Moreover, since $\mathrm{Comp}(C_i) = \{C_i\}$, Lemma 16.14 implies that C_i centralises C_j whenever $i \neq j$. So the map ϕ from Y to G defined by

$$\phi((g_1, g_2, \ldots, g_t, h)) = g_1 g_2 \cdots g_t h$$

for all $(g_1, g_2, \ldots, g_t, h) \in Y$ is a homomorphism.

The definition of $F^*(G)$ implies that $\mathrm{im}\,\phi = F^*(G)$. We prove that the kernel T of ϕ is contained in $Z(Y)$. Thinking of the groups C_i and $F(G)$ as subgroups of Y, we note that ϕ is injective when restricted to C_i (for some i) or to $F(G)$. So $C_i \cap T = F(G) \cap T = \{1\}$. Since T, $F(G)$ and the groups C_i are normal in Y, we find that $[C_i, T] \leqslant C_i \cap T = \{1\}$ and similarly $[F(G), T] = \{1\}$. Thus $T \leqslant Z(Y)$. This shows that $F^*(G)$ is a central product of $F(G)$ and the components C_i of G.

Let $i \in \{1, 2, \ldots, t\}$. By Theorem 16.11, there exists a surjective homomorphism $\alpha_i : \tilde{L}_i \to C_i$. Moreover, $\ker(\alpha_i) \leqslant Z(\tilde{L}_i)$ and $\alpha_i(Z(\tilde{L}_i)) = Z(C_i)$. Let $\pi : X \to Y$ be the map that is equal to α_i when restricted to each factor $\tilde{L}_i$ and is equal to the identity map when restricted to $F(G)$. Then π is a surjective homomorphism, and $\pi(Z(X)) = Z(Y)$. Defining $K = \pi^{-1}(T)$, we find that $K \leqslant Z(X)$. Since the surjective homomorphism $\phi\pi : X \to F^*(G)$ has kernel K, the theorem follows.

We will use the following corollary in Section 16.4:

Corollary 16.17 *Let G be a group of order n and let*

$$\mathrm{Comp}(G) = \{C_1, C_2, \ldots, C_t\}.$$

For $i \in \{1, 2, \ldots, t\}$, define $L_i = C_i/Z(C_i)$. Then

(i) $F^*(G)/Z(F^*(G)) \cong \left(\prod_{i=1}^{t} L_i\right) \times F(G)/Z(F(G))$;

(ii) $|F(G)| \prod_{i=1}^{t} |L_i|$ *divides n; and*

(iii) $t \leqslant \mu/2$.

Proof: Part (i) of the corollary follows from Theorem 16.16. Part (i) implies that $n/|Z(F^*(G))|$ is divisible by $|F(G)/Z(F(G))| \prod_{i=1}^{t} |L_i|$. Since $Z(F(G)) = Z(F^*(G))$, part (ii) follows. The Feit–Thompson Theorem [33] implies that 4 divides $|L_i|$ for all i, and so 2^{2t} divides n. Hence $2t \leqslant \mu$, and part (iii) holds.

Theorem 16.18 *Let G be a (finite) group. Then $C_G(F^*(G)) = Z(F(G))$.*

Proof: By Proposition 16.15, we find that $Z(F(G)) \leqslant C_G(F^*(G))$. So we need to show that $C_G(F^*(G)) \leqslant Z(F(G))$.

Let $H = C_G(F^*(G))$, let $K = C_G(F(G))$ and let $Z = Z(F(G))$. Suppose, for a contradiction, that H strictly contains Z. Then H/Z is a non-trivial normal subgroup of K/Z. Let M/Z be a minimal normal subgroup of K/Z contained in H/Z (for some subgroup M of H containing Z). If M/Z is soluble, then M is a soluble normal subgroup of $C_G(F(G))$, and so $M \leqslant Z$ by Theorem 6.4. This contradicts the fact that M/Z is a minimal normal subgroup of K/Z. So M/Z cannot be soluble, and hence M is a direct product of (isomorphic) non-abelian simple groups. Let R be a subgroup such that $Z \leqslant R \leqslant M$, and such that R/Z corresponds to one of the simple factors of M/Z. By Lemma 16.13, if we write $C = R'$ then $R = CZ$, where C is quasisimple. Moreover, C (being a normal subgroup of the subnormal subgroup R) is subnormal in G, and so $C \in \mathrm{Comp}(G)$. Thus $CZ \leqslant E(G)Z \leqslant F^*(G)$. But $C \leqslant R \leqslant M \leqslant H$, and so $C \leqslant C_G(F^*(G))$. Thus $C \leqslant Z(F^*(G)) = Z$ (the equality following by Proposition 16.15). Hence $R = CZ \leqslant Z$, contradicting the fact that R/Z is non-trivial. This contradiction shows that $H \leqslant Z$, as required.

16.4 The general case of Pyber's theorem

In this section we complete the proof of Pyber's theorem in the general case. We will use the notation defined in Chapter 15 freely. In particular, we suppose G is a group of order n, where $n = p_1^{\alpha_1} p_2^{\alpha_2} \cdots p_k^{\alpha_k}$. We fix the Sylow p_i-subgroups P_i of G for $i \in \{1, 2, \ldots, k\}$. We define

$$F_i = O_{p_i}(G),$$

$$A_i = \mathrm{Aut}\,(F_i) \text{ and}$$

$$A = \mathrm{Aut}\,(F(G)) = A_1 \times A_2 \times \cdots \times A_k.$$

In addition, we define some new notation: the groups A_0 and A^* are defined by

$$A_0 = \mathrm{Aut}\,(E(G)) \text{ and}$$

$$A^* = A_0 \times A.$$

The proof has a very similar structure to the soluble case given in Chapter 15, but the generalised Fitting subgroup $F^*(G)$ replaces the Fitting subgroup $F(G)$ throughout. More precisely, we bound the number of groups G of order n having fixed Sylow subgroups $P_1, P_2, \ldots, P_k$ in three stages. We first count the number of choices for the generalised Fitting subgroup $F^*(G)$. The quotient $G/Z(F^*(G))$ may be realised as a subgroup of $\mathrm{Aut}\,(F^*(G))$.

The natural embedding of $\text{Aut}(F^*(G))$ into A^* realises $G/Z(F^*(G))$ as a subgroup of A^*, and the second stage in our proof counts the number of possibilities for this subgroup. Finally, we use Lemma 15.1 to bound the number of possibilities for G.

Before beginning the proof of the theorem, we prove some results concerning A^*.

Lemma 16.19 *Let G be a group of order n. Define A_0 and A^* as above.*

(i) $|A^*| \leqslant n^{2\mu+1}$.

(ii) *If S is a soluble subgroup of A_0, then $|S| \leqslant n^3$.*

(iii) *There are at most $n^{(41/4)\mu+(25/2)}$ soluble subgroups of A_0.*

(iv) *The number $|\text{Mss}(A^*)|$ of maximal soluble subgroups of A^* is at most $n^{(45/4)\mu+278\,847}$.*

Proof: Since $|E(G)|$ divides n, Corollary 16.7 implies that $E(G)$ can be generated by $\mu+1$ elements. An automorphism of $E(G)$ is determined by the images of these $\mu+1$ elements. So $|A_0| \leqslant n^{\mu+1}$. We showed in the proof of Theorem 15.5 (when giving a bound on $|M|$) that $|A| \leqslant n^\mu$. Therefore, part (i) of the lemma follows.

We now investigate the structure of A_0. Let $\text{Comp}(G) = \{C_1, C_2, \dots, C_t\}$. Then $\text{Comp}(E(G)) = \text{Comp}(G)$, and so A_0 permutes these components amongst themselves. Let B_0 be the kernel of this action. So B_0 is isomorphic to a subgroup of the direct product $\prod_{i=1}^{t} \text{Aut}(C_i)$, and A_0/B_0 is isomorphic to a subgroup of $\text{Sym}(t)$.

For $i \in \{1, 2, \dots, t\}$, define the non-abelian simple group L_i by $L_i = C_i/Z(C_i)$. We claim that the natural map from $\text{Aut}(C_i)$ to $\text{Aut}(L_i)$ is injective. For suppose $\alpha \in \text{Aut}(C_i)$ acts trivially modulo $Z(C_i)$. Then for all $x, y \in C_i$ we have that $\alpha(x) = xz_1$ and $\alpha(y) = yz_2$ for some $z_1, z_2 \in Z(C_i)$ and so

$$\alpha([x, y]) = [xz_1, yz_2] = [x, y],$$

since z_1 and z_2 are central. But C_i is perfect (by definition of a component) and so is generated by its commutators. Hence α is the identity automorphism. So we may regard B_0 as a subgroup of the direct product $\prod_{i=1}^{t} \text{Aut}(L_i)$.

We are now ready to prove part (ii) of the lemma. Let S be a soluble subgroup of A_0. Then S is contained in a subgroup X of A_0 such that $B_0 \leqslant X$ and X/B_0 is a maximal soluble subgroup of A_0/B_0. Clearly, $|S| \leqslant |X|$. To find an upper bound on $|X|$, first note that X/B_0 is a soluble subgroup of

$\mathrm{Sym}(t)$ and so $|X/B_0| \leqslant 24^{t/3}$ by Theorem 10.1. Moreover, one consequence of the Classification is that $|\mathrm{Out}(L)| \leqslant |L|$ for any simple group L, and so

$$|B_0| \leqslant \prod_{i=1}^{t} |\mathrm{Aut}\,(L_i)|$$

$$= \prod_{i=1}^{t} |L_i| \prod_{i=1}^{t} |\mathrm{Out}(L_i)|$$

$$\leqslant \left(\prod_{i=1}^{t} |L_i| \right)^2.$$

Now $E(G)/Z(E(G))$ is a section of G that is isomorphic to $\prod_{i=1}^{t} L_i$. Thus $\prod_{i=1}^{t} |L_i|$ divides n and so $|B_0| \leqslant n^2$. Since $t \leqslant \mu/2$ by Corollary 16.17 (iii),

$$|S| \leqslant |X| = |B_0||X/B_0| \leqslant n^2 24^{t/3} \leqslant n^2 24^{\mu/6} < n^2 2^\mu \leqslant n^3.$$

So part (ii) of the lemma follows.

We now aim to bound the number of soluble subgroups S of A_0. We first bound the number of choices for the subgroup X above, and then bound the number of subgroups S associated with each choice of X. The number of choices for X is at most $|\mathrm{Mss}(A_0/B_0)|$. By Lemma 15.2, $|\mathrm{Mss}(A_0/B_0)| \leqslant |\mathrm{Mss}(\mathrm{Sym}(t))|$. Theorem 11.2 implies that $|\mathrm{Mss}(\mathrm{Sym}(t))| \leqslant t! 2^{16t}$. We observed above that $t \leqslant \mu/2$, and so

$$t! 2^{16t} \leqslant n^{\mu/2} 2^{8\mu} \leqslant n^{\mu/2+8}.$$

So there are at most $n^{\mu/2+8}$ possibilities for X.

We claim that every soluble subgroup S of A_0 can be generated by $\frac{1}{4}(13\mu + 6)$ elements. Proving this claim establishes part (iii) of the lemma; we may see this as follows. Our claim implies that S is determined by at most $\frac{1}{4}(13\mu + 6)$ elements from X, and so there are at most $|X|^{(13\mu+6)/4}$ choices for S once X is fixed. But we observed above that $|X| \leqslant n^3$, and that there are at most $n^{\mu/2+8}$ possibilities for X. Hence the number of soluble subgroups of A_0 is at most $n^{\mu/2+8} n^{(39\mu+18)/4}$, and so part (iii) of the lemma follows from our claim.

We prove the claim as follows. Let S be a soluble subgroup of A_0. Now, SB_0/B_0 is isomorphic to a subgroup of $\mathrm{Sym}(t)$ and so can be generated by $(t+1)/2$ or fewer elements by Theorem 6.11. The group B_0 is isomorphic to a subgroup of the direct product $\prod_{i=1}^{t} \mathrm{Aut}\,(L_i)$, so there exists a subgroup H of $\prod_{i=1}^{t} \mathrm{Aut}\,(L_i)$ isomorphic to $S \cap B_0$. Define the subgroup K of $\prod_{i=1}^{t} \mathrm{Aut}\,(L_i)$ by $K = \prod_{i=1}^{t} L_i$. Then $\prod_{i=1}^{t} \mathrm{Aut}\,(L_i)/K$ is isomorphic to the direct product $\prod_{i=1}^{t} \mathrm{Out}(L_i)$, where $\mathrm{Out}(L_i)$ is the outer automorphism group of L_i. By Corollary 16.17 (ii) we have that $|K|$ divides n. So $|H \cap K|$ divides n and hence $H \cap K$ can be generated by $\mu + 1$ elements, by Corollary 16.7. The Classification implies that if L is simple then $\mathrm{Out}(L)$ has the following structure. There exists a normal series $1 \leqslant D \leqslant E \leqslant \mathrm{Out}(L)$ for $\mathrm{Out}(L)$ such that D and E/D are cyclic, and such that $\mathrm{Out}(L)/E$ is either cyclic or is isomorphic to $\mathrm{Sym}(3)$. In particular, every subgroup of $\mathrm{Out}(L)$ can be generated by 4 or fewer elements. Thus any subgroup of $\prod_{i=1}^{t} \mathrm{Out}(L_i)$ can be generated by $4t$ elements; in particular, this is true of HK/K. Thus S can be generated by $4t + (t+1)/2 + \mu + 1$ elements. Using the fact that $t \leqslant \mu/2$, our claim (and part (iii) of the lemma) is proved.

We now prove part (iv) of the lemma. We have that $|\mathrm{Mss}(A^*)| = |\mathrm{Mss}(A_0 \times A)| = |\mathrm{Mss}(A_0)| \, |\mathrm{Mss}(A)|$. Now $|\mathrm{Mss}(A_0)|$ is bounded above by the number of soluble subgroups of A_0. So the bound of part (iii) of the lemma together with the bound of Corollary 15.4 combine to prove part (iv). This establishes the lemma.

We now state and prove Pyber's theorem.

Theorem 16.20 *The number of groups G of order n with Sylow subgroups $P_1, P_2, \ldots, P_k$ is at most $n^{(97/4)\mu + 278\,852}$.*

Proof: We begin by proving that there are at most $n^{2\mu+3}$ choices for $F^*(G)$ (up to isomorphism). This is clearly true when $n \leqslant 3$, so we may assume that $n \geqslant 4$.

Let $\mathrm{Comp}(G) = \{C_1, C_2, \ldots, C_t\}$, and define $L_i = C_i/Z(C_i)$ for $i \in \{1, 2, \ldots, t\}$. Now $\prod_{i=1}^{t} |L_i|$ divides n by Corollary 16.17 (ii), and so we find that $(|L_1|, |L_2|, \ldots, |L_t|)$ is a multiplicative partition of a divisor r of n. By Lemma 11.1, there are at most r^2 multiplicative partitions of r, and so the number of choices for t and the integers $|L_i|$ is bounded above by

$$\sum_{r|n} r^2 = \sum_{r|n} (n/r)^2 \leqslant n^2 \sum_{r=1}^{\infty} (1/r^2) < 2n^2,$$

since $\pi^2/6 < 2$. The Classification implies that there are at most two simple groups of any order, and so there are at most 2^t choices for the isomorphism classes of the groups L_i once their orders are fixed. But $t \leqslant \mu/2$ by Corollary 16.17 (iii), and so $2^t \leqslant 2^{\frac{1}{2}\mu} \leqslant 2^{\frac{1}{2}\log n} = \sqrt{n}$. Hence the number of choices for the groups L_i is at most n^3, since $2n^2\sqrt{n} \leqslant n^3$.

There are at most n^μ possibilities for the isomorphism class of $F(G)$. This follows by the argument in the proof of Theorem 15.5. To recapitulate, $F(G)$ is a direct product of its Sylow p_i-subgroups F_i. Each F_i is isomorphic to a subgroup of P_i. But P_i has order $p_i^{\alpha_i}$, so P_i has at most $(p_i^{\alpha_i})^{\alpha_i} \leqslant (p_i^{\alpha_i})^\mu$

subgroups. Thus there are at most $\prod_{i=1}^{k}(p_i^{\alpha_i})^{\mu} = n^{\mu}$ possibilities for $F(G)$. Note that by choosing F_i as a normal subgroup of P_i, we determine the action of P_i on F_i by conjugation.

Define the group X and the subgroup K as in Theorem 16.16. We have shown that there are at most $n^{\mu+3}$ possibilities for the isomorphism class of X. To complete the first stage of our proof, we need to show that there are at most n^{μ} choices for K.

We claim that $|Z(X)|$ divides n. To see this, note that $|Z(\tilde{L}_i)|$ divides $|L_i|$ by Proposition 16.12, and clearly $|Z(F(G))|$ divides $|F(G)|$. Since

$$Z(X) = Z(\tilde{L}_1) \times Z(\tilde{L}_2) \times \cdots \times Z(\tilde{L}_t) \times Z(F(G)),$$

our claim now follows by Corollary 16.17 (ii).

Now, K is a subgroup of the abelian group $Z(X)$. So K is abelian of order dividing n, and thus K may be generated by μ elements. Hence there are at most n^{μ} choices for K, as $|Z(X)| \leqslant n$. We have now shown that there are at most $n^{2\mu+3}$ choices for $F^*(G)$ (up to isomorphism), as required.

Let $Z = Z(F^*(G))$. Conjugation induces an embedding of G/Z into $\mathrm{Aut}(F^*(G))$ by Theorem 16.18. Now, there is a natural embedding from $\mathrm{Aut}(F^*(G))$ into A^*, so there exists a natural map $\phi : G \to A^*$ with kernel Z. We aim to count the number of possibilities for $\phi(G)$.

Let $H = \phi(G)$. By Theorem 16.4, there exists a soluble subgroup S of H and an element $x \in H$ such that H is generated by S and xSx^{-1}. By Lemma 16.19 (i), there are at most $n^{2\mu+1}$ possibilities for x. The soluble subgroup S of A^* is contained in a maximal soluble subgroup M. By Lemma 16.19 (iv), there are at most $n^{(45/4)\mu+278\,847}$ such subgroups M. We now aim to give an upper bound for the number of subgroups S of A^* contained in a fixed maximal soluble subgroup M. The argument below is essentially that given in Chapter 15, but with modifications due to the extra group A_0 being involved.

Let $Q_1, Q_2, \ldots, Q_k$ be a Sylow system for S. By replacing the Sylow subgroups P_i of G by appropriate conjugates if necessary, we may assume that $Q_i \leqslant \phi(P_i)$ for $i \in \{1, 2, \ldots, k\}$. By Theorem 6.2, there is (part of) a Sylow system $R_1, R_2, \ldots, R_k$ for M such that $Q_i = S \cap R_i$ for $i \in \{1, 2, \ldots, k\}$. By Theorem 6.2, all Sylow systems of M are conjugate, and so there are at most $|M|$ possibilities for the R_i. Note that $|M| \leqslant |A^*| = n^{2\mu+1}$.

Recall that $A^* = A_0 \times A_1 \times \cdots \times A_k$. For $j \in \{0, 1, \ldots, k\}$, let $\pi_j : A^* \to A_j$ be the natural map. Now $M = M_0 \times M_1 \times \cdots \times M_k$, where $M_j = \pi_j(M)$. Note that M_j is a maximal soluble subgroup of A_j. For each $i \in \{1, 2, \ldots, k\}$,

$$R_i = R_{i0} \times R_{i1} \times \cdots \times R_{ik}$$

where $R_{ij} = \pi_j(R_i)$. Note that R_{ij} is a Sylow p_i-subgroup of M_j and $\pi_j(Q_i) = \pi_j(S) \cap \pi_j(R_i) \leqslant R_{ij}$.

We define an 'approximate Sylow system' $T_1, T_2, \ldots, T_k$ for M as follows. Let $\phi_i = \pi_i \circ \phi$. So ϕ_i is the map from G to $\mathrm{Aut}\,(F_i)$ induced by conjugation. Since $Q_i \leqslant \phi(P_i)$, we have that $\pi_i(Q_i) \leqslant \phi_i(P_i)$. For $i \in \{1, 2, \ldots, k\}$ and $j \in \{0, 1, 2, \ldots, k\}$, define

$$T_{ij} = \begin{cases} R_{ij} & \text{if } i \neq j, \\ \phi_i(P_i) & \text{if } i = j. \end{cases}$$

Our method for enumerating the possibilities for $F^*(G)$ also determined the action of P_i on F_i, so we may consider $\phi_i(P_i)$ as being known for all $i \in \{1, 2, \ldots, k\}$. Thus the subgroups T_{ij} are determined by the Sylow system for M we have chosen. Finally, for $i \in \{1, 2, \ldots, k\}$ define

$$T_i = T_{i0} \times T_{i1} \times \cdots \times T_{ik}.$$

Note that T_i is a p_i-group, and $Q_i \leqslant T_i$.

For $j \in \{1, 2, \ldots, k\}$, define

$$X_j = \prod_{\substack{i=1 \\ i \neq j}}^{k} R_{ij}.$$

Since the subgroups R_i form part of a Sylow system for M, we have that $R_{ij}R_{i'j'} = R_{i'j'}R_{ij}$ and so X_j is well-defined, and is a soluble p'_j-subgroup of A_j. Lemma 15.3 (i) implies that $|X_j| \leqslant p_j^{3\alpha_j}$. But now

$$\prod_{i=1}^{k} |T_i| = \prod_{i=1}^{k} |X_i| \prod_{i=1}^{k} |T_{ii}| \prod_{i=1}^{k} |T_{i0}|$$

$$\leqslant \prod_{i=1}^{k} p_i^{3\alpha_i} \prod_{i=1}^{k} p_i^{\alpha_i} \prod_{i=1}^{k} |T_{i0}|$$

$$= n^4 \prod_{i=1}^{k} |T_{i0}|.$$

But $\prod_{i=1}^{k} |T_{i0}| \leqslant |M_0| \leqslant n^3$, by Lemma 16.19 (ii). Hence $\prod_{i=1}^{k} |T_i| \leqslant n^7$.

Now Q_i is a subgroup of T_i, and is generated by at most α_i elements, so the number of possibilities for Q_i once T_i is fixed is at most $|T_i|^{\alpha_i}$. But

$$\prod_{i=1}^{k} |T_i|^{\alpha_i} \leqslant \prod_{i=1}^{k} |T_i|^{\mu} \leqslant n^{7\mu},$$

and so there are at most $n^{7\mu}$ possibilities for the Q_i once the T_i are chosen.

To summarise, $\phi(G)$ is determined by a soluble subgroup S and an element x. There are at most $n^{2\mu+1}$ choices for x, at most $n^{(45/4)\mu+278\,847}$ choices for M and at most $n^{2\mu+1}$ choices for a Sylow system R_i of M. The approximate Sylow system T_i is then determined; there are at most $n^{7\mu}$ possibilities for the Q_i, and then S is determined. So there are at most $n^{(89/4)\mu+278\,849}$ choices for $\phi(G)$, once $F^*(G)$ is fixed. This completes the second stage of the proof.

To finish the proof, we will apply Lemma 15.1. Now, G is an extension of Z by G/Z. Once $F^*(G)$ is fixed, the isomorphism class of Z is determined. Once $\phi(G)$ is fixed, both the isomorphism class of G/Z and the action of G/Z on Z are determined. Lemma 15.1 implies there are at most n^μ choices for G once $F^*(G)$ and $\phi(G)$ are fixed. There are at most $n^{2\mu+3}$ choices for $F^*(G)$, and then at most $n^{(89/4)\mu+278\,849}$ choices for $\phi(G)$. So we find that there are at most $n^{(97/4)\mu+278\,852}$ choices for G, and so the theorem follows.

Corollary 16.21 *We have that* $f(n) \leqslant n^{\frac{2}{27}\mu^2+O(\mu^{3/2})}$.

Proof: Let $n = p_1^{\alpha_1} p_2^{\alpha_2} \cdots p_k^{\alpha_k}$. We wish to count the number of choices for a group G of order n, up to isomorphism. By Theorem 5.7, there are at most $p_i^{\frac{2}{27}\alpha_i^3+O(\alpha_i^{5/2})}$ choices for the isomorphism class of a Sylow p_i-subgroup of G. Since

$$p_i^{\frac{2}{27}\alpha_i^3+O(\alpha_i^{5/2})} = (p_i^{\alpha_i})^{\frac{2}{27}\alpha_i^2+O(\alpha_i^{3/2})} \leqslant (p_i^{\alpha_i})^{\frac{2}{27}\mu^2+O(\mu^{3/2})},$$

an upper bound for the number of choices for the isomorphism classes of the Sylow subgroups of G is

$$\prod_{i=1}^{k} (p_i^{\alpha_i})^{\frac{2}{27}\mu^2+O(\mu^{3/2})} = n^{\frac{2}{27}\mu^2+O(\mu^{3/2})}.$$

Theorem 16.20 implies that there are at most $n^{O(\mu)}$ choices for the isomorphism class of G once the isomorphism classes of the Sylow subgroups of G have been chosen, and so the corollary follows.

Corollary 16.22 *We have that* $f_A(n) \leqslant n^{(97/4)\mu+278\,853}$.

Proof: The number of choices for the Sylow subgroups $P_1, P_2, \ldots, P_k$ of an A-group of order n is precisely the number of abelian groups of order n. Anticipating an elementary result, Corollary 17.3, from the next chapter, we find that this is at most n and the estimate follows immediately from Theorem 16.20.

As has been noted before (see p. 139), the methods of Chapter 15 ought to yield a result of the form $f_A(n) \leqslant n^{8\mu+b}$ and in fact one should seek to prove that $f_A(n) \leqslant n^{2\mu+o(\mu)}$ or something even better—see Question 22.22 on p. 266.

IV

Other topics

17

Enumeration within varieties of abelian groups

Up to this point the thrust of this work has been the study of the general group enumeration function $f(n)$, although enumeration of special kinds of groups, such as soluble groups and A-groups, has been touched on. We focus now, however, on enumeration within restricted classes of groups and, in particular, enumeration in varieties of groups. Recall that a variety $\mathfrak{V}$ is a class of groups G corresponding to a set of laws $w(x_1, \ldots, x_n)$, that is to say, defined by identical relations of the form $w(g_1, \ldots, g_n) = 1$ required to hold for all choices of $g_1, \ldots, g_n \in G$. For a survey of varieties of groups see Hanna Neumann's book [75]. The varieties of particular interest to us are:

- the variety $\mathfrak{A}$ of abelian groups, defined by the commutator law $[x, y]$;
- the variety $\mathfrak{A}_r$ of abelian groups of exponent dividing r, defined by the laws $[x, y]$ and x^r;
- the product variety $\mathfrak{A}_r \mathfrak{A}_s$ consisting of all groups G with an abelian normal subgroup A of exponent dividing r such that G/A is abelian of exponent dividing s (this variety is defined by all laws $[w_1, w_2]$ and w_3^r, where w_1, w_2, w_3 can be expressed as products of sth powers and commutators);
- the variety $\mathfrak{B}_{p,2}$ of groups of class at most 2 and exponent dividing p, where p is an odd prime number, defined by the laws $[[x, y], z]$ and x^p;
- the Higman variety $\mathfrak{H}_p$ of p-groups of Φ-class 2, defined by the laws x^{p^2}, $[x, y]^p$ and $[[x, y], z]$.

In what follows we shall speak of $\mathfrak{V}$ as being 'a variety of A-groups' if it is generated by finite A-groups. The enumeration functions $f_{\mathfrak{V}}(n)$ exhibit an interesting trichotomy:

$$f_{\mathfrak{V}}(n) \leqslant \begin{cases} n & \text{if } \mathfrak{V} \subseteq \mathfrak{A}, \\ n^{c_1 \mu(n) + c_2} & \text{if } \mathfrak{V} \text{ is a non-abelian variety of A-groups}, \\ n^{c_3 \mu(n)^2 + c_4 \mu(n)^{3/2}} & \text{if } \mathfrak{V} \text{ contains a non-abelian nilpotent group}, \end{cases}$$

163

where c_1, c_2, c_3, c_4 are suitable constants and $\mu(n)$ is as defined on p. 2. In fact the estimate $f_{\mathfrak{V}}(n) \leqslant n$ for subvarieties of $\mathfrak{A}$ is a crude one which, as we shall see below, can be much refined, and, as we have already seen, $c_3 = 2/27$. That the above really is a trichotomy follows from the fact that if $\mathfrak{V}$ is a variety that contains non-abelian finite groups then either it contains a non-abelian finite p-group for some prime number p, in which case if $n = p^m$ then $f_{\mathfrak{V}}(n) \geqslant p^{\frac{2}{27}m^3 - \frac{2}{3}m^2} = n^{\frac{2}{27}\mu(n)^2 - \frac{2}{3}\mu(n)}$ (see Theorems 19.1, 19.2, 19.3 below), or for some distinct prime numbers p, q it contains the variety $\mathfrak{A}_p\mathfrak{A}_q$ (this is an immediate consequence of the theory presented in [75, Chapter 5, Section 3] but it can also be proved easily by direct methods), and in this case there exists $c > 0$ such that $f_{\mathfrak{V}}(n) > n^{c\mu(n)}$ for infinitely many integers n (see Theorem 18.4). In the first three chapters of this last part of the book we examine this phenomenon in more detail.

This chapter is concerned with enumeration results about abelian groups. For a variety $\mathfrak{V}$ that contains only abelian groups, our interest is in proving bounds on the number $f_{\mathfrak{V}}(n)$ of isomorphism classes of groups of order n that are contained in $\mathfrak{V}$. The chapter is divided into two sections. In the first section we review the classification of varieties $\mathfrak{V}$ of abelian groups, and show how $f_{\mathfrak{V}}(n)$ is related to an enumeration function for partitions of a certain type. The second section contains a discussion of the partition enumeration problems that arise and the consequences for our group enumeration problem. Much of the material in this section is drawn from the 1998 Oxford MSc thesis of Duncan Brydon [12].

17.1 Varieties of abelian groups

We are interested in abelian varieties, in other words, subvarieties of the variety $\mathfrak{A}$. By the following theorem due to B. H. Neumann [74], the varieties $\mathfrak{A}$ and $\mathfrak{A}_r$ are the only abelian varieties.

Theorem 17.1 *Let $\mathfrak{V}$ be an abelian variety. Then (using the notation above) either $\mathfrak{V} = \mathfrak{A}$, or there exists a positive integer r such that $\mathfrak{V} = \mathfrak{A}_r$.*

Proof: Let $\mathfrak{V}$ be a variety. Let w be a law in $\mathfrak{V}$. We may write

$$w = x_1^{r_1} x_2^{r_2} \cdots x_k^{r_k} c \tag{17.1}$$

where $r_1, r_2, \ldots, r_k$ are integers and where c is a product of commutators involving the elements x_i only. We claim that if the r_i are not all zero, then w is equivalent to the laws c and x^d, where d is the greatest common divisor of the integers r_i. Clearly the laws c and x^d imply w. To prove the converse, we first observe that substituting $x_i = x$ and $x_j = 1$ whenever $j \neq i$ into w shows that the laws x^{r_i} are consequences of w. By the (extended) Euclidean algorithm, there exist integers $a_1, a_2, \ldots, a_k$ such that

$$a_1 r_1 + a_2 r_2 + \cdots + a_k r_k = d.$$

So the equality

$$(x^{r_1})^{a_1} (x^{r_2})^{a_2} \cdots (x^{r_k})^{a_k} = x^d$$

implies that x^d is a consequence of the laws x^{r_i}. Since the laws $x_i^{r_i}$ are consequences of w, the fact that

$$c = x_k^{-r_k} \cdots x_2^{-r_2} x_1^{-r_1} w$$

implies that c is a consequence of w. So our claim follows.

Now assume that $\mathfrak{V}$ is an abelian variety. Then $[x, y]$ is a law in $\mathfrak{V}$. Firstly, suppose for all positive integers e that x^e is not a law in $\mathfrak{V}$. By our claim above, this implies that any law in $\mathfrak{V}$ may be written as a product c of commutators (for if the r_i are not all zero in the expression (17.1) then x^d is a law in $\mathfrak{V}$). But c is a consequence of $[x, y]$ and hence $\mathfrak{V} = \mathfrak{A}$ in this case.

Suppose that there exists a positive integer e such that x^e is a law in $\mathfrak{V}$. Let r be the smallest positive integer such that x^r is a law in $\mathfrak{V}$. To prove the theorem, we will show that $\mathfrak{V} = \mathfrak{A}_r$, and to do this it suffices to show that any law in $\mathfrak{V}$ is a consequence of x^r and $[x, y]$. Let w be a law in $\mathfrak{V}$, expressed in the form (17.1) as before. If all the integers r_i are zero then $w = c$ and c is a consequence of $[x, y]$, so we may assume that at least one of the integers r_i is non-zero. Let d be the greatest common divisor of the integers r_i. The claim above implies that x^d is a law in $\mathfrak{V}$, and so $x^{\gcd(d, r)}$ is a law in $\mathfrak{V}$. By our choice of r, we must have that $\gcd(d, r) = r$ and so r divides all of the integers $r_1, r_2, \ldots, r_k$. But then $x_i^{r_i}$ is a consequence of x^r, and since c is a consequence of $[x, y]$ we find that w is a consequence of x^r and $[x, y]$. This establishes the theorem.

Define $f_{\mathfrak{A}}(n)$ to be the number of isomorphism classes of abelian groups of order n. For a positive integer r, define $f_{\mathfrak{A}_r}(n)$ to be the number of isomorphism classes of abelian groups of order n and of exponent dividing r. We now show that the enumeration functions $f_{\mathfrak{A}}(n)$ and $f_{\mathfrak{A}_r}(n)$ are closely related to certain enumeration functions $p(m)$ and $p_s(m)$ that arise in the theory of integer partitions. Here, $p(m)$ is the number of (unordered) partitions of a positive integer m, and $p_s(m)$ is the number of partitions π of m such that each part of π is at most s. We will consider these two functions in more detail in the next section.

Theorem 17.2 *Define the functions $f_{\mathfrak{A}}(n)$ and $p(m)$ as above. Let n be a positive integer, and let $n = p_1^{m_1} p_2^{m_2} \cdots p_u^{m_u}$ be the decomposition of n into a product of distinct prime numbers p_i. Then*

$$f_{\mathfrak{A}}(n) = \prod_{i=1}^{u} p(m_i).$$

Proof: An abelian group A may be written uniquely in the form

$$A = P_1 \times P_2 \times \cdots \times P_u$$

where P_i is the (abelian) Sylow p_i-subgroup of A. Since the isomorphism class of A is uniquely determined by the isomorphism classes of the subgroups P_i, we have that

$$f_{\mathfrak{A}}(n) = \prod_{i=1}^{u} f_{\mathfrak{A}}(p_i^{m_i}).$$

So to establish the theorem, it is sufficient to show that $f_{\mathfrak{A}}(p^m) = p(m)$ for any positive integer m and prime number p. The classification theorem for finite abelian groups states that any abelian group P of order p^m is isomorphic to a unique group of the form

$$C_{p^{a_1}} \times C_{p^{a_2}} \times \cdots \times C_{p^{a_k}}$$

where C_r denotes the cyclic group of order r and where $p^{a_{i+1}}$ divides p^{a_i} for $i \in \{1, 2, \ldots, k-1\}$. So $a_1, a_2, \ldots, a_k$ are positive integers such that $a_1 \geqslant a_2 \geqslant \cdots \geqslant a_k$. The condition that $|P| = p^m$ is equivalent to the equality $a_1 + a_2 + \cdots + a_k = m$. But these last two conditions merely state that $a_1, a_2, \ldots, a_k$ form a partition of m, and so $f_{\mathfrak{A}}(p^m) = p(m)$, as required.

This theorem, together with the crude bound on $p(m)$ that we proved in Corollary 5.10, already shows that $f_{\mathfrak{A}}(n)$ grows much more slowly than the group enumeration functions we have studied up to this point.

Corollary 17.3 *Let $\mathfrak{V}$ be an abelian variety. Then $f_{\mathfrak{V}}(n) \leqslant n$ for all positive integers n.*

Proof: Since $\mathfrak{V} \subseteq \mathfrak{A}$, it suffices to prove the corollary when $\mathfrak{V} = \mathfrak{A}$. Corollary 5.10 states that $p(m) \leqslant 2^{m-1}$, and so by Theorem 17.2 (and using the same notation)

$$f_{\mathfrak{A}}(n) \leqslant \prod_{i=1}^{u} p(m_i) \leqslant \prod_{i=1}^{u} 2^{m_i-1} \leqslant \prod_{i=1}^{u} p_i^{m_i} = n.$$

Theorem 17.4 *Let r be a positive integer, and let $r = p_1^{s_1} p_2^{s_2} \cdots p_u^{s_u}$ be the decomposition of r into a product of distinct prime numbers p_i. Let $\pi = \{p_1, p_2, \ldots, p_u\}$. Let n be a positive integer. If n is not a π-number, then $f_{\mathfrak{A}_r}(n) = 0$. If n is a π-number, then $n = p_1^{m_1} p_2^{m_2} \cdots p_u^{m_u}$ for some non-negative integers $m_1, m_2, \ldots, m_u$ and*

$$f_{\mathfrak{A}_r}(n) = \prod_{i=1}^{u} p_{s_i}(m_i).$$

Proof: Suppose that there exists a prime p dividing n that is not one of $p_1, p_2, \ldots, p_u$. Then an abelian group A of order n cannot have exponent dividing r, and so no group of order n can lie in $\mathfrak{A}_r$. Hence $f_{\mathfrak{A}_r}(n) = 0$ in this case.

Suppose now that n is a π-number. Let $m_1, m_2, \ldots, m_u$ be non-negative integers such that $n = p_1^{m_1} p_2^{m_2} \cdots p_u^{m_u}$. Since an abelian group A has exponent dividing r if and only if the same is true of its Sylow subgroups, we may argue as in Theorem 17.2 that

$$f_{\mathfrak{A}_r}(n) = \prod_{i=1}^{u} f_{\mathfrak{A}_r}(p_i^{m_i}).$$

But a p_i-group has exponent dividing r if and only if it has exponent dividing $p_i^{s_i}$, and so

$$f_{\mathfrak{A}_r}(n) = \prod_{i=1}^{u} f_{\mathfrak{A}_{p_i^{s_i}}}(p_i^{m_i}).$$

It remains to show for a prime number p, a positive integer s and a non-negative integer m that $f_{\mathfrak{A}_{p^s}}(p^m) = p_s(m)$. To see this, we use the correspondence given in the proof of Theorem 17.2 between isomorphism classes of abelian groups P of order p^m and partitions $a_1, a_2, \ldots, a_k$ of m. It is clear that an abelian group P of order p^m has exponent dividing p^s if and only if all the parts of the corresponding partition are at most s. Hence there is a one-to-one correspondence between isomorphism classes of abelian groups of order p^m in $\mathfrak{A}_{p^s}$ and partitions of m whose parts are all less than or equal to s. Thus $f_{\mathfrak{A}_{p^s}}(p^m) = p_s(m)$, and the theorem follows.

17.2 Enumerating partitions

Our next aim is to provide bounds on the functions $p(m)$ and $p_s(m)$ that arise in our enumeration problems. We first consider $p(m)$, the number of partitions of m. As we mentioned in earlier chapters, a theorem of Hardy and Ramanujan [44] (see Andrews [3]) gives an excellent bound on the size of $p(m)$.

Theorem 17.5 *Let* $p(m)$ *be the number of partitions of the integer m. Then*

$$p(m) \sim_m \frac{1}{4\sqrt{3}m} e^{\pi\sqrt{(2/3)m}}.$$

Here '$\sim_m$' means that the ratio of the two sides approaches 1 as $m \to \infty$. The proof of this theorem uses a fair amount of analysis, and is beyond the scope of this book. However, it is possible to say something about the rate of growth of $p(m)$ using elementary arguments, and we will do this below. Before doing this, we note that Theorem 17.5 may be combined with Theorem 17.2 to give one of the enumeration results we are seeking. Let n be a positive integer. We may write $n = p_1^{m_1} p_2^{m_2} \cdots p_u^{m_u}$ where the p_i are distinct prime numbers and the m_i are positive integers. We then define

$$\eta(n) = \min\{m_1, m_2, \ldots, m_u\}.$$

Theorem 17.6 *Let* u *be a fixed positive integer. If* $n = p_1^{m_1} p_2^{m_2} \cdots p_u^{m_u}$ *where the* p_i *are distinct prime numbers and the* m_i *are positive integers, then*

$$f_{\mathfrak{A}}(n) \sim_\eta \prod_{i=1}^{u} \frac{1}{4\sqrt{3}m_i} e^{\pi\sqrt{(2/3)m_i}}.$$

Before turning to the enumeration of partitions with parts of bounded size, we include a proposition that establishes the approximate rate of growth of $p(m)$ using elementary arguments. The proof we use to establish the upper bound of the proposition is taken from a 1942 paper of Erdős [27], who notes that Hardy and Ramanujan [44] mention the existence of such a proof.

Proposition 17.7 *For all sufficiently large integers m,*

$$2^{\sqrt{m}} \leqslant p(m) < e^{\pi\sqrt{(2/3)m}}.$$

Clearly (given that we know that Theorem 17.5 is in fact true) the upper bound of Theorem 17.7 is closer to the truth than the lower bound. In fact, an improved lower bound of the form $e^{(\pi\sqrt{2/3}-\epsilon)\sqrt{m}} < p(m)$ for any $\epsilon > 0$ can be proved using the same techniques used in the proof of the upper bound we give below.

Proof of Proposition 17.7: We begin by establishing the lower bound. Let k be the largest integer such that $\frac{1}{2}(k+1)(k+2) \leqslant m$. This bound on k implies that $m - \sum_{i=1}^{k} i \geqslant k+1$. Note that if $m \geqslant 28$ then $k \geqslant \sqrt{m}$. Let S be any subset of $\{1, 2, \ldots, k\}$. Then there exists a unique integer s such that $S \cup \{s\}$ is a partition of m. Moreover, $s \geqslant k+1$ and so partitions of m arising from distinct subsets S are distinct. There are 2^k subsets of $\{1, 2, \ldots, k\}$ and so $p(m) \geqslant 2^k \geqslant 2^{\sqrt{m}}$ as required.

We now establish the upper bound. We first observe that the function $p(m)$ satisfies the identity

$$mp(m) = \sum_{v=1}^{m} \sum_{k=1}^{\lfloor m/v \rfloor} vp(m - kv). \tag{17.2}$$

To see this, we sum all the parts of all the partitions of m in two different ways. The parts of each of the $p(m)$ partitions of m clearly sum to m, giving the left-hand side of (17.2). Now, each integer v occurs in exactly $p(m - v)$ partitions of m. More generally, v occurs k or more times in exactly $p(m - kv)$ partitions of m. Hence the parts equal to v contribute $\sum_{k=1}^{\lfloor m/v \rfloor} vp(m - kv)$ to our sum. This gives the right-hand side of (17.2).

Let $c = \pi \sqrt{2/3}$. We prove that $p(m) < e^{c\sqrt{m}}$ by induction on m. The following two facts are needed in our proof. The first fact is that $\sqrt{1-t} < 1 - \frac{1}{2}t$ when $0 < t \leqslant 1$, which follows from the fact that $1 - t < 1 - t + \frac{1}{4}t^2 = \left(1 - \frac{1}{2}t\right)^2$. The second fact is that $\frac{e^{-x}}{(1-e^{-x})^2} < \frac{1}{x^2}$ for any non-zero real number x. This may be established by elementary calculus: the function f defined by $f(x) = x^2 e^{-x}/(1 - e^{-x})^2$ has no stationary points in $\mathbb{R} \setminus \{0\}$, and by L'Hôpital's rule we see that $f(x) \to 1$ as $x \to 0$.

Clearly, $p(m) < e^{c\sqrt{m}}$ holds when $m = 1$. Assume that $m > 1$ and that $p(r) < e^{c\sqrt{r}}$ whenever $r < m$. Then

$$mp(m) < \sum_{v=1}^{m} \sum_{k=1}^{\lfloor m/v \rfloor} v e^{c\sqrt{m-kv}} < \sum_{v=1}^{m} \sum_{k=1}^{\lfloor m/v \rfloor} v e^{c\sqrt{m} - (ckv)/2\sqrt{m}},$$

by the first fact above, since $kv/m \leqslant 1$. Hence

$$mp(m) < \sum_{v=1}^{\infty} \sum_{k=1}^{\infty} v e^{c\sqrt{m} - (ckv)/2\sqrt{m}}$$

$$= e^{c\sqrt{m}} \sum_{k=1}^{\infty} \frac{e^{-ck/2\sqrt{m}}}{(1 - e^{-ck/2\sqrt{m}})^2}.$$

Since $\dfrac{e^{-x}}{(1-e^{-x})^2} < \dfrac{1}{x^2}$ for any non-zero real number x, we find that

$$mp(m) < e^{c\sqrt{m}} \sum_{k=1}^{\infty} \frac{4m}{c^2 k^2} = me^{c\sqrt{m}},$$

where we are using the well-known fact that $\sum_{k=1}^{\infty} \frac{1}{k^2} = \frac{\pi^2}{6}$. The upper bound now follows by induction on m, and so the proposition is established.

We now turn our attention to the enumeration function $p_s(m)$ that counts the number of partitions of m whose parts are all at most s.

Theorem 17.8 *Let s be a fixed positive integer. Then*

$$\frac{1}{s!}\binom{m+s-1}{s-1} \leqslant p_s(m) \leqslant \frac{1}{s!}\binom{m+s-1}{s-1} + \binom{m+s-2}{s-2}. \qquad (17.3)$$

In particular,

$$p_s(m) \sim_m \frac{1}{s!(s-1)!}m^{s-1}. \qquad (17.4)$$

Proof: The theorem is trivial when $s = 1$, so we may assume that $s \geqslant 2$. We claim that $p_s(m)$ is equal to the number of partitions with at most s parts. To see this, we use the notion of a partition diagram. If $a_1, a_2, \ldots, a_k$ is a partition of m with $a_1 \geqslant a_2 \geqslant \cdots \geqslant a_k$, the partition diagram is a picture consisting of an array of squares. There are a_i squares in row i, and the rows are left justified. For example, the left-hand portion of Figure 17.1 shows the partition diagram for the partition $4, 3, 3, 1, 1$ of 12. Reflecting a partition diagram in a 'north-west to south-east' diagonal produces another partition diagram: see the right-hand portion of Figure 17.1 (which is the diagram for the partition $5, 3, 3, 1$). Our claim follows, since the process of reflection provides a bijection between the partitions of m with each part at most s and the partitions of m with at most s parts.

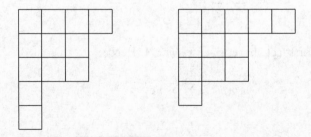

Figure 17.1   A partition diagram and its reflection

We count the partitions of m with at most s parts as follows. Let Ω_s be the set defined by

$$\Omega_s = \left\{ (a_1, a_2, \ldots, a_s) \in \mathbb{Z}^s \mid \sum_{i=1}^{s} a_i = m \text{ and } a_i \geqslant 0 \text{ for all } i \in \{1, 2, \ldots, s\} \right\}.$$

The group $\mathrm{Sym}(s)$ acts on Ω_s by permuting the components of each vector in Ω_s, and there is a one-to-one correspondence between the orbits of $\mathrm{Sym}(s)$ on Ω_s and the partitions of m with at most s parts. So to provide bounds on $p_s(m)$ it suffices to provide bounds on this number of orbits.

Let Λ be the set

$$\left\{ (b_1, b_2, \ldots, b_s) \in \mathbb{Z}^s \mid \sum_{i=1}^{s} b_i = m + s \text{ and } b_i \geqslant 1 \text{ for all } i \in \{1, 2, \ldots, s\} \right\}$$

of ordered partitions of $m + s$ with s parts. There is a bijection between Ω_s and Λ, where we set $b_i = a_i + 1$ for all i. We observed in the proof of Lemma 5.9 that ordered partitions of $m + s$ correspond to choices of a subset of the $m + s - 1$ '+' signs in the expression $(1 + 1 + \cdots + 1) = m + s$ that we change to ')+('. An ordered partition has s parts if and only if exactly $s - 1$ '+' signs are changed, and so $|\Omega_s| = |\Lambda| = \binom{m+s-1}{s-1}$.

Each orbit of $\mathrm{Sym}(s)$ on Ω_s contains at most $|\mathrm{Sym}(s)|$ elements of Ω_s, and so the number of orbits of $\mathrm{Sym}(s)$ on Ω_s is at least

$$\frac{|\Omega_s|}{|\mathrm{Sym}(s)|} = \frac{1}{s!} \binom{m+s-1}{s-1}.$$

This proves the lower bound of (17.3).

To prove our upper bound on $p_s(m)$, we use the orbit counting lemma (often erroneously attributed to Burnside) which states that when a finite group G acts on a set Ω, the number of orbits of G on Ω is equal to the mean number of fixed points of an element $g \in G$. (The lemma is proved by counting in two ways pairs (g, ω) where $\omega \in \Omega$ is fixed by $g \in G$, and then using the orbit stabiliser theorem.) In our case, the orbit counting lemma states that

$$p_s(m) = \frac{1}{s!} \sum_{g \in \mathrm{Sym}(s)} F(g),$$

where $F(g)$ is the number of points in Ω_s fixed by g.

An element $(a_1, a_2, \ldots, a_s) \in \Omega_s$ is fixed by $g \in \mathrm{Sym}(s)$ if and only if the integers a_i are equal as i runs through each cycle of g. This implies that $F(g)$ is determined by the lengths of the cycles in g. Moreover, if $h \in \mathrm{Sym}(s)$ is such that the set partition of $\{1, 2, \ldots, s\}$ induced by the cycle structure of h is a refinement of that induced by the cycle structure of g, then every $\omega \in \Omega_s$ fixed by g is also fixed by h and so $F(h) \geqslant F(g)$. Thus for a non-identity element $g \in \text{Sym}(s)$, we have that $F(g)$ is bounded above by $F(\tau)$, where $\tau \in \text{Sym}(s)$ is a transposition. Without loss of generality, define $\tau = (12)$. Since the identity element of $\text{Sym}(s)$ has $|\Omega_s|$ fixed points, we therefore have

$$
\begin{aligned}
p_s(m) &= \frac{|\Omega_s|}{s!} + \frac{1}{s!} \sum_{g \in \text{Sym}(s) \setminus \{1\}} F(g) \\
&\leqslant \frac{1}{s!} \binom{m+s-1}{s-1} + \frac{s!-1}{s!} F(\tau) \\
&\leqslant \frac{1}{s!} \binom{m+s-1}{s-1} + F(\tau).
\end{aligned}
\tag{17.5}
$$

The set F of points in Ω_s that are fixed by τ is given by

$$
F = \{(a_1, a_2, \ldots, a_s) \in \Omega_s \mid a_1 = a_2\}.
$$

But the map from F to Ω_{s-1} defined by

$$
(a_1, a_2, a_3, \ldots, a_s) \mapsto (a_1 + a_2, a_3, \ldots, a_s)
$$

is injective (since $a_1 = a_2$), and so

$$
F(h) = |F| \leqslant |\Omega_{s-1}| = \binom{m+s-2}{s-2}.
$$

This bound and (17.5) combine to establish the upper bound of (17.3). Equation (17.4) now follows by observing that for any constants a and b the binomial coefficient $\binom{m+a}{b}$ is a polynomial of degree b in m whose leading coefficient is $1/b!$ and whose remaining coefficients depend only on a and b. This establishes the theorem.

Theorem 17.8 and Theorem 17.4 together have the following immediate corollary. Recall the definition of $\eta(n)$ stated just before Theorem 17.6.

Theorem 17.9 *Let r be a fixed positive integer, and write $r = p_1^{s_1} p_2^{s_2} \cdots p_u^{s_u}$ where the p_i are distinct prime numbers and the s_i are positive integers. If $n = p_1^{m_1} p_2^{m_2} \cdots p_u^{m_u}$ where the m_i are non-negative integers, then*

$$
f_{\mathfrak{A}_r}(n) \sim_\eta \left(\prod_{i=1}^u 1/s_i!(s_i - 1)! \right) \prod_{i=1}^u m_i^{s_i - 1}.
$$

17.3 Further results on abelian groups

There is a major area concerned with the enumeration of abelian groups that we have not touched on. For $x > 0$ define

$$F_{\mathfrak{A}}(x) = \sum_{m \leqslant x} f_{\mathfrak{A}}(m),$$

the function that enumerates abelian groups of order at most x. The asymptotic behaviour of $F_{\mathfrak{A}}(x)$ as $x \to \infty$ was first considered by Paul Erdős and George Szekeres in 1935 (see [30]). Define $C_1 = \zeta(2)\zeta(3)\zeta(4) \cdots = 2.29485\ldots$, where $\zeta(s)$ is the Riemann zeta function, definable for $s > 1$ by $\zeta(s) = \sum_{n \geqslant 1} n^{-s}$ or as $\prod_{p \text{ prime}}(1 - p^{-s})^{-1}$. Erdős and Szekeres proved that $F_{\mathfrak{A}}(x) = C_1 x + O(\sqrt{x})$. This was improved by D. G. Kendall and R. A. Rankin, who proved in [54] (*Math. Reviews* **12** (1951) p. 316 contains a small correction to this paper) that

$$F_{\mathfrak{A}}(x) = C_1 x - C_2 x^{1/2} + O(x^{1/3} \log^2 x),$$

where $C_2 = \zeta(1/2)\zeta(3/2)\zeta(2)\zeta(5/2) \cdots = 14.6\ldots$, and further improved by later authors to

$$F_{\mathfrak{A}}(x) = C_1 x - C_2 x^{1/2} + C_3 x^{1/3} + R(x),$$

where $C_3 = \prod_{r \geqslant 1, \ r \neq 3} \zeta(r/3)$ and $R(x) = o(x^{1/3})$. The remainder term $R(x)$ was shown by H.-E. Richert in 1953 to be $O(x^{3/10} \log^{9/10} x)$ and over the years this has gradually been improved by various mathematicians, with the most recent estimate due to Liu [60] being that

$$R(x) = O(x^{50/199+\varepsilon}) \quad \text{for any } \varepsilon > 0.$$

These results rely on careful study of the generating function

$$\sum_{n=1}^{\infty} f_{\mathfrak{A}}(n) n^{-s} = \prod_{r=1}^{\infty} \zeta(rs)$$

and are a long way beyond the scope of this book.

Abelian groups may be viewed as $\mathbb{Z}$-modules. Analogues of enumeration theorems for finite abelian groups have been proved for categories of modules over other rings. They have also been proved for certain categories of finite rings. Much of this work can be placed in a very general setting. The interested reader is referred to Chapters 1 and 5 of Knopfmacher's book [58].

18

Enumeration within small varieties of A-groups

We have seen in the previous chapter that varieties of abelian groups contain relatively few groups of any given order. Indeed, Corollary 17.3 shows that $f_{\mathfrak{V}}(n) \leqslant n$ for any variety $\mathfrak{V}$ of abelian groups. Once a variety contains non-abelian finite groups, the enumeration function starts behaving differently. This chapter considers the enumeration functions of certain small varieties that contain non-abelian groups. More precisely, let p and q be distinct prime numbers. Let $\mathfrak{U}$ be the variety $\mathfrak{A}_p\mathfrak{A}_q$ consisting of extensions of a group in $\mathfrak{A}_p$ by a group in $\mathfrak{A}_q$. (Recall that for an integer r, the variety $\mathfrak{A}_r$ consists of the abelian groups of exponent dividing r. So $\mathfrak{A}_p$ and $\mathfrak{A}_q$ consist of the elementary abelian p- and q-groups respectively.) Let $\mathfrak{V}$ be the variety $\mathfrak{A}_p\mathfrak{A}_q \vee \mathfrak{A}_q\mathfrak{A}_p$, the smallest variety containing both $\mathfrak{A}_p\mathfrak{A}_q$ and $\mathfrak{A}_q\mathfrak{A}_p$. This chapter investigates the enumeration functions $f_{\mathfrak{U}}(n)$ and $f_{\mathfrak{V}}(n)$ for the varieties $\mathfrak{U}$ and $\mathfrak{V}$. The results in this chapter are taken from, or are slight refinements of, the DPhil thesis of Geetha Venkataraman [93] and a subsequent technical report [95].

We will see that both $\mathfrak{U}$ and $\mathfrak{V}$ contain only A-groups. In fact, $\mathfrak{U}$ is minimal in the sense that it contains non-abelian groups but every proper subvariety of $\mathfrak{U}$ is abelian. As both $\mathfrak{U}$ and $\mathfrak{V}$ contain only soluble groups, the main result of Chapter 12 implies there exist constants $c_{p,q}, d_{p,q}, c'_{p,q}$ and $d'_{p,q}$ such that

$$f_{\mathfrak{U}}(n) \leqslant n^{c_{p,q}\mu+d_{p,q}} \text{ and } f_{\mathfrak{V}}(n) \leqslant n^{c'_{p,q}\mu+d'_{p,q}}$$

for all positive integers n. (Recall that $\mu(n)$ is the largest power of a prime that divides n.) However, in contrast to the class of all A-groups we are able to provide constants $c_{p,q}$ and $c'_{p,q}$ that are best possible. In fact, $c'_{p,q} = \max\{c_{p,q}, c_{q,p}\}$ so the leading term of $f_{\mathfrak{V}}(n)$ is determined by the enumeration function of one of its two minimal non-abelian subvarieties $\mathfrak{A}_p\mathfrak{A}_q$ and $\mathfrak{A}_q\mathfrak{A}_p$. It is an interesting question to ask how far this is true in general: for which varieties of A-groups is the leading term of the corresponding enumeration function determined by the minimal non-abelian subvarieties that arise?

This chapter contains two sections. The first investigates the enumeration function $f_{\mathfrak{U}}(n)$, and the second the enumeration function $f_{\mathfrak{V}}(n)$.

18.1 A minimal variety of A-groups

Throughout this section, p and q will be distinct prime numbers and $\mathfrak{U}$ will be the variety $\mathfrak{A}_p\mathfrak{A}_q$ of extensions of a group in $\mathfrak{A}_p$ by a group in $\mathfrak{A}_q$. Clearly, a finite group $G \in \mathfrak{U}$ has order $p^{\alpha}q^{\beta}$ for some non-negative integers α and β. We aim to provide good bounds on $f_{\mathfrak{U}}(n)$, where $n = p^{\alpha}q^{\beta}$.

We define $c_{p,q}$ to be the constant

$$c_{p,q} = \frac{1}{d} - 2\sqrt{\left(\frac{\log p}{\log q}\right)^2 + \frac{\log p}{d\log q} + 2\frac{\log p}{\log q}}, \qquad (18.1)$$

where d is the order of p modulo q. We will prove in this section that $f_{\mathfrak{U}}(n) \leqslant n^{c_{p,q}\mu+1}$, and that the constant $c_{p,q}$ is the best possible. The constant $c_{p,q}$ appears because of the following lemma.

Lemma 18.1 *Let p and q be distinct primes, and let d be the order of p modulo q. Then the constant $c_{p,q}$ defined above is the maximum value attained by the function*

$$g(t) = \frac{(\frac{1}{d} - t)t}{(\log p/\log q) + t} \qquad (18.2)$$

on the interval $0 \leqslant t \leqslant \frac{1}{d}$.

The lemma is proved by elementary calculus, and so we omit the proof.

Let P and Q be groups. Let $\theta : Q \to \mathrm{Aut}\,(P)$ be a homomorphism, and write θ_b for the automorphism of P corresponding to the image of $b \in Q$ under θ. We define the semidirect product $P \rtimes_{\theta} Q$ of P and Q associated with θ to be the group whose underlying set is $P \times Q$ and whose multiplication is defined by

$$(a_1, b_1)(a_2, b_2) = (a_1 \theta_{b_1}(a_2), b_1 b_2).$$

We claim that every finite group in $\mathfrak{U}$ is isomorphic to such a semidirect product, where P and Q are elementary abelian p- and q-groups respectively. To see this, let G be a finite group in $\mathfrak{U}$ and suppose G has order $p^{\alpha}q^{\beta}$. Then G is an extension of an elementary abelian group P of order p^{α} by an elementary abelian group of order q^{β}. Any Sylow q-subgroup Q of G provides a complement of P in G, and so G is a semidirect product of P

by Q. Since Q is clearly elementary abelian, our claim follows. So to count the number of isomorphism classes of groups of order $p^\alpha q^\beta$ in $\mathfrak{U}$, it suffices to count the number of isomorphism classes of groups $P \rtimes_\theta Q$ where P is a fixed elementary abelian group of order p^α, Q is a fixed elementary abelian group of order q^β and θ runs over all homomorphisms from Q to Aut (P). Our strategy to provide an upper bound on the number of these isomorphism classes is as follows. We first establish (see Proposition 18.2) a bijection between the set of isomorphism classes we wish to count and the set of orbits of a certain group acting on a set. We then establish the upper bound we need by finding a small set Δ that intersects every orbit non-trivially and analysing its orbit structure.

Let $X_{\alpha,\beta}$ be the group Aut $(P) \times$ Aut (Q). Let $\Gamma_{\alpha,\beta}$ be the set of homomorphisms $\theta : Q \to$ Aut (P). There is an action of $X_{\alpha,\beta}$ on $\Gamma_{\alpha,\beta}$ defined as follows. Let $(\kappa, \lambda) \in X_{\alpha,\beta}$, where $\kappa \in$ Aut (P) and $\lambda \in$ Aut (Q). Let $\theta \in \Gamma_{\alpha,\beta}$. Then we define $(\kappa, \lambda)\theta = \theta'$, where θ' is defined by

$$\theta'_b(a) = \kappa \theta_{\lambda^{-1}(b)} \kappa^{-1}(a) \tag{18.3}$$

for all $a \in P$ and $b \in Q$.

Proposition 18.2 *In the notation defined above, there is a one-to-one correspondence between the isomorphism classes of groups in $\mathfrak{U}$ of order $p^\alpha q^\beta$ and the set of orbits of $X_{\alpha,\beta}$ on $\Gamma_{\alpha,\beta}$.*

Proof: We need to prove that $P \rtimes_\theta Q \cong P \rtimes_{\theta'} Q$ if and only if θ and θ' lie in the same orbit of $X_{\alpha,\beta}$.

Suppose that θ and θ' lie in the same orbit of $X_{\alpha,\beta}$. Let $\kappa \in$ Aut (P) and $\lambda \in$ Aut (Q) be such that $(\kappa, \lambda)\theta = \theta'$, so θ and θ' satisfy (18.3). Define the map $\phi : P \rtimes_\theta Q \to P \rtimes_{\theta'} Q$ by $\phi((a, b)) = (\kappa(a), \lambda(b))$ for all $a \in P$ and $b \in Q$. Clearly ϕ is bijective. It is easy to check that

$$\phi((a_1, b_1))\phi((a_2, b_2)) = (\kappa(a_1)\, \theta'_{\lambda(b_1)}\kappa(a_2), \lambda(b_1)\lambda(b_2)) \text{ and}$$

$$\phi((a_1, b_1)(a_2, b_2)) = (\kappa(a_1)\, \kappa\theta_{b_1}(a_2), \lambda(b_1)\lambda(b_2))$$

for all $a_1, a_2 \in P$ and $b_1, b_2 \in Q$. So ϕ is a homomorphism if and only if $\theta'_{\lambda(b_1)}\kappa(a_2) = \kappa\theta_{b_1}(a_2)$ for all $a_2 \in P$ and $b_1 \in Q$. But substituting $a = \kappa(a_2)$ and $b = \lambda(b_1)$ into this expression gives precisely the condition (18.3), and so ϕ is an isomorphism as required.

To prove the converse, suppose that $\phi : P \rtimes_\theta Q \to P \rtimes_{\theta'} Q$ is an isomorphism. Now, $\phi(1 \times Q)$ is a Sylow q-subgroup of $P \rtimes_{\theta'} Q$. Since all Sylow q-subgroups are conjugate, there exists an inner automorphism γ of $P \rtimes_{\theta'} Q$ such that $\gamma(\phi(1 \times Q)) = 1 \times Q$. By replacing ϕ by $\gamma\phi$ if necessary, we may therefore assume that $\phi(1 \times Q) = 1 \times Q$. Clearly, since $P \times 1$ is the unique Sylow subgroup of $P \rtimes_\theta Q$ and $P \rtimes_{\theta'} Q$, we have that $\phi(P \times 1) = P \times 1$. Define $\kappa \in \text{Aut}(P)$ and $\lambda \in \text{Aut}(Q)$ by

$$\phi((a, 1)) = (\kappa(a), 1) \text{ for all } a \in P \text{ and}$$

$$\phi((1, b)) = (1, \lambda(b)) \text{ for all } b \in Q.$$

Thus, for all $a \in P$ and $b \in Q$,

$$\phi((a, b)) = \phi((a, 1)(1, b))$$

$$= \phi((a, 1))\phi((1, b))$$

$$= (\kappa(a), 1)(1, \lambda(b))$$

$$= (\kappa(a), \lambda(b)).$$

Now,

$$\phi((1, b))\phi((a, 1)) = (1, \lambda(b))(\kappa(a), 1) = (\theta'_{\lambda(b)}(\kappa(a)), \lambda(b)) \text{ and}$$

$$\phi((1, b)(a, 1)) = \phi(\theta_b(a), b) = (\kappa\theta_b(a), \lambda(b)).$$

Thus $\theta'_{\lambda(b)}(\kappa(a)) = \kappa\theta_b(a)$ for all $a \in P$ and $b \in Q$. But replacing a by $\kappa^{-1}(a)$ and b by $\lambda^{-1}(b)$ in this last equality shows that (18.3) holds, and so $\theta' = (\kappa, \lambda)\theta$ and the proposition is established.

We will now investigate the structure of the orbits of $X_{\alpha,\beta}$ on $\Gamma_{\alpha,\beta}$ more closely. As P is elementary abelian, we may regard P as a vector space V of dimension α over $\mathbb{F}_p$. Now, a typical element $\theta \in \Gamma_{\alpha,\beta}$ makes V into a Q-module in a natural way. We may therefore regard $X_{\alpha,\beta}$ as acting on a set of Q-modules of dimension α over $\mathbb{F}_p$. By definition of Q-module isomorphism, $\theta, \theta' \in \Gamma_{\alpha,\beta}$ give rise to isomorphic Q-modules if and only if there exists $\kappa \in \text{Aut}(P) = \text{GL}(V)$ such that $\theta'_b = \kappa\theta_b\kappa^{-1}$ for all $b \in Q$. But this condition is exactly equivalent to the condition that $(\kappa, 1)\theta = \theta'$. Hence the $(\text{Aut}(P) \times 1)$-orbits on $\Gamma_{\alpha,\beta}$ are in one-to-one correspondence with the isomorphism classes of Q-modules of dimension α over $\mathbb{F}_p$.

We recap some facts about Q-modules over $\mathbb{F}_p$, where Q is an elementary abelian group of order q^β. Since p and q are coprime, every Q-module is a direct sum of irreducible Q-modules. Let $Y_1, Y_2, \ldots, Y_\ell$ be a complete set of non-isomorphic non-trivial irreducible Q-modules. Let d be the order of p modulo q. Then $\ell = (q^\beta - 1)/d$, the modules Y_i all have dimension d and the kernel of the action of Q on Y_i has index q in Q. Moreover, for any of the $(q^\beta - 1)/(q - 1)$ subgroups K of index q in Q, there are exactly $(q - 1)/d$ non-trivial irreducible modules Y_i with kernel K. These $(q - 1)/d$ modules are closely related. Indeed, let Y be one such module. Let $x \in Q \setminus K$, so every element of Q may be written uniquely in the form $x^r k$ for some $k \in K$ and $r \in \{0, 1, \ldots, q-1\}$. For an integer $e \in \{1, 2, \ldots, q-1\}$, define the Q-module $Y[e]$ to have the same underlying set as Y, but with the action of $x^r k$ on $v \in Y$ given by $x^{er} k v$. As e runs through the set $\{1, 2, \ldots, q-1\}$, every isomorphism class of irreducible Q-modules with kernel K occurs exactly d times as $Y[e]$.

Let $y_1, y_2, \ldots, y_\beta$ be a minimal generating set for Q. By reordering the Y_i if necessary, we may suppose that for all $i \in \{1, 2, \ldots, \beta\}$ the kernel of the action of Q on Y_i is the subgroup K_i generated by $y_1, y_2, \ldots, y_{i-1}, y_{i+1}, \ldots, y_\beta$. For the rest of this section, we will stick to this notation for irreducible Q-modules. Moreover, we will denote the trivial irreducible Q-module by Y_0.

Theorem 18.3 *Let p and q be distinct primes. Let $\mathfrak{U}$ be the variety $\mathfrak{U}_p \mathfrak{U}_q$ of groups that are extensions of an elementary abelian p-group by an elementary abelian q-group. Define $c_{p,q}$ as in Equation (18.1). Then for all positive integers n, the number $f_\mathfrak{U}(n)$ of isomorphism classes of groups in $\mathfrak{U}$ of order n satisfies*

$$f_\mathfrak{U}(n) \leqslant n^{c_{p,q}\mu + 1}.$$

Proof: When n is divisible by primes other than p and q, $f_\mathfrak{U}(n) = 0$ and so the theorem is trivially true. So we may suppose that $n = p^\alpha q^\beta$ for some non-negative integers α and β. Proposition 18.2 shows that it suffices to give a good upper bound on the number of orbits of $X_{\alpha,\beta}$ on $\Gamma_{\alpha,\beta}$.

We observed above that the $(\mathrm{Aut}\,(P) \times 1)$-orbits on $\Gamma_{\alpha,\beta}$ correspond to isomorphism classes of Q-modules of dimension α over $\mathbb{F}_p$. For a Q-module V over $\mathbb{F}_p$, there exist unique non-negative integers $m_0, m_1, \ldots, m_\ell$ such that

$$V \cong m_0 Y_0 \oplus m_1 Y_1 \oplus \cdots \oplus m_\ell Y_\ell. \tag{18.4}$$

Since $\dim Y_i = d$ whenever $i \geqslant 1$, and since $\dim Y_0 = 1$, the module V has dimension α if and only if

$$m_0 + d(m_1 + m_2 + \cdots + m_\ell) = \alpha.$$

For an integer s such that $0 \leqslant s \leqslant \beta$, define Δ_s to be the subset of $\Gamma_{\alpha,\beta}$ corresponding to those modules V such that $Y_1, Y_2, \ldots, Y_s$ all occur with positive multiplicity in the decomposition (18.4), and such that the kernel of the action of Q on V has index exactly q^s in Q. We aim to show that the set

$$\Delta = \Delta_0 \cup \Delta_1 \cup \cdots \cup \Delta_\beta$$

contains a representative of every orbit of $X_{\alpha,\beta}$ on $\Gamma_{\alpha,\beta}$. The theorem will then follow once we give a good upper bound on the number of $(\operatorname{Aut}(P) \times 1)$-orbits passing through Δ (as this number is itself an upper bound on the number of $X_{\alpha,\beta}$-orbits passing through Δ).

Suppose $\theta \in \Gamma_{\alpha,\beta}$, and let L be the kernel of the action of Q on the corresponding Q-module U. Suppose L has index q^s in Q. We will show that there exists an element of Δ_s in the same $X_{\alpha,\beta}$ orbit as θ. Let $Z_1, Z_2, \ldots, Z_c$ be a minimal set of irreducible submodules of U such that $\cap_{i=1}^{c} L_i = L$, where L_i is the kernel of the action of Q restricted to Z_i. (When $s = 0$ we have that $c = 0$, since we use the convention that the intersection of an empty set of submodules of U is equal to U.) Clearly none of the Z_i are trivial, and so each L_i has index q in Q. It is not difficult to show that, by the minimality of the set of Z_i, the intersection of any j of the subgroups L_i has index exactly q^j in Q. In particular, we find that $c = s$.

Let $z_1, z_2, \ldots, z_s \in Q$ be chosen so that

$$z_j \in \left(L_1 \cap \cdots \cap L_{j-1} \cap L_{j+1} \cap \cdots \cap L_s\right) \setminus L.$$

It is not difficult to see that $Q = \langle z_1 \rangle \times \cdots \times \langle z_s \rangle \times L$. Moreover,

$$L_i = \langle z_1 \rangle \times \cdots \times \langle z_{i-1} \rangle \times \langle z_{i+1} \rangle \times \cdots \times \langle z_s \rangle \times L.$$

Recall that we have chosen a standard basis $y_1, y_2, \ldots, y_\beta$ for Q, and a set of irreducible modules Y_i and subgroups K_i. Let $\lambda' \in \operatorname{Aut}(Q)$ be an automorphism mapping z_i to y_i when $1 \leqslant i \leqslant s$, and mapping a minimal generating set for L onto $\{y_{s+1}, y_{s+2}, \ldots, y_\beta\}$. Let W be the Q-module corresponding to $(1, \lambda')\theta$. For any $i \in \{1, 2, \ldots, s\}$, the subset Z_i of our original Q-module U is also an irreducible submodule Z_i' of W, but now the kernel of this action is $\lambda'(L_i) = K_i$. Hence there exists $e_i \in \{1, 2, \ldots, q-1\}$ such that $Y_i[e_i]$ is isomorphic to Z_i'. Let $\lambda \in \operatorname{Aut}(Q)$ be an automorphism such that $\lambda(z_i) = y_i^{e_i}$ when $1 \leqslant i \leqslant s$, and mapping a minimal generating set of L onto $\{y_{s+1}, y_{s+2}, \ldots, y_\beta\}$. Then it is not difficult to check that for all $i \in \{1, 2, \ldots, s\}$ the submodule consisting of the underlying vector space Z_i whose module structure is given by the appropriate restriction of $(1, \lambda)\theta$ is isomorphic to Y_i. Thus the module associated with $(1, \lambda)\theta$ contains submodules isomorphic to $Y_1, Y_2, \ldots, Y_s$. Since the index of the kernel of an element in $\Gamma_{\alpha,\beta}$ is preserved by the action of $X_{\alpha,\beta}$, we find that $(1, \lambda)\theta \in \Delta_s$, as required.

We claim that the number of orbits in Δ_s is at most $n^{c_{p,q}\mu}$. Let $\theta \in \Delta_s$ and let V be the associated Q-module. Now, V is isomorphic to a direct sum of the form (18.4), where $m_1, m_2, \ldots, m_s$ are all positive. This implies that V contains at least s non-trivial non-isomorphic irreducible submodules, each of dimension d and so $sd \leqslant \alpha$. Thus when $s > \alpha/d$ we find that $\Delta_s = \emptyset$ and our claim follows trivially. We may therefore assume that $0 \leqslant s \leqslant \alpha/d$.

The intersection M of $K_1, K_2, \ldots, K_s$ is a subgroup of index q^s in Q, and so the kernel of the action of any irreducible submodule of V must contain M. Let I be defined to be the subset of $\{0, 1, 2, \ldots, \ell\}$ such that $i \in I$ if and only if the kernel associated with Y_i contains M. Then whenever $i \notin I$ we find that $m_i = 0$. Note that there are exactly $(q^s - 1)/(q - 1)$ subgroups of Q of index q that contain M, and each such subgroup is the kernel of $(q - 1)/d$ non-trivial irreducible Q-modules. Hence

$$|I| = 1 + ((q^s - 1)/(q - 1))((q - 1)/d) = 1 + (q^s - 1)/d.$$

Since the orbit containing θ depends only on the isomorphism class of the associated Q-module, we find that the number of orbits containing elements of Δ_s is at most equal to the number of vectors $(m_0, m_1, \ldots, m_\ell)$ of non-negative integers with the property that $m_0 + d(m_1 + \cdots + m_\ell) = \alpha$, and such that m_i is positive whenever $1 \leqslant i \leqslant s$ and $m_i = 0$ whenever $i \notin I$.

Define $a_1, a_2, \ldots, a_{(q^s-1)/d}$ by setting $a_i = m_j$, where j is the ith element of $\{1, 2, \ldots, \ell\}$ that is contained in I. Note that m_0 is defined uniquely by $m_1, m_2, \ldots, m_\ell$ when their sum is at most α/d, and there is no appropriate value for m_0 otherwise. So the number of orbits in Δ_s is at most the size of the set

$$\left\{ (a_1, a_2, \ldots, a_{(q^s-1)/d}) \;\Big|\; \sum_{i=1}^{(q^s-1)/d} a_i \leqslant \alpha/d \text{ and } a_i \geqslant 1 \text{ for } i \in \{1, 2, \ldots, s\} \right\},$$

where the a_i are non-negative integers. There is an injection mapping this set into the ordered partitions $(b_0, b_1, b_2, \ldots, b_{(q^s-1)/d})$ of $\lfloor \alpha/d \rfloor + (q^s - 1)/d - s + 1$ with $(q^s - 1)/d + 1$ parts:

$$b_i = \begin{cases} 1 + \lfloor \alpha/d \rfloor - \sum_{j=1}^{(q^s-1)/d} a_j & \text{if } i = 0, \\ a_i & \text{if } 1 \leqslant i \leqslant s \text{ and} \\ a_i + 1 & \text{if } s+1 \leqslant i \leqslant (q^s - 1)/d. \end{cases}$$

Hence the number of orbits in Δ_s is at most

$$\binom{\lfloor \alpha/d \rfloor + (q^s - 1)/d - s}{(q^s - 1)/d} = \binom{\lfloor \alpha/d \rfloor + (q^s - 1)/d - s}{\lfloor \alpha/d \rfloor - s}$$

$$= \prod_{j=0}^{\lfloor \alpha/d \rfloor - s - 1} \frac{\lfloor \alpha/d \rfloor + (q^s - 1)/d - s - j}{\lfloor \alpha/d \rfloor - s - j}$$

$$\leqslant \prod_{j=0}^{\lfloor \alpha/d \rfloor - s - 1} q^s$$

$$\leqslant q^{s(\alpha/d - s)}$$

$$= n^{(\log q / \log n) s (\alpha/d - s)}.$$

Now, $s \leqslant \beta$ (as Q has a subgroup of index q^s) and so

$$\log n = \alpha \log p + \beta \log q \geqslant \alpha \log p + s \log q. \tag{18.5}$$

Set $t = s/\alpha$, so $0 \leqslant t \leqslant 1/d$. Then (18.5) shows that the exponent of our upper bound is at most

$$\frac{\log q}{\alpha \log p + \alpha t \log q} \alpha t (\alpha/d - \alpha t) = \frac{t(1/d - t)}{(\log p / \log q) + t} \alpha \leqslant c_{p,q} \mu,$$

the inequality following by Lemma 18.1 and since $\alpha \leqslant \mu$. So there are at most $n^{c_{p,q}\mu}$ orbits passing through Δ_s, and our claim follows.

Since $f_{\mathfrak{U}}(n)$ is at most the sum of the number of $X_{\alpha,\beta}$-orbits passing through each of the sets Δ_s, we find that

$$f_{\mathfrak{U}}(n) \leqslant \sum_{s=0}^{\beta} n^{c_{p,q}\mu} \leqslant \beta n^{c_{p,q}\mu} \leqslant n^{c_{p,q}\mu + 1},$$

as required.

We will now show that the constant $c_{p,q}$ in Theorem 18.3 is the best possible.

Theorem 18.4 *Let p and q be distinct primes. Let $\mathfrak{U}$ be the variety $\mathfrak{A}_p \mathfrak{A}_q$ of groups that are extensions of an elementary abelian p-group by an elementary abelian q-group. Define $c_{p,q}$ by Equation (18.1). For every positive real number ϵ, there exist infinitely many positive integers n such that*

$$f_{\mathfrak{U}}(n) > n^{(c_{p,q} - \epsilon)\mu}.$$

So the constant $c_{p,q}$ in Theorem 18.3 is the best possible.

Proof: Let $n = p^\alpha q^\beta$, and suppose that α is divisible by d, where d is the order of p modulo q. Furthermore, suppose that $\alpha > d$. We aim to provide a lower bound on the number of $X_{\alpha,\beta}$-orbits on $\Gamma_{\alpha,\beta}$, which will give a good lower bound for $f_{\mathfrak{U}}(n)$ for infinitely many values of n.

We first find a lower bound on the number of $(\mathrm{Aut}\,(P) \times 1)$-orbits on $\Gamma_{\alpha,\beta}$. Before the statement of Theorem 18.3, we observed that the number of $(\mathrm{Aut}\,(P) \times 1)$-orbits on $\Gamma_{\alpha,\beta}$ is equal to the number of Q-modules of dimension α over $\mathbb{F}_p$ up to isomorphism. Moreover, each such Q-module V may be written in the form (18.4) for unique non-negative integers $m_0, m_1, \ldots, m_\ell$ where $m_0 + d(m_1 + m_2 + \cdots + m_\ell) = \alpha$. If we consider just those isomorphism classes with $m_0 = 0$, we find that the number of such modules (up to isomorphism) is equal to the size of the set

$$\{(m_1, m_2, \ldots, m_\ell) \mid m_i \geqslant 0 \text{ for all } i \text{ and } \sum_{i=1}^{\ell} m_i = \alpha/d\}.$$

Hence the number of $(\mathrm{Aut}\,(P) \times 1)$-orbits is at least

$$|\{(m_1, m_2, \ldots, m_\ell) \mid m_i \geqslant 0 \text{ for all } i \text{ and } \sum_{i=1}^{\ell} m_i = \alpha/d\}|$$

$$= |\{(a_1, a_2, \ldots, a_\ell) \mid a_i \geqslant 1 \text{ for all } i \text{ and } \sum_{i=1}^{\ell} a_i = \ell + \alpha/d\}|$$

$$= \binom{\ell + \alpha/d - 1}{\ell - 1}$$

$$= \binom{\ell + \alpha/d - 1}{\alpha/d}.$$

Now, for any positive integers u and v with $v \leqslant u$,

$$\binom{u}{v} = \prod_{i=0}^{v-1} (u-i)/(v-i) \geqslant \prod_{i=0}^{v-1} u/v = (u/v)^v.$$

Hence the number of $(\mathrm{Aut}\,(P) \times 1)$-orbits on $\Gamma_{\alpha,\beta}$ is at least

$$\left(\frac{\ell + \alpha/d - 1}{\alpha/d}\right)^{\alpha/d} = \left(\frac{(q^\beta - 1)/d + \alpha/d - 1}{\alpha/d}\right)^{\alpha/d}$$

$$= \left(\frac{q^\beta + \alpha - 1 - d}{\alpha}\right)^{\alpha/d}$$

$$\geqslant q^{\beta\alpha/d}/(\alpha)^{\alpha/d},$$

the last step following since $\alpha > d$.

Now, $\mathrm{Aut}\,(P) \times 1$ has index $|\mathrm{Aut}\,(Q)|$ in $X_{\alpha,\beta}$, and so each $X_{\alpha,\beta}$-orbit is the union of at most $|\mathrm{Aut}\,(Q)|$ of the $(\mathrm{Aut}\,(P) \times 1)$-orbits. Since $|\mathrm{Aut}\,(Q)| = |\mathrm{GL}(\beta, q)| \leqslant q^{\beta^2}$, we find that the number $f_{\mathfrak{U}}(p^\alpha q^\beta)$ of $X_{\alpha,\beta}$ orbits is at least

$$q^{(\beta\alpha/d) - \beta^2}/(\alpha)^{\alpha/d} = n^e,$$

where

$$e = \frac{\beta(\alpha/d - \beta)\log q}{\alpha \log p + \beta \log q} - \frac{(\alpha \log \alpha)/d}{\alpha \log p + \beta \log q}. \tag{18.6}$$

We now use this lower bound on the number of orbits to prove Theorem 18.3. Let $g(t)$ be the function defined by the formula (18.2), and suppose that the maximum of $g(t)$ on the interval $0 \leqslant t \leqslant 1/d$ is attained at the point $t = t_1$. Let ϵ be a positive real number. We wish to show that $f_{\mathfrak{U}}(n) \geqslant n^{(c_{p,q} - \epsilon)\mu}$ for infinitely many values of n. Let t' be a rational number such that $0 \leqslant t' < 1/d$ and $g(t_1) - g(t') < \epsilon$. Consider integers n of the form $n = p^\alpha q^\beta$, where α is divisible by both d and the denominator of t', where $\alpha > d$ and where $\beta = t'\alpha$. Since $t' < 1$, we have that $\alpha = \mu(n)$ for integers n of this form. Substituting $\beta = t'\alpha$ and $\alpha = \mu$ into (18.6), we find that $f_{\mathfrak{U}}(n) \geqslant n^e$, where

$$e = \frac{t'(1/d - t')}{(\log p / \log q) + t'} \mu - \frac{(\log \mu)/d}{\log p + t' \log q}$$

$$= g(t')\mu - \frac{1}{d(\log p + t' \log q)} \log \mu.$$

But $g(t') > c_{p,q} - \epsilon$ and $\log \mu = O(\mu)$. So for all sufficiently large n of the form we are considering, we find that $f_{\mathfrak{U}}(n) \geqslant n^{(c_{p,q} - \epsilon)\mu}$, and the theorem is proved.

We remark that, just as is the case when enumerating soluble A-groups in Chapter 12, we have no hope of finding a corresponding lower bound for $f_{\mathfrak{U}}(n)$ of the form $n^{c\mu}$ as the integers n of the form p^μ (for example) all have the property that $f_{\mathfrak{U}}(n) \leqslant n$.

In Chapter 12 we proved that the number $f_{A,\,\mathrm{sol}}(n)$ of (isomorphism classes of) soluble A-groups of order n was at most $n^{11\mu + 13}$. So if we define

$$\ell = \limsup_{n \to \infty} \log f_{A,\,\mathrm{sol}}(n) / (\mu(n) \log n)$$

we find that $\ell \leqslant 11$. We may reduce this upper bound on ℓ further by more careful arguments, but we know of no proved upper bounds that seem realistic. Since all finite groups in the variety $\mathfrak{A}_p \mathfrak{A}_q$ are soluble A-groups, Theorem 18.4 implies that $\ell \geqslant c_{p,q}$ for any primes p and q. Let q be a prime such that $p = 2q + 1$ is also prime. (Primes q of this form are known as *Sophie Germain* primes.) The multiplicative order of p modulo q is 1. Suppose that there are infinitely many Sophie Germain primes q. (This is thought likely to be true, but is not currently known.) Then as $q \to \infty$ over Sophie Germain primes, we find that $\log p / \log q \to 1$, and so (18.1) shows that $c_{p,q} \to 3 - 2\sqrt{2} = 0.17157\ldots$ Therefore, $\ell \geqslant 3 - 2\sqrt{2}$ if there are infinitely many Sophie Germain primes. (For a lower bound that does not rely on this assumption, we may take q to be the largest currently known Sophie Germain prime, namely $q = 2\,540\,041\,185 \cdot 2^{114\,729} - 1$. Then $c_{2q+1,q}$ already agrees with $3 - 2\sqrt{2}$ to several decimal places.) It would be very interesting to know whether this lower bound on ℓ is in fact an equality.

18.2 The join of minimal varieties

This section considers the variety $\mathfrak{V} = \mathfrak{A}_p\mathfrak{A}_q \vee \mathfrak{A}_q\mathfrak{A}_p$, the smallest variety containing both $\mathfrak{A}_p\mathfrak{A}_q$ and $\mathfrak{A}_q\mathfrak{A}_p$.

Proposition 18.5 *Let p and q be distinct primes, and let $\mathfrak{V} = \mathfrak{A}_p\mathfrak{A}_q \vee \mathfrak{A}_q\mathfrak{A}_p$. Let G be a finite group in the variety $\mathfrak{V}$. Then $G \cong X \times Y$, where $X \in \mathfrak{A}_p\mathfrak{A}_q$ and $Y \in \mathfrak{A}_q\mathfrak{A}_p$.*

Proof: The varieties $\mathfrak{A}_p$ and $\mathfrak{A}_q$ are clearly locally finite and soluble (since their k-generator free groups are abelian of orders p^k and q^k respectively). Subgroups and images of locally finite groups are locally finite. Moreover, any group that is the extension of a locally finite group by a locally finite group is itself locally finite. The same properties are true for soluble groups. Thus the varieties $\mathfrak{U}_1 = \mathfrak{A}_p\mathfrak{A}_q$ and $\mathfrak{U}_2 = \mathfrak{A}_q\mathfrak{A}_p$ are locally finite and soluble, and so the variety $\mathfrak{U}_1\mathfrak{U}_2$ is also locally finite and soluble. Since $\mathfrak{V} \subseteq \mathfrak{U}_1\mathfrak{U}_2$, we find that $\mathfrak{V}$ is locally finite and soluble. Since $\mathfrak{U}_1$ and $\mathfrak{U}_2$ have exponent pq, the same is true for $\mathfrak{V}$. Hence a finite group $G \in \mathfrak{V}$ is soluble of order $p^\alpha q^\beta$ for some integers α and β.

Suppose that G is generated by k elements, and let F be a free group on k generators. Let K_1 and K_2 be the fully invariant subgroups of F corresponding to the varieties $\mathfrak{U}_1$ and $\mathfrak{U}_2$ respectively. Then $K_1 \cap K_2$ is the fully invariant subgroup of F corresponding to $\mathfrak{V}$. Since $\mathfrak{V}$ is locally finite, $F/(K_1 \cap K_2)$ is finite. Now F/K_1 lies in $\mathfrak{U}_1$, and so all its p-subgroups and q-subgroups are elementary abelian. The same is true for F/K_2, since this group lies in $\mathfrak{U}_2$. Since $F/(K_1 \cap K_2)$ is isomorphic to a subgroup of $(F/K_1) \times (F/K_2)$, we find that all p-subgroups and q-subgroups of $F/(K_1 \cap K_2)$ are elementary abelian. But the finite group $G \in \mathfrak{V}$ is isomorphic to a quotient of $F/(K_1 \cap K_2)$, and so its Sylow p-subgroups and Sylow q-subgroups are elementary abelian.

Define $G_1 = G/O_p(G)$. Now, $O_p(G_1) = \{1\}$ and so $F(G_1) = O_q(G_1) \times O_p(G_1) = O_q(G_1)$. Any Sylow q-subgroup Q_1 of G_1 centralises $O_q(G_1)$, since G_1 (being a quotient of the A-group G) is an A-group. Now G_1 is soluble (as G is soluble) and so Corollary 6.5 shows that

$$Q_1 \leqslant C_{G_1}(O_q(G_1)) = C_{G_1}(F(G_1)) \leqslant F(G_1) = O_q(G_1).$$

Hence $Q_1 = O_q(G_1)$ and the Sylow q-subgroup Q_1 of G_1 is normal. Clearly, the quotient G_1/Q_1 is a p-group. Moreover, Q_1 and G_1/Q_1 are elementary abelian since the Sylow subgroups of G are elementary abelian.

Define $G_2 = G/O_q(G)$. Then we may show, just as above, that G_2 is an extension of an elementary abelian p-group P_2 by an elementary abelian q-group.

Let π_1 be the natural homomorphism from G to G_1. Now, G_1/Q_1 acts on Q_1 by conjugation, and so Q_1 may be thought of as a G_1/Q_1-module. Since G_1/Q_1 is a p-group, the module is completely reducible by Maschke's theorem. Now, $O_q(G)O_p(G)/O_p(G)$ is a submodule of Q_1, and so there exists a submodule S that is a complement to $\pi_1(O_q(G))$ in Q_1. Let $X = \pi_1^{-1}(S)$. Since S is a submodule of Q_1, we have that S is normal in G_1 and so X is normal in G. By construction, X contains $O_p(G)$ and $X/O_p(G)$ is a complement to $O_q(G)O_p(G)/O_p(G)$ in Q_1.

Similarly, we may find a normal subgroup Y of G that contains $O_q(G)$ and is such that $Y/O_q(G)$ is a complement to $O_p(G)O_q(G)/O_q(G)$ in P_2.

The image of XY under π_1 contains $O_q(G)O_p(G)/O_p(G)$ (as $O_q(G) \leqslant Y$) and its complement S in Q_1 (by definition of X). Hence $XY/O_p(G)$ contains a Sylow q-subgroup of G_1 and so XY contains a Sylow q-subgroup of G. Similarly, by examining $XY/O_q(G)$ as a subgroup of G_2, we find that XY contains a Sylow p-subgroup of G. Hence $XY = G$. It is not difficult to verify that $|G| = |X||Y|$, and so G is isomorphic to $X \times Y$. Since X is an extension of the elementary abelian p-group $O_p(G)$ by an elementary abelian q-group, we find that $X \in \mathfrak{A}_p\mathfrak{A}_q$. Similarly, $Y \in \mathfrak{A}_q\mathfrak{A}_p$. Hence G is isomorphic to the direct product of a group in $\mathfrak{A}_p\mathfrak{A}_q$ and a group in $\mathfrak{A}_q\mathfrak{A}_p$, and so the proposition follows.

Theorem 18.6 *Let p and q be distinct primes. Let $\mathfrak{V}$ be the variety $\mathfrak{A}_p\mathfrak{A}_q \vee \mathfrak{A}_q\mathfrak{A}_p$. Define the constants $c_{p,q}$ and $c_{q,p}$ as in Equation (18.1). Let $c'_{p,q} = \max\{c_{p,q}, c_{q,p}\}$. Then for all positive integers n,*

$$f_{\mathfrak{V}}(n) \leqslant n^{c'_{p,q}\mu+2}.$$

Moreover, for all positive real numbers ϵ, there exist an infinite number of positive integers n such that $f_{\mathfrak{V}}(n) > n^{(c'_{p,q}-\epsilon)\mu}$, and so the constant $c'_{p,q}$ is the best possible.

Proof: We begin by establishing the upper bound of the theorem. By Proposition 18.5, every finite member G of $\mathfrak{V}$ is isomorphic to $X \times Y$, where

$X \in \mathfrak{A}_p\mathfrak{A}_q$ has order n_1 and $Y \in \mathfrak{A}_q\mathfrak{A}_p$ has order n_2. There are at most n choices for n_1. The value of n_2 is determined by n_1. Once n_1 and n_2 are fixed, the number of choices for G is at most

$$f_{\mathfrak{A}_p\mathfrak{A}_q}(n_1)f_{\mathfrak{A}_q\mathfrak{A}_p}(n_2) \leqslant n_1^{c_{p,q}\mu(n_1)+1} n_2^{c_{q,p}\mu(n_2)+1}$$

by Theorem 18.3. But

$$
\begin{aligned}
n_1^{c_{p,q}\mu(n_1)+1} n_2^{c_{q,p}\mu(n_2)+1} &\leqslant n_1^{c_{p,q}\mu+1} n_2^{c_{q,p}\mu+1} \qquad \text{(since } \mu(n_1) \leqslant \mu \text{ and } \mu(n_2) \leqslant \mu\text{)} \\
&\leqslant (n_1 n_2)^{\max\{c_{p,q},c_{q,p}\}\mu+1} \\
&= n^{c_{p,q}\mu+1}.
\end{aligned}
$$

Hence

$$f_{\mathfrak{V}}(n) \leqslant n^{c_{p,q}\mu+2},$$

and the upper bound of the theorem follows.

Since $\mathfrak{V}$ contains $\mathfrak{A}_p\mathfrak{A}_q$, we have that $f_{\mathfrak{V}}(n) \geqslant f_{\mathfrak{A}_p\mathfrak{A}_q}(n)$, and so Theorem 18.4 shows that

$$f_{\mathfrak{V}}(n) > n^{(c_{p,q}-\epsilon)\mu}$$

for an infinite number of positive integers n. Similarly, since $\mathfrak{V}$ contains $\mathfrak{A}_q\mathfrak{A}_p$,

$$f_{\mathfrak{V}}(n) > n^{(c_{q,p}-\epsilon)\mu}$$

for an infinite number of positive integers n. Since $c'_{p,q} = c_{p,q}$ or $c'_{p,q} = c_{q,p}$, one of these two inequalities implies the second statement of the theorem. Hence the theorem is proved.

19

Enumeration within small varieties of p-groups

The enumeration function $f_{\mathfrak{V}}(n)$ of a variety $\mathfrak{V}$ grows rather slowly if $\mathfrak{V}$ contains only abelian groups (Corollary 17.3 states that $f_{\mathfrak{V}}(n) \leqslant n$ in this case). If $\mathfrak{V}$ contains non-abelian groups, but all these groups are A-groups, the results of the previous chapter and of Chapter 12 show that the enumeration function of $\mathfrak{V}$ grows more rapidly (something like $n^{c\mu}$), but still much more slowly than the general group enumeration function. However, once $\mathfrak{V}$ contains finite groups that are not A-groups (and so contains a variety of non-abelian p-groups), the enumeration function of $\mathfrak{V}$ behaves much more like the general group enumeration function. Indeed, a lower bound for $f_{\mathfrak{V}}(n)$ grows like $n^{\frac{2}{27}\mu^2}$ for infinitely many values of n, matching the leading term of the general group enumeration function.

In this chapter, we will investigate two small non-abelian varieties of p-groups: the Higman variety $\mathfrak{H}_p$ of p-groups of Φ-class 2 defined by the laws x^{p^2}, $[x, y]^p$ and $[x, y, z]$; and the subvariety $\mathfrak{B}_{p,2}$ of $\mathfrak{H}_p$ consisting of groups of exponent p and nilpotency class 2 defined by the laws x^p and $[x, y, z]$. (Since any group of exponent 2 is abelian, $\mathfrak{B}_{p,2}$ is an abelian variety when $p = 2$. Since this chapter is concerned with non-abelian varieties, we will not consider $\mathfrak{B}_{p,2}$ when $p = 2$.) The significance of these two varieties lies in the following theorem.

Theorem 19.1 *Any variety containing a finite non-abelian p-group contains* $\mathfrak{B}_{p,2}$ *when p is odd, and contains* $\mathfrak{H}_p$ *when p = 2.*

Proof: Let $\mathfrak{V}$ be a variety containing a finite non-abelian p-group P. Since $\mathfrak{V}$ is closed under taking quotients, we may assume that the derived subgroup of P has order p (since we may replace P by the quotient by a normal subgroup of index p in P' if necessary). Thus P satisfies the laws $[x, y, z]$ and $[x, y]^p$. Since P is a non-trivial finite p-group, P has exponent p^ℓ for some positive integer ℓ; moreover, $\ell \geqslant 2$ when $p = 2$ since P is non-abelian. Thus $\mathfrak{V}$ contains a subvariety $\mathfrak{U}$ (namely the subvariety generated by P) in which the laws x^{p^ℓ}, $[x, y]^p$ and $[x, y, z]$ hold.

To prove the theorem, it is sufficient to prove that $\mathfrak{U}$ is the variety defined by the laws x^{p^ℓ}, $[x, y]^p$ and $[x, y, z]$, for then it is clear that $\mathfrak{U}$ contains $\mathfrak{H}_p$ or $\mathfrak{B}_{p,2}$ as appropriate. Let w be a law in $\mathfrak{U}$. (So the law w is an element of a free group F on a countable generating set.) We aim to show that w is a consequence of the laws x^{p^ℓ}, $[x, y]^p$ and $[x, y, z]$. We may write

$$w = x_1^{r_1} x_2^{r_2} \cdots x_k^{r_k} w'$$

where r_i are positive integers and where $w' \in F'$ is a product of commutators. As in the proof of Theorem 17.1, we may show that the laws w and x^{p^ℓ} are equivalent to the laws w' and x^d, where $d = \gcd(r_1, r_2, \ldots, r_k, p^\ell)$. But $P \in \mathfrak{U}$ has exponent p^ℓ, and so we must have that $d = p^\ell$. Hence the laws x^{p^ℓ} and w are equivalent to the laws x^{p^ℓ} and w'. We may write

$$w' = \left(\prod_{(i,j) \in S} [x_i, x_j]^{s_{ij}} \right) w''$$

where the integers s_{ij} are positive, where the product is over some finite set S of pairs of integers (i, j) such that $i < j$ and where w'' is a product of commutators that lie in the third term F_3 of the lower central series for F. Again we may argue as in the proof of Theorem 17.1 to show that the laws w' and $[x, y]^p$ are equivalent to the laws w'' and $[x, y]^{d'}$, where

$$d' = \gcd(\{s_{ij} : (i, j) \in S\} \cup \{p\}).$$

Since p is a prime, $d' = 1$ or $d' = p$. The former case cannot occur since this would imply that $\mathfrak{U}$ is abelian, contradicting the fact that $P \in \mathfrak{U}$. Hence, w' and $[x, y]^p$ are equivalent to w'' and $[x, y]^p$. To summarise, we have shown that the laws w, x^{p^ℓ}, $[x, y]^p$ and $[x, y, z]$ are equivalent to the laws w'', x^{p^ℓ}, $[x, y]^p$ and $[x, y, z]$. But $w'' \in F_3$ and thus w'' is a consequence of $[x, y, z]$. So w is a consequence of x^{p^ℓ}, $[x, y]^p$ and $[x, y, z]$, and the theorem follows.

So Theorem 19.1 implies that lower bounds on the enumeration functions of $\mathfrak{B}_{p,2}$ and $\mathfrak{H}_p$ give rise to lower bounds on the enumeration function of any variety containing non-abelian p-groups. Once we have proved these lower bounds, the statement made at the end of the first paragraph of the chapter will be justified.

The chapter is organised as follows. Section 19.1 contains bounds on the enumeration function $f_{\mathfrak{H}_p}(p^m)$, and on the function $f_{\mathfrak{B}_{p,2}}(p^m)$ in the case when p is odd. The leading terms of these two enumeration functions are equal, and so it is interesting to ask how far the growth rates of these functions differ. In Section 19.2, we provide a partial answer to this question by proving an upper bound (which we conjecture is tight) for the leading term of the exponent of the ratio $f_{\mathfrak{H}_p}(p^m)/f_{\mathfrak{B}_{p,2}}(p^m)$.

19.1 Enumerating two small varieties

This section contains upper and lower bounds on the functions $f_{\mathfrak{H}_p}(p^m)$ and $f_{\mathfrak{B}_{p,2}}(p^m)$. All the arguments in this section are either contained in Higman [45] or are small modifications of the arguments given there.

Theorem 19.2 *Let p be a prime number, and let $\mathfrak{H}_p$ be the variety of p-groups of Φ-class 2 defined by the laws x^{p^2}, $[x, y]^p$ and $[x, y, z]$. Then*

$$p^{\frac{2}{27}m^3 - \frac{4}{9}m^2} \leqslant f_{\mathfrak{H}_p}(p^m) \leqslant m p^{\frac{2}{27}m^3 + \frac{11}{24}m}.$$

Proof: We recall the notation that was introduced in Chapter 4, especially Section 4.1. We define F_r to be the free group on r generators $x_1, x_2, \ldots, x_r$, and G_r to be the quotient of F_r by the subgroup generated by all words of the form x^{p^2}, $[x, y]^p$ or $[x, y, z]$. We identify the elements x_i with their images in G_r. Lemma 4.2 implies that G_r is a finite p-group. Moreover, the Frattini subgroup $\Phi(G_r)$ of G_r is central and elementary abelian, of order $p^{\frac{1}{2}r(r+1)}$ and index p^r in G_r.

By construction, G_r satisfies the laws x^{p^2}, $[x, y]^p$ and $[x, y, z]$, and so G_r and any quotient of G_r lie in $\mathfrak{H}_p$. In the proof of the lower bound for $f(p^m)$ given in Proposition 4.4 and Theorem 4.5, we constructed at least $p^{\frac{2}{27}m^2(m-6)}$ isomorphism classes of groups of order p^m by taking quotients of G_r for some r. Since all these groups lie in $\mathfrak{H}_p$, the lower bound of the theorem follows immediately.

We now prove the upper bound of the theorem. Let H be a group in $\mathfrak{H}_p$ of order p^m, and suppose that H has a minimal generating set consisting of r elements. Let $s = m - r$. Lemma 3.12 implies that $|H/\Phi(H)| = p^r$, and so $|\Phi(H)| = p^s$. By Lemma 4.1, there exists a surjective homomorphism γ from G_r to H. Let $N = \ker \gamma$, so we have that H is isomorphic to G_r/N. Because G_r and H are p-groups, their Frattini subgroups are generated by the elements of the form x^p and $[x, y]$ (by Lemma 3.12) and so $\gamma(\Phi(G_r)) = \Phi(H)$. Now, $\Phi(G_r)$ has index p^r in G_r and $\Phi(H)$ has index p^r in H, and so $N \leqslant \Phi(G_r)$. Since $\Phi(H)$ has order p^s, we find that N has index p^s in $\Phi(G_r)$. Thus the number of isomorphism classes of groups of order p^m in $\mathfrak{H}_p$ having a minimal generating set of r elements is at most the number of subgroups N of $\Phi(G_r)$ of index p^s. Now, $\Phi(G_r)$ is elementary abelian of order $p^{\frac{1}{2}r(r+1)}$, so we may regard $\Phi(G_r)$ as a vector space over $\mathbb{F}_p$ of dimension $\frac{1}{2}r(r+1)$ and N as a subspace of codimension s. By duality, the number of choices for N is the number of choices for a subspace of dimension s in a space of dimension $\frac{1}{2}r(r+1)$. Hence, by Proposition 3.16 there are at most

$$p^{s\left(\frac{1}{2}r(r+1)-s+1\right)} = p^{\frac{1}{2}r(r+1)s-s(s-1)}$$

choices for a subgroup N of $\Phi(G_r)$ of the correct form. Thus

$$f_{\mathfrak{H}_p}(p^m) \leqslant \sum p^{\frac{1}{2}r(r+1)s-s(s-1)}, \qquad (19.1)$$

where the sum is over pairs of integers r and s such that $r+s=m$, such that $r \geqslant 1$ and such that $s \geqslant 0$. It is not difficult to show that the maximum value of $\frac{1}{2}r(r+1)s - s(s-1)$ occurs when $r = (2/3)m + \delta$ and $s = (1/3)m - \delta$, where $0 \leqslant \delta \leqslant \frac{1}{2}$. By substituting these expressions for r and s into the expression $\frac{1}{2}r(r+1)s - s(s-1)$, it is easy to show that the maximum value of $\frac{1}{2}r(r+1)s - s(s-1)$ is at most $\frac{2}{27}m^3 + \frac{11}{24}m$. This bound, together with the fact that there are m terms on the right-hand side of (19.1), establishes the upper bound of the theorem.

Theorem 19.3 *Let p be an odd prime number, and let $\mathfrak{B}_{p,2}$ be the variety of groups of exponent p and nilpotency class at most 2 defined by the laws x^p and $[x, y, z]$. Then*

$$p^{\frac{2}{27}m^3 - \frac{2}{3}m^2} \leqslant f_{\mathfrak{B}_{p,2}}(p^m) \leqslant m p^{\frac{2}{27}m^3 - \frac{2}{9}m^2 + \frac{49}{72}m}.$$

Proof: As $\mathfrak{B}_{p,2}$ is a subvariety of $\mathfrak{H}_p$, any group H in $\mathfrak{B}_{p,2}$ with a minimal generating set of size r is isomorphic to a quotient of the group G_r by a subgroup N of $\Phi(G_r)$. Indeed, since a quotient G_r/N is in $\mathfrak{B}_{p,2}$ if and only if N contains $(G_r)^p$, the isomorphism classes of groups H in $\mathfrak{B}_{p,2}$ which have order p^{r+s} and have a minimal generating set of size r are in one-to-one correspondence with the orbits of $\operatorname{Aut}(G_r)$ on the set of subgroups N of $\Phi(G_r)$ of index p^s that contain $(G_r)^p$.

Let $\pi: G_r \to G_r$ be defined by $\pi(x) = x^p$ for all $x \in G_r$. Since G_r has nilpotency class 2, Lemma 3.3 implies that

$$\pi(xy) = (xy)^p = x^p y^p [y, x]^{\frac{1}{2}p(p-1)}.$$

Since p is odd, p divides $\frac{1}{2}p(p-1)$, and since G_r' has exponent p we find that $[y, x]^{\frac{1}{2}p(p-1)} = 1$. Thus π is a homomorphism. Now, $(G_r)^p = \operatorname{im}\pi = \langle \pi(x_1), \pi(x_2), \ldots, \pi(x_r) \rangle$, the last equality following since the x_i generate

G_r and π is a homomorphism. Hence $(G_r)^p$ is the subgroup of $\Phi(G_r)$ of order p^r generated by $x_1^p, x_2^p, \ldots, x_r^p$. Since (by the proof of Lemma 4.2) there are no redundant elements amongst these generators, and since $(G_r)^p$ is an elementary abelian p-group, we find that $(G_r)^p$ has order p^r.

The number of subgroups N of $\Phi(G_r)$ of index p^s containing $(G_r)^p$ is equal to the number of subgroups of $\Phi(G_r)/(G_r)^p$ of index p^s. Since $\Phi(G_r)/(G_r)^p$ is elementary abelian of order $p^{\frac{1}{2}r(r+1)-r} = p^{\frac{1}{2}r(r-1)}$, the number of subgroups N of the form we require is equal to the number of subspaces of codimension s of a vector space of dimension $\frac{1}{2}r(r-1)$.

We first establish the lower bound of the theorem. The lower bound is trivial when $m \leqslant 9$, so we may assume that $m \geqslant 10$. We note (just as in Proposition 4.4) that each $\mathrm{Aut}\,(G_r)$-orbit has length at most p^{r^2}. Moreover, Proposition 3.16 implies that there are at least $p^{(\frac{1}{2}r(r-1)-s)s}$ subgroups N containing $(G_r)^p$, and so there are at least $p^{(\frac{1}{2}r(r-1)-s)s-r^2}$ isomorphism classes of groups H of order p^{r+s} in $\mathfrak{B}_{p,2}$ with $|\Phi(H)| = p^s$. If we substitute $r = \frac{2}{3}m - \delta$ and $s = \frac{1}{3}m + \delta$ into the above formula, where $\delta \in \{0, \frac{1}{3}, \frac{2}{3}\}$ is chosen so that r and s are integers, we find that we have established the lower bound of the theorem.

To prove the upper bound, we use exactly the same argument as in the previous theorem. The number of subspaces of codimension s in a vector space of dimension $\frac{1}{2}r(r-1)$ is at most p to the power $(\frac{1}{2}r(r-1)-s+1)s$, and so

$$f_{\mathfrak{B}_{p,2}}(p^m) \leqslant \sum p^{(\frac{1}{2}r(r-1)-s+1)s}, \tag{19.2}$$

where the sum is over integers r and s satisfying the conditions $r \geqslant 1$, $s \geqslant 0$ and $r + s = m$. It is not difficult to show that

$$\left(\frac{1}{2}r(r-1)-s+1\right)s \leqslant \frac{2}{27}m^3 - \frac{2}{9}m^2 + \frac{49}{72}m.$$

This bound, together with the fact that there are m terms on the right-hand side of (19.2), establishes the upper bound we require.

We comment that there is a more natural proof of Theorem 19.3 that uses relatively free groups in $\mathfrak{B}_{p,2}$; we use relatively free groups in $\mathfrak{H}_p$ because we have already developed the relevant theory in Chapter 4.

19.2 The ratio of two enumeration functions

In this section, we establish an upper bound, due to Blackburn [7], on the ratio $f_{\mathfrak{H}_p}(p^m)/f_{\mathfrak{B}_{p,2}}(p^m)$. Note that Theorems 19.2 and 19.3 already combine to show that this ratio is at most $p^{\frac{2}{3}m^2+O(m)}$. However, we will provide an improved upper bound, which has a leading term that we conjecture to be correct.

For a variety $\mathfrak{V}$ of p-groups, we will write $\mathfrak{V}[r,s]$ for the set of (isomorphism classes of) groups $G \in \mathfrak{V}$ with $\Phi(G)$ of index p^r and of order p^s. Thus

$$f_{\mathfrak{V}}(p^m) = \sum_{r+s=m} |\mathfrak{V}[r,s]|.$$

Theorem 19.4 *Let p be an odd prime, and let $\mathfrak{H}_p$ and $\mathfrak{B}_{p,2}$ be the varieties defined in Theorems 19.2 and 19.3. Then*

$$f_{\mathfrak{H}_p}(p^m)/f_{\mathfrak{B}_{p,2}}(p^m) \leqslant p^{\frac{2}{9}m^2+O(m^{3/2})}.$$

Proof: We begin by showing that for any non-negative integers r and s,

$$|\mathfrak{H}_p[r,s]| \leqslant p^{rs} \sum_{j=0}^{s} |\mathfrak{B}_{p,2}[r,j]|.$$

We will again use the group G_r defined in Chapter 4 and used in the previous section. By Lemma 4.2, $\Phi(G_r)$ is an elementary abelian p-group of order $p^{\frac{1}{2}r(r+1)}$. Moreover, the proof of Lemma 4.2 showed that if $x_1, x_2, \ldots, x_r$ is a minimal generating set for G_r then the set

$$\{x_i^p : 1 \leqslant i \leqslant r\} \cup \{[x_j, x_i] : 1 \leqslant i < j \leqslant r\}$$

is a minimal generating set for $\Phi(G_r)$. We have seen in the proof of Theorem 19.3 that the set $\{x_i^p : 1 \leqslant i \leqslant r\}$ generates $(G_r)^p$ (we are using the fact that p is odd at this point). Since G_r has nilpotency class 2, the set $\{[x_j, x_i] : 1 \leqslant i < j \leqslant r\}$ generates G_r'. Hence $\Phi(G_r) = (G_r)^p \times G_r'$.

Let X_s be the set of subgroups of $\Phi(G_r)$ of index p^s in $\Phi(G_r)$, and let $Y_s \subseteq X_s$ consist of those subgroups containing $(G_r)^p$. We have already seen that the group $\operatorname{Aut}(G_r)$ acts naturally on X_s (and so on Y_s, since $(G_r)^p$ is a characteristic subgroup of G_r). Moreover, $|\mathfrak{H}_p[r,s]|$ is equal to the number of $\operatorname{Aut}(G_r)$-orbits on X_s, and $|\mathfrak{B}_{p,2}[r,s]|$ is equal to the number of $\operatorname{Aut}(G_r)$-orbits on Y_s. Define $\Gamma : X_s \to \bigcup_{i=0}^{s} Y_i$ by

$$\Gamma(N) = (N \cap G_r')(G_r)^p$$

for any $N \in X_s$. Since N has index p^s in $\Phi(G_r)$, we have that $N \cap G_r'$ has index at most p^s in G_r', and so $\Gamma(N) \in Y_i$ for some $i \in \{0, 1, \ldots, s\}$. Thus Γ is well-defined. Since G_r' and $(G_r)^p$ are both characteristic subgroups of G_r, the function Γ respects the action of $\operatorname{Aut}(G_r)$ on the sets X_s and $\bigcup_{i=0}^{s} Y_i$.

Let $i \in \{0, 1, \ldots, s\}$ and let $M \in Y_i$. We claim that M has at most p^{rs} preimages under Γ. For suppose that $N \in X_s$ is such that $\Gamma(N) = M$. By definition of Γ, we have that $(N \cap G'_r)(G_r)^p/(G_r)^p = M/(G_r)^p$. Since G'_r and $(G_r)^p$ intersect trivially, $N \cap G'_r$ is determined by this equality. Since $N \cap G'_r$ has index p^i in G'_r, we find that NG'_r/G'_r has index p^{s-i} in $\Phi(G_r)/G'_r$. But $\Phi(G_r)/G'_r$ is elementary abelian of order p^r, and so the number of choices for NG'_r/G'_r is at most the number of subspaces of codimension $s - i$ in a vector space over $\mathbb{F}_p$ of dimension r; this number is bounded above by $p^{r(s-i)}$. Since $G'_r/(N \cap G'_r)$ has order p^i, there are at most $(p^i)^{r-(s-i)}$ choices for $N/(N \cap G'_r)$ once $N \cap G'_r$ and NG'_r/G'_r are fixed. But N is determined by $N/(N \cap G'_r)$ and $N \cap G'_r$, and so the number of preimages of M is at most $p^{r(s-i)+(r-(s-i))i}$. Since $r(s-i) + (r-(s-i))i \leqslant rs$, our claim follows.

Since every element of $\bigcup_{i=0}^{s} Y_i$ has at most p^{rs} preimages under Γ, and since Γ respects the action of $\mathrm{Aut}\,(G_r)$ on the sets X_s and $\bigcup_{i=0}^{s} Y_i$, we have shown that

$$|\mathfrak{H}_p[r, s]| \leqslant \sum_{i=0}^{s} p^{rs} |\mathfrak{B}_{p,2}[r, i]|.$$

Writing a sum over all integers r and s such that $r \geqslant 1$, $s \geqslant 0$ and $r + s = m$ as an unlabelled sum, we find that

$$f_{\mathfrak{H}_p}(p^m) = \sum |\mathfrak{H}_p[r, s]| = \sum \sum_{i=0}^{s} p^{rs} |\mathfrak{B}_{p,2}[r, i]|$$

$$\leqslant m^2 p^{r_0 s_0} |\mathfrak{B}_{p,2}[r_0, i_0]|,$$

where the maximum value of $p^{rs} |\mathfrak{B}_{p,2}[r, i]|$ over the region $r \geqslant 1$, $s \geqslant 0$, $r + s = m$ and $0 \leqslant i \leqslant s$ occurs when $r = r_0$, $s = s_0$ and $i = i_0$. Thus

$$f_{\mathfrak{H}_p}(p^m) \leqslant m^2 p^{r_0 s_0} f_{\mathfrak{B}_{p,2}}(p^{r_0+i_0})$$

$$\leqslant m^2 p^{r_0 s_0} f_{\mathfrak{B}_{p,2}}(p^m),$$

the last equality following since $f_{\mathfrak{B}_{p,2}}(p^m)$ is a non-decreasing function of m. Hence

$$f_{\mathfrak{H}_p}(p^m)/f_{\mathfrak{B}_{p,2}}(p^m) \leqslant m^2 p^{r_0 s_0}.$$

To prove the theorem, it remains to show that $r_0 s_0 \leqslant \frac{2}{9} m^2 + O(m^{3/2})$.

From the proof of Theorem 19.3, we have that

$$\left(\frac{1}{2} r(r-1) - i\right) i - r^2 + rs \leqslant \log_p \left(p^{rs} |\mathfrak{B}_{p,2}[r, i]|\right) \leqslant \left(\frac{1}{2} r(r-1) - i + 1\right) i + rs. \tag{19.3}$$

The maximum value of $\log_p\left(p^{rs}|\mathfrak{B}_{p,2}[r,i]|\right)$ is at least $\frac{2}{27}m^3 - \frac{4}{9}m^2$. (This bound is trivial when $m \leqslant 2$. When $m > 2$ we may set $r = \frac{2}{3}m - \delta$ and $s = i = \frac{1}{3}m + \delta$ in (19.3), where $\delta \in \{0, 1/3, 2/3\}$ is chosen so as to make r, s and i integers.) Using the fact that the upper bound in (19.3) is at most $\frac{1}{2}r(r+1)s$, it is not difficult to verify that whenever $r = \frac{2}{3}m - x$ and $s = \frac{1}{3}m + x$ where $|x| \geqslant 2\sqrt{m}$, we have that $\log_p\left(p^{rs}|\mathfrak{B}_{p,2}[r,i]|\right)$ is less than $\frac{2}{27}m^3 - \frac{4}{9}m^2$ for all sufficiently large m. Hence $r_0 = \frac{2}{3}m - x$ and $s_0 = \frac{1}{3}m + x$ for some x such that $|x| \leqslant 2\sqrt{m}$. But this means that $r_0 s_0 \leqslant \frac{2}{9}m^2 + O(m^{3/2})$, and so the theorem is proved.

20

Miscellanea

This chapter is divided into several sections, in each of which we emphasise a separate topic in the theory of group enumeration. Section 20.1 is devoted to the problem of enumerating groups which are generated by at most d elements, for a fixed integer d. Section 20.2 contains a proof that groups with many non-abelian factors in their composition series are rare. Section 20.3 contains an enumeration of the graded Lie rings of order p^m. This enumeration is analogous to the enumeration of p-groups given in Chapters 4 and 5, but uses more elementary methods. In Section 20.4 we show that the error term of $O(m^{5/2})$ in the upper bound in the enumeration of p-groups can be improved to $O(m^2)$ if we restrict our attention to counting p-groups of nilpotency class at most 3.

20.1 Enumerating d-generator groups

In this section we consider the problem of estimating the number $f_d(n)$ of isomorphism classes of groups of order n that can be generated by d generators. We first give a brief history of progress on this problem.

The function $f_d(n)$ was first considered by Peter Neumann [76] in 1969, in the special case where n has the form p^m for a fixed prime p. In their 1987 paper, McIver and Neumann [67] showed that $f_d(p^m) \leqslant p^{\frac{1}{2}(d+1)m^2 + O(m)}$.

Extrapolating from this result, Laslo Pyber [83] conjectured that $f_d(n) \leqslant n^{c \log n}$ for some constant c depending only on d. Avinoam Mann [63] proved that the number of soluble d-generator groups of order n is at most $n^{(d+1)\lambda(n)}$ (recall the definition of $\lambda(n)$ from Chapter 1). In fact, he proved that the number of d-generator groups of order n that do not have any composition factors contained within three specific families of simple groups is at most $n^{c\lambda(n)}$. Finally, in 2001, Alex Lubotzky [61] proved that $f_d(n) \leqslant n^{cd \log n}$ for

195

some constant c. More precisely, he showed that $f_d(n) \leqslant n^{2(d+1)\lambda(n)}$ (see [61, p. 198]).

The methods of Mann and Lubotzky involve bounding the number of relations required to present a d-generator group of order n. In his paper [65], Mann focuses on the function $h(n, r)$ which he defines to be the number of groups of order n (up to isomorphism of course) that can be defined by r relations. If such a group is generated by d generators then $d \leqslant r$, so Lubotzky's theorem implies that $h(n, r) \leqslant n^{2(r+1)\lambda(n)}$, but this is not optimal. Confirming a conjecture proposed by Mann in the cited paper, A. Jaikin-Zapirain and L. Pyber (see [52]) have proved the following:

Theorem 20.1 *There exists* $c > 0$ *such that* $h(n, r) \leqslant n^{cr}$ *for all* n, r.

For nilpotent groups Mann proves the stronger result that $h_{\mathrm{nil}}(n, r) = o(n^r)$. For soluble groups he proves that $h_{\mathrm{sol}}(n, r) \leqslant n^{r+2\lambda(n)}$ and he proves that in general $h(n, r) \leqslant n^{r \log \log n + 2\lambda(n) + 3}$.

We return to $f_d(n)$: Avinoam Mann [63] provided a lower bound by constructing many d-generator groups of order p^m; his construction shows that the upper bound is reasonable when n is a power of a prime. In a recent paper [51], Andrei Jaikin-Zapirain has tightened both the upper and lower bounds in the case when $n = p^m$ for a fixed prime p. More precisely, he has proved that

$$p^{\frac{1}{4}(d-1)m^2 + o(m^2)} \leqslant f_d(p^m) \leqslant p^{\frac{1}{2}(d-1)m^2 + o(m^2)}.$$

We now summarise the contents of this section. We will not prove the upper bound for $f_d(n)$ in the general case, since Lubotzky's proof uses profinite methods which are beyond the scope of this book. We will reproduce Avinoam Mann's upper bound in the soluble case here, as it only uses elementary methods. We then turn our attention to the case when n is a power of a fixed prime p. Mann's theorem has as a corollary an upper bound for $f_d(p^m)$. We show that this upper bound is reasonable by proving Jaikin-Zapirain's lower bound for $f_d(p^m)$. (In fact, for the sake of simplicity, we prove a slightly weaker bound than Jaikin-Zapirain. The proof of Jaikin-Zapirain's lower bound uses some standard material on the lower exponent p central series of a group. The material we need is summarised in Lemma 20.6 below.)

Mann's proof begins by bounding the number of relations needed in a presentation of a soluble d-generator group.

Theorem 20.2 *Let G be a soluble group of order n, and suppose that G is generated by $x_1, x_2, \ldots, x_d$. Then there exists a presentation for G in these generators that consists of at most $(d+1)\lambda(n)$ relations.*

Proof: We prove the theorem by induction on $\lambda(n)$. When $\lambda(n) = 1$, the group G has prime order and so is cyclic. Then G is generated by one of the generators x_i, and a presentation for G consists of the relation $x_i^n = 1$ and the relations $x_j = x_i^{k_j}$ for $j \in \{1, 2, \ldots, d\} \setminus \{i\}$ and for some integers k_j. Thus the theorem holds in this case.

Assume (as an inductive hypothesis) that $\lambda(n) > 1$ and that the theorem is true for proper divisors of n. Let N be a minimal normal subgroup of G. Since G is soluble, N is an elementary abelian group of order p^k for some prime number p and positive integer k. By the inductive hypothesis, there exists a presentation for G/N in the generators $x_1 N, x_2 N, \ldots, x_d N$ consisting of r relations where $r \leqslant (d+1)\lambda(n/p^k) = (d+1)(\lambda(n) - k)$. For $i \in \{1, 2, \ldots, r\}$, let $u_i(x_1 N, x_2 N, \ldots, x_d N)$ be the ith relation in this presentation.

Let $y_1 \in N$ be a non-identity element. Since N is a minimal normal subgroup, N is generated by the conjugates of y_1. So we may choose elements $y_2, y_3, \ldots, y_k$ of N and words $v_2, v_3, \ldots, v_k$ in $x_1, x_2, \ldots, x_d$ such that $y_i = v_i^{-1} y_1 v_i$ for all $i \in \{2, 3, \ldots, k\}$ and such that N is generated by $y_1, y_2, \ldots, y_k$. We may express y_1 as a word w_1 in $x_1, x_2, \ldots, x_d$. For all $i \in \{2, 3, \ldots, k\}$, y_i may be expressed as the word $w_i = v_i^{-1} w_1 v_i$ in $x_1, x_2, \ldots, x_d$. Finally, we find words $t_i(y_1, y_2, \ldots, y_k)$ such that $u_i(x_1, x_2, \ldots, x_d) = t_i(y_1, y_2, \ldots, y_k)$ for all $i \in \{1, 2, \ldots, r\}$ and words $s_{ij}(y_1, y_2, \ldots, y_k)$ such that $x_j^{-1} y_i x_j = s_{ij}(y_1, y_2, \ldots, y_k)$ for all $i \in \{1, 2, \ldots, k\}$ and $j \in \{1, 2, \ldots, d\}$. For the sake of brevity, we write t_i and s_{ij} for the words $t_i(w_1, w_2, \ldots, w_k)$ and $s_{ij}(w_1, w_2, \ldots, w_k)$ in $x_1, x_2, \ldots, x_d$.

We claim that the relations

$$x_j^{-1} w_i x_j = s_{ij}, \quad w_1^p = 1, \quad [w_1, w_\ell] = 1 \quad \text{and} \quad u_m = t_m$$

form a presentation for G, where i, j, ℓ and m range over the sets $\{1, 2, \ldots, k\}$, $\{1, 2, \ldots, d\}$, $\{2, 3, \ldots, k\}$ and $\{1, 2, \ldots, r\}$ respectively.

Let the group X be presented by the generators $x_1, x_2, \ldots, x_d$ and the relations given above. Clearly G is a homomorphic image of X, and so to prove our claim it suffices to show that X has order at most n. Identifying words with their images in X, the subgroup Y of X generated by $w_1, w_2, \ldots, w_k$ is abelian. (To see this, note that the w_i are all conjugates of w_1, and so for some word z we find that $[w_i, w_j] = 1$ if and only if $[w_1, z^{-1} w_j z] = 1$. This last identity follows from Lemma 3.1, the relations $[w_1, w_\ell] = 1$ and the fact that $\langle w_1, w_2, \ldots, w_k \rangle$ is normal in $\langle x_1, x_2, \ldots, x_d \rangle$.) Moreover the elements

w_i all have order p in X, since the relations imply that w_1 has order p and that $w_2, w_3, \ldots, w_k$ are conjugates of w_1. Hence Y is an elementary abelian group of order at most p^k. Moreover, the relations $x_j^{-1} w_i x_j = s_{ij}$ and the fact that Y is finite imply that $x_j^{-1} Y x_j = Y$ for all $j \in \{1, 2, \ldots, d\}$. Hence Y is normal in X. But $t_m \in Y$ for all $m \in \{1, 2, \ldots, r\}$, and so the relations $u_m = 1$ hold in X/Y. Thus X/Y is a quotient of G/N, and so $|X/Y| \leqslant n/p^k$. Hence $|X| \leqslant (n/p^k)p^k = n$, and our claim is established.

We have constructed a presentation for G in $x_1, x_2, \ldots, x_d$ consisting of $kd + k + r$ relations. But $(d+1)k + r \leqslant (d+1)\lambda(n)$ and so we have established our inductive step. The theorem now follows by induction.

Theorem 20.3 *There are at most $n^{(d+1)\lambda(n)}$ soluble d-generator groups of order n.*

Proof: We argue using induction on $\lambda(n)$. The theorem is trivially true when $\lambda(n) = 1$, and so we assume that $\lambda(n) > 1$ and that the theorem is true for all smaller values of $\lambda(n)$.

Let G be a soluble group of order n, and let N be a minimal normal subgroup of G. Then (since G is soluble) N is elementary abelian of order p^k for some prime number p and for some positive integer k. There are $\lambda(n)$ choices for p^k, after which the isomorphism class of N is fixed. Moreover, by our inductive hypothesis, there are at most $(n/p^k)^{(d+1)\lambda(n/p^k)}$ choices for the isomorphism class of G/N. Let $y_1, y_2, \ldots, y_k$ be a generating set for N. By Theorem 20.2, there exists a presentation for G/N having r relations $u_m(x_1 N, x_2 N, \ldots, x_d N)$ where $r \leqslant (d+1)\lambda(n/p^k)$. Writing u_m for $u_m(x_1, x_2, \ldots, x_m)$, we have that $u_m \in N$. Now (as in Theorem 20.2) G is determined by the presentation for G/N, by a presentation for N and by $kd + r$ equations of the form

$$x_j^{-1} y_i x_j = s_{ij} \quad \text{and} \quad u_m = t_m$$

for $i \in \{1, 2, \ldots, k\}$, $j \in \{1, 2, \ldots, d\}$ and $m \in \{1, 2, \ldots, r\}$, where $s_{ij}, t_m \in N$. Therefore there are at most $|N|^{kd+r}$ choices for G once the isomorphism classes of N and G/N have been chosen. Note that

$$|N|^{kd+r} \leqslant (p^k)^{kd+(d+1)(\lambda(n/p^k))} = (p^k)^{d\lambda(n)+\lambda(n/p^k)}.$$

Hence the number of possibilities for G is at most

$$\lambda(n)(n/p^k)^{(d+1)\lambda(n/p^k)}(p^k)^{d\lambda(n)+\lambda(n/p^k)} \leqslant \lambda(n)n^{(d+1)\lambda(n)-1} \leqslant n^{(d+1)\lambda(n)}.$$

The theorem now follows by induction on $\lambda(n)$.

We now specialise to the case when n is a power of a prime. We aim to prove the following theorem:

Theorem 20.4 *Let p be a fixed prime. Let $f_d(p^m)$ be the number of (isomorphism classes of) d-generator groups of order p^m. Then*

$$p^{\frac{1}{4}(d-2)m^2+o(m^2)} \leqslant f_d(p^m) \leqslant p^{(d+1)m^2}.$$

The upper bound of Theorem 20.4 is a corollary of Theorem 20.3, and so we concentrate our efforts on proving the lower bound.

To show there are many d-generator groups of order p^m, it suffices to show that there are many normal groups N of index p^m in the free group F on d generators $x_1, x_2, \ldots, x_d$. For let G be a d-generator group of order p^m. If a quotient F/N is isomorphic to G, the isomorphism induces a homomorphism from F to G with kernel N. There are $(p^m)^d$ homomorphisms from F to G (as there are p^m choices for the image of each free generator x_i of F, and the homomorphism is determined once the images of these d generators are fixed). Thus, at most p^{dm} normal subgroups N of F have the property that F/N is isomorphic to G. So $f_d(p^m)$ is at least p^{-dm} times the number of normal groups of F of index p^m.

To find many subgroups N that are normal in F and of index p^m, we will find certain normal subgroups H of F, of index less than p^m in F, such that $H/([H,F]H^p)$ is large. Any subgroup N such that $[H,F]H^p \leqslant N \leqslant H$ is normal in F, and there are many choices for N since $H/[H,F]H^p$ is elementary abelian. Counting such subgroups N that are of index p^m in F will give us the lower bound we need.

For a normal subgroup H of the free group F on d generators, define $r_p(H)$ by

$$|H/([H,F]H^p)| = p^{r_p(H)}.$$

Lemma 20.5 *Let d be a positive integer, and let p be a prime. Let H_1 and H_2 be normal subgroups of the free group F on d generators $x_1, x_2, \ldots, x_d$. Suppose that the index of H_1 in F is a power of p and suppose that H_2 is a subgroup of H_1 of index p. Then*

$$r_p(H_2) \leqslant r_p(H_1) + d.$$

Moreover, if there exists $z \in H_1 \setminus H_2$ such that $z^p \in [H_2, F]H_2^p$, then

$$r_p(H_2) \leqslant r_p(H_1) + d - 1.$$

Proof: For $i \in \{1, 2\}$ define the subgroup K_i by $K_i = [H_i, F]H_i^p$. So $|H_i/K_i| = p^{r_p(H_i)}$. Since $H_2 \leqslant H_1$ we have that $K_2 \leqslant K_1$. Since H_2 has index p in H_1, we have that $K_1 \leqslant H_2$. Now

$$p^{r_p(H_2)} = |H_2/K_2| = p^{-1}|H_1/K_2|$$
$$= p^{-1}(|H_1/K_1|)(|K_1/K_2|) = p^{r_p(H_1)-1}|K_1/K_2|. \tag{20.1}$$

To provide an upper bound on $r_p(H_2)$, we aim to provide an upper bound on $|K_1/K_2|$.

Let $z \in H_1 \setminus H_2$. We claim that K_1 is generated by K_2 together with the elements

$$[z, x_1], [z, x_2], \ldots, [z, x_d] \text{ and } z^p. \tag{20.2}$$

It is clear that all the elements above lie in K_1. A typical element of H_1 has the form wz^i, where $w \in H_2$ and $i \in \{0, 1, \ldots, p-1\}$. So $[H_1, F]$ is generated by elements of the form $[wz^i, x]$ where $x \in F$. The subgroup H_2/K_2 is central in F/K_2 and so

$$[wz^i, x]K_2 = [z^i, x]K_2 = [z, x]^i K_2,$$

the last step following from Lemma 3.3 since $[z, x] \in H_2$. For any $j \in \{1, 2, \ldots, d\}$, we have that $[z, x_j] \in [H_1, F] \leqslant H_2$, and so $[z, x_j]K_2$ is central in F/K_2. So writing x as a word in the generators $x_1, x_2, \ldots, x_d$ and their inverses, we may express $[z, x]K_2$ as a product of elements $[z, x_j]K_2$ and their inverses using part (2) of Lemma 3.1. Thus $[H_1, F]$ is contained in the subgroup generated by K_2 and the elements (20.2) above. Similarly, H_1^p is contained in this subgroup, since if $w \in H_2$ and $i \in \{0, 1, \ldots, p-1\}$ then

$$(wz^i)^p K_2 = w^p z^{ip} K_2 = z^{ip} K_2 = (z^p)^i K_2.$$

Since $K_1 = [H_1, F]H_1^p$, our claim follows.

Now, K_1 is generated by K_2 and $d+1$ extra elements. Since K_1/K_2 is a subgroup of the elementary abelian p-group H_2/K_2, this means that $|K_1/K_2| \leqslant p^{d+1}$. So, using (20.1),

$$p^{r_p(H_2)} = p^{r_p(H_1)-1}|K_1/K_2| \leqslant p^{r_p(H_1)+d}.$$

Thus $r_p(H_2) \leqslant r_p(H_1) + d$ and the first statement of the lemma is proved.

Suppose now that $z^p \in K_2$. Then K_1 is generated by K_2 together with the d elements $[z, x_j]$ where $j \in \{1, 2, \ldots, d\}$. Thus $|K_1/K_2| \leqslant p^d$ and so, using (20.1) as before, we find that $r_p(H_2) \leqslant r_p(H_1) + d - 1$. This proves the lemma. $\quad\blacksquare$

Let F be a group. The *lower exponent p central series* $L_1 \geqslant L_2 \geqslant L_3 \geqslant \cdots$ of F is defined by $L_1 = F$ and $L_{i+1} = [L_i, F]L_i^p$ for $i \geqslant 1$. The standard facts about this series that we need are contained in the following lemma:

Lemma 20.6 *Let p be a prime. Let F be a group, and let $L_1 \geqslant L_2 \geqslant \cdots$ be its lower exponent p central series.*

(i) *Let i and e be integers such that $i \geqslant 2$ and $e \geqslant 1$. Then for any $g \in L_i$ and any $x \in F$*

$$[g, x]^{p^e} \in [g^{p^e}, x] L_{i+e+2}.$$

(ii) *Suppose that F is generated by a set X. Let i be a positive integer. Then L_i is generated modulo L_{i+1} by the elements of the form*

$$[a_1, a_2, \ldots, a_s]^{p^{i-s}}$$

where $a_1, a_2, \ldots, a_s \in X$ and where $1 \leqslant s \leqslant i$.

(iii) *Let d be a positive integer. Define $M_d(i)$ by*

$$M_d(i) = (1/i) \sum \mu(r) d^{i/r}$$

where the sum is over the positive divisors r of i and where μ is the Möbius function. (The Möbius function μ is defined by $\mu(r) = 0$ whenever a square of a prime divides r, otherwise $\mu(p_1 p_2 \ldots p_s) = (-1)^s$ where $p_1, p_2, \ldots, p_s$ are distinct primes.) Let F be the free group on d generators. Then for $i \geqslant 1$,

$$p^{r_p(L_i)} = |L_i / L_{i+1}| = p^{\sum_{j=1}^{i} M_d(j)}.$$

Proof: See Vaughan-Lee [92, Lemmas 2.2.2 and 2.2.3] for proofs of the first two parts of the lemma. Part (iii) of the lemma follows from Bryant and Kovács [11, Section 3] (who relate the order of L_i / L_{i+1} to the dimension of the ith grade of a free graded Lie algebra) together with standard results in the theory of free Lie algebras (which determine this dimension); see, for example, [41, Section 11.2].

Lemma 20.7 *Let d be a fixed integer and let p be a prime. Let F be the free group on d generators $x_1, x_2, \ldots, x_d$. Let $L_1 \geqslant L_2 \geqslant \cdots$ be the lower exponent p central series of F. Then, as $i \to \infty$,*

$$r_p(L_i) \sim \frac{d^{i+1}}{i(d-1)},$$

and the index of L_i in F is $p^{s(i)}$, where

$$s(i) \sim \frac{d^{i+1}}{i(d-1)^2}.$$

Proof: Let $M_d(i)$ be the function defined in Lemma 20.6. We claim that $M_d(i) \sim d^i/i$. The function $M_d(i)$ is defined as a sum over the positive divisors r of i. The term corresponding to $r = 1$ is d^i/i, and all the remaining terms have absolute value at most $d^{i/2}/i$. Since i has at most i positive divisors, we find that

$$d^i/i - d^{i/2} \leqslant M_d(i) \leqslant d^i/i + d^{i/2}.$$

The ratio of $d^{i/2}$ to d^i/i tends to 0 as $i \to \infty$, and so $M_d(i) \sim d^i/i$, as claimed.

To complete the proof of Lemma 20.7, we use the following elementary result. Let (a_i) and (b_i) be two real sequences such that $a_i \to \infty$ and $b_i \to \infty$ as $i \to \infty$. Suppose that $a_i - a_{i-1} \sim b_i - b_{i-1}$ as $i \to \infty$. Then $a_i \sim b_i$ as $i \to \infty$.

By Lemma 20.6,

$$r_p(L_i) = \sum_{j=1}^{i} M_d(j).$$

By the estimate above, $r_p(L_i) - r_p(L_{i-1}) = M_d(i) \sim d^i/i$. But we also have that

$$\frac{d^{i+1}}{i(d-1)} - \frac{d^i}{(i-1)(d-1)} = \frac{d^i}{i(d-1)}\left(d - \frac{i}{i-1}\right) \sim \frac{d^i}{i}.$$

The first assertion of the lemma now follows, by applying the elementary result referred to above (in the case $a_i = r_p(L_i)$ and $b_i = d^{i+1}/(i(d-1))$).

To prove the second statement of the lemma, first note that

$$p^{s(i)} = |F/L_i| = |L_1/L_i| = \prod_{j=1}^{i-1} |L_j/L_{j+1}| = \prod_{j=1}^{i-1} p^{r_p(L_j)}.$$

Hence $s(i) = \sum_{j=1}^{i-1} r_p(L_j)$. The elementary result referred to above now implies the second assertion of the lemma, once we have observed that

$$s(i) - s(i-1) = r_p(L_{i-1}) \sim \frac{d^i}{(i-1)(d-1)} \sim \frac{d^i}{i(d-1)}$$

and

$$\frac{d^{i+1}}{i(d-1)^2} - \frac{d^i}{(i-1)(d-1)^2} = \frac{d^i(d-i/(i-1))}{i(d-1)^2} \sim \frac{d^i}{i(d-1)}.$$

So Lemma 20.7 is proved.

Lemma 20.8 *Let d be a positive integer, and let p be a prime. Let F be the free group on d generators $x_1, x_2, \ldots, x_d$. There exists a chain of normal subgroups $F = H_0 \geqslant H_1 \geqslant H_2 \geqslant \cdots$ of F such that for each k the subgroup H_k has index p^k in F and*

$$r_p(H_k) \geqslant (d-1)k(1+o(1))$$

as $k \to \infty$.

Proof: Let $L_1, L_2, \ldots$ be the lower exponent p central series of F. For each positive integer i, define $M_{i,0}, M_{i,1}, \ldots, M_{i,i}$ as follows. Let $M_{i,j}$ be the subgroup generated by L_{i+1} and all elements of the form

$$[a_1, a_2, \ldots, a_s]^{p^{i-s}}$$

where $a_1, a_2, \ldots, a_s \in \{x_1, x_2, \ldots, x_d\}$ and where $1 \leqslant s \leqslant i - j$. By part (ii) of Lemma 20.6,

$$L_i = M_{i,0} \geqslant M_{i,1} \geqslant \cdots \geqslant M_{i,i-1} \geqslant M_{i,i} = L_{i+1},$$

and so we may think of the series of subgroups $M_{i,j}$ as refining the lower exponent p central series of F.

Note that $M_{i,i-1}$ is generated by L_{i+1} and the d elements

$$x_1^{p^{i-1}}, x_2^{p^{i-1}}, \ldots, x_d^{p^{i-1}}.$$

Since $M_{i,i-1}/L_{i+1}$ is a subgroup of the elementary abelian p-group L_i/L_{i+1}, this means that $M_{i,i-1}/L_{i+1}$ has order at most p^d. (In fact $|M_{i,i-1}/L_{i+1}| = p^d$, though we shall not need this.) Similarly, $M_{i,i-2}$ is generated modulo $M_{i,i-1}$ by the elements

$$[x_a, x_b]^{p^{i-2}}$$

where $a, b \in \{1, 2, \ldots, d\}$. Since $[x_a, x_b]^{p^{i-2}}$ is the inverse of $[x_b, x_a]^{p^{i-2}}$, we may restrict the generating set to those elements where $a < b$. Thus $M_{i,i-2}$ is generated modulo $M_{i,i-1}$ by $\frac{1}{2}d(d-1)$ elements, and so

$$|M_{i,i-2}/L_{i+1}| = |M_{i,i-2}/M_{i,i-1}||M_{i,i-1}/L_{i+1}| \leqslant p^{\frac{1}{2}d(d-1)+d} = p^{\frac{1}{2}d(d+1)}.$$

(Again, in fact, $|M_{i,i-2}/L_{i+1}| = p^{\frac{1}{2}d(d+1)}$.)

Let $H_0 \geqslant H_1 \geqslant \cdots$ be a refinement of the series

$$M_{1,0} \geqslant M_{1,1} \geqslant M_{2,1} \geqslant M_{2,2} \geqslant M_{3,1} \geqslant M_{3,2} \geqslant M_{3,3} \geqslant M_{4,1} \geqslant \cdots$$

where H_{k+1} has index p in H_k for all non-negative integers k. Now $M_{i,j}$ is generated modulo $M_{i,j+1}$ by elements of the form

$$[a_1, a_2, \ldots, a_{i-j}]^{p^j} \tag{20.3}$$

where $a_1, a_2, \ldots, a_{i-j} \in \{x_1, x_2, \ldots, x_d\}$. So since $M_{i,j}/M_{i,j+1}$ is elementary abelian, we may insist that H_k is generated by H_{k+1} and an element of the above form whenever $M_{i,j} \geqslant H_k > M_{i,j+1}$.

Each subgroup H_k has the property that $L_i \geqslant H_k > L_{i+1}$ for some integer i. Since F acts trivially by conjugation on L_i/L_{i+1}, any subgroup lying between L_i and L_{i+1} is normal. Hence the subgroups H_k are normal in F. So it remains to establish the lower bound on $r_p(H_k)$. We do this in three stages.

We first relate $r_p(H_k)$ to $r_p(H_{k+1})$. We show that in most cases, $r_p(H_{k+1}) \leqslant r_p(H_k) + d - 1$. In the second stage, we use this result to provide a bound on $r_p(L_{i+1})$ in terms of $r_p(H_k)$, where $L_i \geqslant H_k > L_{i+1}$. In the final stage, we use our knowledge of $r_p(L_{i+1})$ to provide the lower bound on $r_p(H_k)$ we are seeking.

Let k be an integer, $k \geqslant 0$, and let i and j be the integers such that $M_{i,j} \geqslant H_k > M_{i,j+1}$. Let $\ell = \log_p |H_k/L_{i+1}|$ (and so $H_{k+\ell} = L_{i+1}$). As the first stage in deriving our bound on $r_p(H_k)$, we claim that

$$r_p(H_{k+1}) \leqslant r_p(H_k) + d - 1 \text{ whenever } \ell > \frac{1}{2}d(d+1). \qquad (20.4)$$

For suppose that $\ell > \frac{1}{2}d(d+1)$. Then since $|M_{i,i-2}/L_{i+1}| \leqslant p^{\frac{1}{2}d(d+1)}$, we have that $j \leqslant i - 3$. By our choice of H_k, there exists an element $z \in H_k \setminus H_{k+1}$ of the form (20.3). Our claim will follow by Lemma 20.5, provided that we can show that $z^p \in [H_{k+1}, F]H_{k+1}^p$.

Since $j \leqslant i - 3$, we have that $i - j - 1 \geqslant 2$. Since $[a_1, a_2, \ldots, a_{i-j-1}] \in L_{i-j-1}$, part (i) of Lemma 20.6 implies that

$$z^p = [[a_1, a_2, \ldots, a_{i-j-1}], a_{i-j}]^{p^{j+1}}$$

$$\in [[a_1, a_2, \ldots, a_{i-j-1}]^{p^{j+1}}, a_{i-j}]L_{(i-j-1)+(j+1)+2}.$$

But

$$L_{(i-j-1)+(j+1)+2} = L_{i+2} = [L_{i+1}, F]L_{i+1}^p \leqslant [H_{k+1}, F]H_{k+1}^p,$$

since $H_{k+1} \geqslant L_{i+1}$. Moreover, $[a_1, a_2, \ldots, a_{i-j-1}]^{p^{j+1}} \in M_{i,j+1}$ by the definition of the subgroup $M_{i,j+1}$. Since $M_{i,j+1} \leqslant H_{k+1}$,

$$z^p \in [M_{i,j+1}, F]L_{i+2} \leqslant [H_{k+1}, F]L_{i+2} \leqslant [H_{k+1}, F]H_{k+1}^p,$$

and so our claim is established.

As the second stage in deriving our lower bound on $r_p(H_k)$, we claim that

$$r_p(L_{i+1}) \leqslant r_p(H_k) + (d-1)\ell + (1/2)d(d+1), \qquad (20.5)$$

where i and ℓ are defined as before. By Lemma 20.5, $r_p(H_{s+1}) \leqslant r_p(H_s) + d$ for any non-negative integer s. So an easy argument using induction shows that $r_p(H_{k+t}) \leqslant r_p(H_k) + dt$ for any non-negative integer t. This bound in the case when $t = \ell$ shows that

$$r_p(L_{i+1}) \leqslant r_p(H_k) + d\ell = r_p(H_k) + (d-1)\ell + \ell.$$

So the claim follows in the case when $\ell \leqslant \frac{1}{2}d(d+1)$. Now suppose that

$\ell > \frac{1}{2}d(d+1)$. Applying our bound (20.4) a total of $\ell - \frac{1}{2}d(d+1)$ times, we find that

$$r_p(H_{k+\ell-\frac{1}{2}d(d+1)}) \leqslant r_p(H_k) + (d-1)(\ell - \frac{1}{2}d(d+1)).$$

Hence, using Lemma 20.5,

$$r_p(L_{i+1}) = r_p(H_{k+\ell})$$

$$\leqslant r_p(H_{k+\ell-\frac{1}{2}d(d+1)}) + d\left(\frac{1}{2}d(d+1)\right)$$

$$\leqslant r_p(H_k) + (d-1)(\ell - \frac{1}{2}d(d+1)) + d\left(\frac{1}{2}d(d+1)\right)$$

$$= r_p(H_k) + (d-1)\ell + \frac{1}{2}d(d+1).$$

This proves our claim.

We are now ready for the last stage of deriving our lower bound on $r_p(H_k)$: proving the bound itself. Suppose, for a contradiction, that there exists a positive real number ϵ and an infinite strictly increasing sequence of positive integers $k_1, k_2, \ldots$ such that

$$r_p(H_{k_j}) \leqslant (1-\epsilon)(d-1)k_j$$

for $j \geqslant 1$. Define integers $i_1, i_2, \ldots$ by

$$L_{i_j} \geqslant H_{k_j} > L_{i_j+1},$$

and define $\ell_1, \ell_2, \ldots$ by $\ell_j = \log_p |H_{k_j}/L_{i_j+1}|$. By Equation (20.5),

$$r_p(L_{i_j+1}) \leqslant r_p(H_{k_j}) + (d-1)\ell_j + (1/2)d(d+1)$$

$$\leqslant (1-\epsilon)(d-1)k_j + (d-1)\ell_j + (1/2)d(d+1)$$

$$= (d-1)\log_p(|F/L_{i_j+1}|) + (1/2)d(d+1) - \epsilon(d-1)k_j,$$

the last step following since the definition of ℓ_j and the fact that H_{k_j} has index p^{k_j} in F together show that $k_j + \ell_j = \log_p(|F/L_{i_j+1}|)$. Since $L_{i_j} \geqslant H_{k_j}$, we have that $k_j \geqslant \log_p(|F/L_{i_j}|)$ and so

$$r_p(L_{i_j+1}) \leqslant (d-1)\log_p(|F/L_{i_j+1}|) + (1/2)d(d+1) - \epsilon(d-1)\log_p(|F/L_{i_j}|).$$
$$(20.6)$$

Lemma 20.7 tells us that

$$r_p(L_{i_j+1}) \sim \frac{d^{i_j+2}}{(i_j+1)(d-1)} \qquad (20.7)$$

as $j \to \infty$. However, if we write $g(j)$ for the right-hand side of (20.6), Lemma 20.7 implies that

$$g(j) \sim \frac{d^{i_j+2}}{(i_j+1)(d-1)} - \epsilon \frac{d^{i_j+1}}{i_j(d-1)}$$

$$\sim (1 - \epsilon/d) \frac{d^{i_j+2}}{(i_j+1)(d-1)} \qquad (20.8)$$

as $j \to \infty$. But the two approximations (20.7) and (20.8) contradict the inequality (20.6). This contradiction proves the lemma.

Proof of Theorem 20.4: As remarked after the statement of Theorem 20.4, the upper bound of Theorem 20.4 is a corollary of Theorem 20.3. So we need to prove the lower bound of the theorem.

Let m be a positive integer. Define $k = \lceil m/2 \rceil$ and $s = \lfloor m/2 \rfloor$. By Lemma 20.8, there exists a normal subgroup H_k of F of p^k such that

$$r_p(H_k) \geqslant (d-1)k(1+o(1)).$$

Define $K = [H_k, F]H_k^p$, so $|H_k/K| = p^{r_p(H_k)}$. Every subgroup N of F such that $H_k \geqslant N \geqslant K$ is normal, since F acts trivially by conjugation. Since H_k/K is an elementary abelian p-group, there are at least $p^{s(r_p(H_k)-s)}$ subgroups N containing K that have index p^s in H_k, by Proposition 3.16 and Corollary 3.17. Using the definition of s and our bound on $r_p(H_k)$, we see that

$$p^{s(r_p(H_k)-s)} \geqslant p^{\frac{1}{2}m(\frac{1}{2}m(d-1)-\frac{1}{2}m)+o(m^2)} = p^{\frac{1}{4}(d-2)m^2+o(m^2)}.$$

Since any subgroup of index p^s in H_k has index p^m in F, we have found $p^{\frac{1}{4}(d-2)m^2+o(m^2)}$ normal subgroups N of F having index p^m in F. Each of these subgroups N corresponds to a d-generator group F/N of order p^m. At most p^m quotients F/N can be isomorphic to a given group G (since an isomorphism from F/N to G induces a homomorphism from F to G with kernel N, and there are at most p^{md} homomorphisms from F to G). Hence we have found $p^{\frac{1}{4}(d-2)m^2+o(m^2)}$ d-generator groups of order p^m, and the theorem follows.

20.2 Groups with few non-abelian composition factors

This section contains a proof of a theorem on the enumeration of non-soluble groups due to A. R. Camina, G. R. Everest and T. M. Gagen [15]. The theorem was conjectured by John G. Thompson [91], and attempts to capture the feeling that non-soluble groups are rare. The proof uses two facts about simple groups that follow from the Classification: firstly that there are at most two simple groups of any order, and secondly that $|\text{Out}(L)| \leqslant |L|$ for any simple group L.

For a finite group G, define

$$\tau(G) = \frac{1}{|G|} \prod |H/K|,$$

where the product runs over the non-abelian factors in a composition series for G. So $1/|G| \leqslant \tau(G) \leqslant 1$, and $\tau(G) = 1/|G|$ if and only if G is soluble.

For a positive integer n, define $F(n)$ to be the number of (isomorphism classes of) groups of order at most n; so $F(n) = \sum_{i=1}^{n} f(i)$. For a positive real number ϵ, define $F_\epsilon(n)$ to be the number of (isomorphism classes of) groups G of order at most n such that $\tau(G) \geqslant \epsilon$. For a fixed value of ϵ, the function $F_\epsilon(n)$ counts only finitely many soluble groups (those of order at most $1/\epsilon$). One can think of $F_\epsilon(n)$ as counting those groups either having many non-abelian composition factors, or having a few large non-abelian composition factors.

Theorem 20.9 *Let ϵ be a positive real number. Then*

$$\lim_{n \to \infty} F_\epsilon(n)/F(n) = 0.$$

This theorem has recently been improved by Benjamin Klopsch in [57]:

Theorem 20.10 *For every $\epsilon > 0$ there exist $b, c > 0$ such that if $n \geqslant 60$ then*

$$n^{b \log \log n} \leqslant F_\epsilon(n) \leqslant n^{c \log \log n}.$$

To give some idea of what is involved we give a proof of Theorem 20.9, leaving the reader to consult [57] for the proof of the stronger result. Theorem 20.9 requires an upper bound on $F_\epsilon(n)$. Our arguments will be fairly crude, as we only aim to show that $F_\epsilon(n)$ grows significantly more slowly than $F(n)$.

Lemma 20.11 *The number of groups G of order n having no non-trivial soluble normal subgroups is at most $n^{6 \log n \log \log n}$.*

In this lemma, and for the remainder of this section, we will use the convention that $\log n \log \log n = 0$ when $n = 1$ (to circumvent the problem that $\log \log 1$ is not defined). Note that using this convention we have that $\log n \log \log n$ is a non-decreasing function on the set of positive integers n.

Note also that in [65] Avinoam Mann proves a refinement of this estimate, showing that the number of d-generator groups G of order n that have no non-trivial soluble normal subgroup is at most $n^{d \log \log n + 3}$. Since $d \leqslant \lambda(n)$ this is not only a refinement but also an improvement.

Proof of Lemma 20.11 The lemma holds trivially when $n \leqslant 4$, so we may assume that $n > 4$. Let G be a group of order n having no non-trivial soluble normal subgroups. Let S be the socle of G (so S is the subgroup generated by all minimal normal subgroups of G). Our condition on G implies that all minimal normal subgroups of G are direct products of non-abelian simple groups. So there exist an integer t and non-abelian simple groups $L_1, L_2, \ldots, L_t$ such that

$$S \cong L_1 \times L_2 \times \cdots \times L_t.$$

In particular, the centre $Z(S)$ of S is trivial. The centraliser C of S in G is a normal subgroup, so if C is non-trivial then C must have a non-trivial intersection with S (by definition of the socle). But $C \cap S = Z(S) = \{1\}$, and so $C = \{1\}$. So G acts faithfully by conjugation on S. In particular, we may regard G as a subgroup of Aut (S).

We saw at the start of the proof of Theorem 16.20 that there are at most n^3 possibilities for the isomorphism class of a product of non-abelian simple groups, where the product has order dividing n. (The fact that there are at most two simple groups of any fixed order is used here.) So there are at most n^3 possibilities for S.

Let S be fixed. Since G has order n, it can be generated by $\log n$ elements. So the number of possibilities for G once S is fixed is at most the number of subgroups of Aut (S) that can be generated by $\log n$ elements. This number is bounded above by $|\text{Aut}(S)|^{\log n}$, and so we require an upper bound on $|\text{Aut}(S)|$. An automorphism of S permutes the minimal normal subgroups $L_1, L_2, \ldots, L_t$ of S, so there is a natural homomorphism from Aut (S) into Sym(t). Let K be the kernel of this homomorphism. Now, K is isomorphic to $\prod_{i=1}^{t} \text{Aut}(L_i)$. The Classification implies that $|\text{Out}(L_i)| \leqslant |L_i|$, and so

$$|K| = \prod_{i=1}^{t} |\text{Aut}(L_i)| = \prod_{i=1}^{t} |L_i||\text{Out}(L_i)| \leqslant \prod_{i=1}^{t} |L_i|^2 = |S|^2 \leqslant n^2.$$

Now $t \leqslant \log n$, since $n \geqslant |S| = \prod_{i=1}^{t} |L_i| \geqslant 2^t$. Hence

$$|\text{Aut}(S)| \leqslant |K||\text{Sym}(t)| \leqslant n^2 t! \leqslant n^2 (\log n)! \leqslant n^2 (\log n)^{\log n}.$$

Hence the number of possibilities for G is bounded above by

$$n^3 \left(n^2 (\log n)^{\log n} \right)^{\log n}.$$

It is easy to see that

$$(\log n)^{(\log n)^2} n^{2\log n+3} = n^{\log n \log\log n + 2\log n + 3} \leqslant n^{6\log n \log\log n},$$

and so the lemma follows.

Lemma 20.12 *Let n be a positive integer, and let q be a prime power dividing n. Let A be an elementary abelian group of order q and let H be a group of order n/q. Then there are at most $q^{2(\log n)^2}$ isomorphism classes of groups G that are extensions of A by H.*

Proof: An extension G of A by H imposes an H-module structure M on A. Theorem 7.4 implies that an extension of A by H is determined by M together with an element of the second cohomology group $H^2(H, M)$.

Let $q = p^k$, where p is a prime and k is a positive integer. Since A is generated by k elements, $|\text{Aut}(A)| \leqslant (p^k)^k = p^{k^2}$. The H-module structure of A is determined by the actions of each element of some generating set for H on A, and H can certainly be generated by $\log(n/q)$ elements. Hence there are at most $p^{k^2 \log(n/q)}$ possibilities for the module structure M. It remains to bound the number of choices for an element of $H^2(H, M)$.

Since A is a p-group, so is $H^2(H, M)$. Let P be a Sylow p-subgroup of H. Then $|P| = p^\ell$ for some integer ℓ such that $k + \ell \leqslant \log n$. By Corollary 7.16, $H^2(H, M) \leqslant H^2(P, M)$. We need to find an upper bound on $|H^2(P, M)|$. Now, $|H^2(P, M)|$ is the number of equivalence classes of exact sequences of the form

$$1 \to M \to E \to P \to 1.$$

Since M, P and E are p-groups, they are soluble. In particular, since P is an ℓ-generator group, P may be defined by $\ell(\ell+1)$ relations $u_1, u_2, \ldots, u_{\ell(\ell+1)}$ in a generating set $x_1, x_2, \ldots, x_\ell$, by Theorem 20.2. A presentation for E may be constructed from three ingredients: a presentation for M; some relations of the form $u_i = t_i$ where t_i is an element of M; and relations determining the action of the generators x_i on M. The proof of this fact is very similar to the proof of Theorem 20.3, so we omit the details. The only freedom we have in constructing this presentation for E is the choice of the elements t_i. Hence there are at most $(p^k)^{\ell(\ell+1)}$ choices for the isomorphism class of E in the above exact sequence. The maps from M to E and from E to P are determined by the process of constructing a presentation for E of this form, so

$$|H^2(H, M)| \leqslant |H^2(P, M)| \leqslant (p^k)^{\ell(\ell+1)}.$$

Since a group G of the form we are counting is determined by the action of H on M together with an element of $H^2(H, M)$, the lemma follows once we observe that

$$p^{k^2 \log(n/q)} (p^k)^{\ell(\ell+1)} = q^{k \log(n/q) + \ell + \ell^2} \leqslant q^{2(\log n)^2}.$$

Proof of Theorem 20.9: For a positive integer n, let $f_\epsilon(n)$ be the number of (isomorphism classes of) groups of order n such that $\tau(G) \geqslant \epsilon$. We claim that

$$f_\epsilon(n) \leqslant n^{6 \log n \log \log n} + \sum q^{2(\log n)^2} f_{q\epsilon}(n/q) \tag{20.9}$$

where the sum runs over prime power divisors q of n such that $2 \leqslant q \leqslant 1/\epsilon$.

Let G be a group of order n such that $\tau(G) \geqslant \epsilon$. Suppose that G has a non-trivial soluble normal subgroup, and let A be a minimal soluble normal subgroup of G. Then A is elementary abelian of order q, where q is a prime power dividing n. Now,

$$1 \geqslant \tau(G/A) = q\tau(G) \geqslant q\epsilon,$$

the equality following from the definition of τ and the fact that A is soluble. Thus $q \leqslant 1/\epsilon$. Note also that the quotient G/A has order n/q, and the above inequality shows that $\tau(G/A) \geqslant q\epsilon$.

For a fixed value of q, there are at most $f_{q\epsilon}(n/q)$ possibilities for G/A, the isomorphism class of A is determined completely and by Lemma 20.12 there are at most $q^{2(\log n)^2}$ extensions of A by G/A. So the number of groups G of order n having a non-trivial soluble normal subgroup and such that $\tau(G) \geqslant \epsilon$ is at most

$$\sum q^{2(\log n)^2} f_{q\epsilon}(n/q),$$

where the sum runs over prime power divisors q of n such that $2 \leqslant q \leqslant 1/\epsilon$. The number of groups G of order n with no non-trivial soluble normal subgroup is bounded above by $n^{6 \log n \log \log n}$, by Lemma 20.11. Hence our claim (20.9) follows.

We claim that for all positive integers n and positive real numbers ϵ,

$$f_\epsilon(n) \leqslant 2n^{6 \log n \log \log n} (1/\epsilon)^{2(\log n)^2}. \tag{20.10}$$

Clearly this is true when $\epsilon > 1$ as then $f_\epsilon(n) = 0$. It is also easy to check that the result holds when $n \leqslant 4$. Assume, as an inductive hypothesis, that (20.10) holds for smaller values of n. Then

$$f_\epsilon(n) \leqslant n^{6\log n \log \log n} + \sum q^{2(\log n)^2} f_{q\epsilon}(n/q)$$

$$\leqslant n^{6\log n \log \log n} + \sum 2q^{2(\log n)^2}(n/q)^{6\log(n/q)\log\log(n/q)}(1/q\epsilon)^{2(\log(n/q))^2}$$

$$\leqslant n^{6\log n \log \log n} + \sum 2q^{2(\log n)^2}(n/q)^{6\log n \log \log n}(1/q\epsilon)^{2(\log n)^2}$$

(since $q\epsilon \leqslant 1$ for all prime powers q in our sum)

$$= n^{6\log n \log \log n} + 2\sum (n/q)^{6\log n \log \log n}(1/\epsilon)^{2(\log n)^2}$$

$$= \left((1/\epsilon)^{-2(\log n)^2} + 2\sum q^{-6\log n \log \log n}\right) n^{6\log n \log \log n}(1/\epsilon)^{2(\log n)^2}.$$

Now,

$$2\sum q^{-6\log n \log \log n} < 2\sum_{q=2}^{\infty} q^{-6} \leqslant 2\int_1^{\infty} x^{-6}dx = 2/5.$$

Moreover $(1/\epsilon)^{-2(\log n)^2} \leqslant 1$, and so

$$f_\epsilon(n) \leqslant (1+2/5)n^{6\log n \log \log n}(1/\epsilon)^{2(\log n)^2} < 2n^{6\log n \log \log n}(1/\epsilon)^{2(\log n)^2},$$

and our claim (20.10) follows.

Now let ϵ be fixed. Since $f_\epsilon(n) \leqslant 2^{O((\log n)^2 \log \log n)}$ by (20.10),

$$F_\epsilon(n) = \sum_{i=1}^n f_\epsilon(i) \leqslant n 2^{O((\log n)^2 \log \log n)} = 2^{O((\log n)^2 \log \log n)}.$$

Theorem 4.5 shows that $f(2^m) \geqslant 2^{\frac{2}{27}m^2(m-6)}$. Hence

$$F(n) \geqslant f(2^{\lfloor \log n \rfloor}) \geqslant 2^{\frac{2}{27}(\log n-1)^2(\log n-7)} = 2^{\frac{2}{27}(\log n)^3 + O(\log n)^2}.$$

These bounds on $F_\epsilon(n)$ and $F(n)$ show that $\lim_{n\to\infty} F_\epsilon(n)/F(n) = 0$, and the theorem follows.

20.3 Enumerating graded Lie rings

A *Lie ring* is an abelian group L, written additively, together with a mapping $[,]: L \times L \to L$ such that

$$[x, x] = 0, \tag{20.11}$$

$$[x, y+z] = [x, y] + [x, z], \tag{20.12}$$

$$[x+y, z] = [x, z] + [y, z], \tag{20.13}$$

$$[[x, y], z] + [[y, z], x] + [[z, x], y] = 0, \tag{20.14}$$

for all $x, y, z \in L$. These identities imply in particular that $[x, y] = -[y, x]$ (to see this, expand $[x + y, x + y]$ and use the fact that $[-x, y] = -[x, y]$). The conditions on the mapping $[\,,\,]$ are analogous to the properties of commutators in a group (see Equations (3.2), (3.3) and (3.4) in Chapter 3). Indeed, the well-known technique (see Gorenstein [36, Section 5.6]) of associating a Lie ring with a group shows that groups and Lie rings are closely related. For subsets $X, Y \subseteq L$, we write $[X, Y]$ for the subgroup generated by all elements of the form $[x, y]$ where $x \in X$ and $y \in Y$. In exactly the same way as for groups, we define the lower central series $R_1, R_2, R_3, \ldots$ of L by setting $R_1 = L$ and $R_i = [R_{i-1}, L]$ for $i > 1$. We say that L is nilpotent (of class at most c) if $R_{c+1} = \{0\}$ for some integer c.

The techniques of Chapter 5 can be used to show that the number of nilpotent Lie rings with p^m elements is at most $p^{\frac{2}{27} m^3 + O(m^{5/2})}$, and the techniques of Chapter 4 can be used to show that the exponent in this bound has the correct leading term. The aim of this section is to show that a similar upper bound (but with a better error term) can be derived in a much more elementary fashion if we restrict ourselves to a smaller class of Lie rings, namely the graded Lie rings generated by the first term of their grading.

A Lie ring L is *graded* if L is expressed as a direct sum

$$L = L_1 \oplus L_2 \oplus \cdots \oplus L_c$$

of subgroups L_i, such that $[L_i, L_j] \subseteq L_{i+j}$ for all integers $i, j \in \{1, 2, \ldots, c\}$ (where we set $L_k = \{0\}$ for integers k such that $k > c$). Clearly, a graded Lie ring is nilpotent of class at most c. We are interested in those graded Lie rings L that are generated by the first term L_1 of their grading. (This condition on L is equivalent to the condition that $[L_i, L_1] = L_{i+1}$ for all positive integers i. The equivalence of these conditions follows easily from the following fact. Let X generate L as a Lie ring. Then L is generated as an abelian group by the set of 'left-normed products from X', namely the set of elements that can be written in the form $[[\cdots[[x_1, x_2], x_3], \ldots, x_{k-1}], x_k]$ for some $x_1, x_2, \ldots, x_k \in X$. See, for example, Vaughan-Lee [92, Section 1.2] for a proof of this fact.) Graded Lie rings that are generated by the first term of their grading are often those that arise in practice. For example, the associated Lie ring of a group is of this type. For the rest of this section, for the sake of brevity, we will refer to a graded Lie ring that is generated by the first term of its grading simply as a graded Lie ring.

Lemma 20.13 *Let L be a graded Lie ring with p^m elements. Then the isomorphism class of L is determined by the isomorphism classes of the groups*

$L_1, L_2, \ldots, L_c$ *of its grading, together with the maps* $\gamma_1, \gamma_2, \ldots, \gamma_{c-1}$, *where*

$$\gamma_i : L_i \times L_1 \to L_{i+1}$$

is formed by restricting the map [,] *to* $L_i \times L_1$.

Proof: Clearly the isomorphism class of L as a group is determined by the isomorphism classes of $L_1, L_2, \ldots, L_c$, and so to prove the lemma it is sufficient to prove that the map [,] is determined by the maps γ_i.

Let $\gamma : L \times L_1 \to L_2 \oplus L_3 \oplus \cdots \oplus L_c$ be the restriction of [,] to $L \times L_1$. We claim that γ is determined by the maps γ_i. To see this, let $x \in L$ and let $y \in L_1$. Then $x = x_1 + x_2 + \cdots + x_c$, where $x_i \in L_i$. But then

$$[x, y] = [x_1 + x_2 + \cdots + x_c, y]$$
$$= [x_1, y] + [x_2, y] + \cdots + [x_c, y] \qquad \text{(by the equality (20.13))}$$
$$= \gamma_1(x_1, y) + \gamma_2(x_2, y) + \cdots + \gamma_{c-1}(x_{c-1}, y) + 0,$$

and so our claim follows.

We now show that [,] is determined by γ. Let $x, y \in L$. We may write $y = y_1 + y_2 + \cdots + y_c$ where $y_i \in L_i$, and (just as above) we may use the identity (20.12) to express $[x, y]$ as a sum of terms of the form $[x, y_j]$. So to show that [,] is determined by γ, it is sufficient to prove that $[x, y]$ is determined by γ for all $x \in L$ and $y \in L_j$ where $j \in \{1, 2, \ldots, c\}$. Indeed, since $[L, L_c] = 0$, we may assume that $j \neq c$.

We claim that $[x, y]$ is determined by γ for all $x \in L$ and $y \in L_j$. We prove our claim by using induction on j. When $j = 1$, $[x, y] = \gamma(x, y)$ and so our claim is correct in this case. Suppose that $1 < j < c$ and that the claim is true for all smaller values of j. Since L_1 generates L, we have that $L_j = [L_{j-1}, L_1]$ and so we may write

$$y = \sum_\alpha [z_\alpha, w_\alpha]$$

where $z_\alpha \in L_{j-1}$, $w_\alpha \in L_1$ and where α runs over some finite indexing set. Hence

$$[x, y] = [x, \sum_\alpha [z_\alpha, w_\alpha]]$$
$$= \sum_\alpha [x, [z_\alpha, w_\alpha]]$$

$$= - \sum_\alpha [[z_\alpha, w_\alpha], x]$$

$$= \sum_\alpha [[w_\alpha, x], z_\alpha] + \sum_\alpha [[x, z_\alpha], w_\alpha]$$

$$= - \sum_\alpha [[x, w_\alpha], z_\alpha] + \sum_\alpha [[x, z_\alpha], w_\alpha].$$

Since $w_\alpha \in L_1$ and $z_\alpha \in L_{j-1}$ our inductive hypothesis implies that $[x, w_\alpha]$ and then $[[x, w_\alpha], z_\alpha]$ are determined by γ. Similarly, $[[x, z_\alpha], w_\alpha]$ is determined by γ. Hence $[x, y]$ is determined by γ and so our claim follows by induction on j. Thus the lemma is established.

Theorem 20.14 *Let p be a prime number. The number of isomorphism classes of graded Lie rings with p^m elements that are generated by their first grading is at most*

$$p^{\frac{2}{27} m^3 + O(m)}.$$

Proof: Let L be a graded Lie ring with p^m elements, and suppose that L is generated by its first grading. Let $L_1, L_2, \ldots, L_c$ be the terms in its grading. By Lemma 20.13, to give an upper bound on the number of possibilities for L it suffices to bound the number of ways of choosing $L_1, L_2, \ldots, L_c$ up to group isomorphism and the number of ways of choosing the maps $\gamma_1, \gamma_2, \ldots, \gamma_{c-1}$ defined in the statement of the lemma.

Define the integer s_i by $|L_i| = p^{s_i}$. Since $p^m = |L| = \prod_{i=1}^c |L_i| = p^{\sum_{i=1}^c s_i}$, we find that $s_1, s_2, \ldots, s_c$ form an ordered partition of m. The number of ordered partitions of m is 2^{m-1} by Lemma 5.9, and so there are at most 2^{m-1} possibilities for the integer c and the integers $s_1, s_2, \ldots, s_c$. Once s_i is fixed, the number of possibilities for the isomorphism class of L_i as an abelian group is at most the number of partitions of s_i, by Theorem 17.2. This number is bounded above by $2^{s_i - 1}$, since there are $2^{s_i - 1}$ ordered partitions of s_i. Now $\prod_{i=1}^c 2^{s_i - 1} = 2^{m-c} \leqslant 2^{m-1}$, and so there are at most 2^{m-1} choices for the abelian groups $L_1, L_2, \ldots, L_c$ once the integers s_i have been fixed. Thus there are at most 2^{2m-2} choices for c, the sequence $s_1, s_2, \ldots, s_c$ and the abelian groups $L_1, L_2, \ldots, L_c$.

Suppose now that the group isomorphism classes of $L_1, L_2, \ldots, L_c$ are fixed. Let $\gamma_1, \gamma_2, \ldots, \gamma_{c-1}$ be the sequence of maps defined in the statement of Lemma 20.13. So $\gamma_i : L_i \times L_1 \to L_{i+1}$ is the restriction of $[,]$ to $L_i \times L_1$. We will show that there are at most $p^{\frac{2}{27} m^3}$ possibilities for the maps γ_i. This is trivial when $c = 1$, so from now on we assume that $c > 1$.

We claim that there are at most $p^{\frac{1}{2}s_1^2 s_2}$ possibilities for γ_1. To see this, let $g_1, g_2, \ldots, g_{r_1}$ be a minimal generating set for L_1. Note that since $|L_1| = p^{s_1}$, we have that $r_1 \leqslant s_1$. It is not difficult to see (using Equations (20.12) and (20.13) and the fact that $\gamma_1(g_j, g_i) = -\gamma_1(g_i, g_j)$) that the map γ_1 is determined by the $\frac{1}{2}r_1(r_1 - 1)$ values $\gamma_1(g_i, g_j)$ where $1 \leqslant i < j \leqslant r_1$. So the number of possibilities for γ_1 is at most $(p^{s_2})^{\frac{1}{2}r_1(r_1-1)}$. Since $r_1 \leqslant s_1$, there are at most $p^{\frac{1}{2}s_1^2 s_2}$ possibilities for γ_1, as claimed.

Let i be an integer such that $2 \leqslant i \leqslant c - 1$. We claim that there are at most $p^{s_1 s_i s_{i+1}}$ possibilities for γ_i. Let r_1 and the elements g_i be defined as in the previous paragraph. Let $h_1, h_2, \ldots, h_{r_i}$ be a generating set for L_i, where we may take $r_i \leqslant s_i$ since $|L_i| = p^{s_i}$. Then, as before, we may use the identities (20.12) and (20.13) to show that γ_i is determined by the $r_1 r_i$ images $\gamma_i(h_j, g_k)$ where $1 \leqslant j \leqslant r_i$ and $1 \leqslant k \leqslant r_1$. Since there are at most $p^{s_{i+1}}$ possibilities for each of these images, we find that the number of possibilities for γ_i is at most $p^{r_1 r_i s_{i+1}}$. Since $r_1 \leqslant s_1$ and $r_i \leqslant s_i$, there are at most $p^{s_1 s_i s_{i+1}}$ possibilities for γ_i, and our claim follows.

By Lemma 20.13, the map $[\,,\,]$ is determined by the maps $\gamma_1, \gamma_2, \ldots, \gamma_{c-1}$, and so there are at most $p^{M_c(s_1, s_2, \ldots, s_c)}$ possibilities for $[\,,\,]$ where

$$M_c(s_1, s_2, \ldots, s_c) = \frac{1}{2}s_1^2 s_2 + \sum_{i=2}^{c-1} s_1 s_i s_{i+1}.$$

It remains to show that $M_c(s_1, s_2, \ldots, s_c) \leqslant \frac{2}{27}m^3$. Setting $x_i = s_i/m$, we see that it is sufficient to show that $M_c(x_1, x_2, \ldots, x_c) \leqslant \frac{2}{27}$ for all real numbers x_i such that $x_1 + x_2 + \cdots + x_c = 1$ and $x_i \geqslant 0$ for all $i \in \{1, 2, \ldots, c\}$. We show this by induction on c. When $c = 2$ the result is easy to prove, so we assume that $c > 2$ and that the result holds for all smaller values of c. At an internal maximum, the theory of Lagrange multipliers shows that the partial derivatives of M_c with respect to each of the variables x_i must be equal. Now,

$$\frac{\partial M_c}{\partial x_c} = x_1 x_{c-1}$$

and

$$\frac{\partial M_c}{\partial x_{c-2}} = \begin{cases} x_1 x_2 + x_2 x_3 & \text{when } c = 3, \\ \frac{1}{2}x_1^2 + x_1 x_3 & \text{when } c = 4, \\ x_1 x_{c-3} + x_1 x_{c-1} & \text{when } c \geqslant 5. \end{cases}$$

In all of these cases, using the fact that $\frac{\partial M_c}{\partial x_c} = \frac{\partial M_c}{\partial x_{c-2}}$ at an internal maximum, it is easy to derive the contradiction that $x_i = 0$ for some i. And so M_c has no internal maxima. Finally, we check the values of M_c on the boundary of the region. The function M_c is zero when $x_1 = 0$. But when $x_i = 0$ for some $i \in \{2, 3, \ldots, c\}$ we find that

$$M_c(x_1, x_2, \ldots, x_{i-1}, 0, x_{i+1}, \ldots, x_c) \leqslant M_{c-1}(x_1, x_2, \ldots, x_{i-1}, x_{i+1}, \ldots, x_c),$$

and the right-hand side is at most $\frac{2}{27}$ by our inductive hypothesis. Hence, by induction on c, the function M_c is bounded above by $\frac{2}{27}$ on our region and so there are at most $p^{\frac{2}{27} m^3}$ choices for the map [,]. Since there were at most 2^{2m-2} choices for the isomorphism classes of $L_1, L_2, \ldots, L_c$ as groups, the theorem follows by Lemma 20.13.

20.4 Groups of nilpotency class 3

In this section we show that the error term in the Sims bound for the number of p-groups of order p^m may be improved if we restrict our enumeration to p-groups of nilpotency class at most 3. Indeed, our aim in this section is to show that the number of such groups is at most $p^{\frac{2}{27} m^3 + O(m^2)}$. The material in this section is taken from Simon Blackburn's DPhil thesis [7]. We need an upper bound on the number of graded Lie rings $L = L_1 \oplus L_2 \oplus L_3$ where L_1, L_2 and L_3 are specified groups, but the techniques of the last section do not provide a good enough bound. So our initial aim is to establish an improved bound for the number of Lie algebras of this type.

Lemma 20.15 *Let p be a prime number. Let m and b be integers such that $0 \leqslant b \leqslant m$. Let G be an abelian group of order p^m. The number of subgroups H of G of order p^b is at most the number $n_{m,b}$ of subspaces of dimension b in a vector space of dimension m over $\mathbb{F}_p$. In particular, the number of subgroups H is at most $p^{b(m-b+1)}$.*

Proof: We will use the identity

$$n_{m,b} = n_{m-1,b} + p^{m-b} n_{m-1,b-1}. \tag{20.15}$$

To see why this is true, let V be a vector space of dimension m over $\mathbb{F}_p$, let U be a fixed subspace of V of dimension $m - 1$ and let x be a fixed element of $V \setminus U$. There are $n_{m-1,b}$ subspaces W of dimension b that are contained in U. A subspace W of dimension b that is not contained in U may be written in the form $W = \langle W_0, x + u \rangle$ where W_0 is a subspace of U of dimension $b - 1$ and where $u \in U$. Note that distinct choices for W_0 give rise to distinct subspaces W, since $W_0 = W \cap U$. Note, however, that $\langle W_0, x + u_1 \rangle = \langle W_0, x + u_2 \rangle$ if and only if $u_1 - u_2 \in W_0$. Thus W is determined by a subspace W_0 of U

of dimension $b-1$ together with a coset of W_0 in U. Thus the number of subspaces W of dimension b not contained in U is $p^{m-b}n_{m-1,b-1}$, and so the identity (20.15) is established.

We prove the lemma by induction on m. If $m=0$ or $m=1$ the assertion is trivial, and so we assume (as an inductive hypothesis) that $m>1$ and the assertion is true for all groups of order less than p^m. Let M be a subgroup of G of index p and let x be a fixed element of $G\setminus M$. Let H be a subgroup of G of order p^b. The number of such subgroups H which are contained in M is at most $n_{m-1,b}$ by our inductive hypothesis. Let $H_0=H\cap M$. If H is not contained in M, then H_0 has order p^{b-1} and so there are at most $n_{m-1,b-1}$ possibilities for H_0, again by our inductive hypothesis. Now $H=\langle H_0, x+y\rangle$ for some element $y\in M$. Moreover $\langle H_0, x+y_1\rangle = \langle H_0, x+y_2\rangle$ when $y_1-y_2\in H_0$, and so there are at most $|M/H_0|$ choices for H once H_0 is fixed. Therefore there are at most $p^{m-b}n_{m-1,b-1}$ subgroups H not contained in M. Hence, by (20.15), there are at most $n_{m,b}$ subgroups in all. So the lemma follows by induction on m.

Proposition 20.16 *Let p be a prime number and let m be a positive integer. Let L_1, L_2 and L_3 be abelian groups of orders p^{s_1}, p^{s_2} and p^{s_3} respectively, where $s_1+s_2+s_3=m$. Then the number of graded Lie rings $L=L_1\oplus L_2\oplus L_3$ generated by L_1 is at most $(s_2+1)p^{M+O(m^2)}$, where*

$$M=\max_{0\leqslant b\leqslant \min\{s_2,s_3\}}\{bs_1(s_2+s_3-b)+\tfrac{1}{2}(s_2-b)s_1^2\}$$

Proof: By Lemma 20.13, we need to find an upper bound for the number of choices of γ_1 and γ_2, where $\gamma_1:L_1\times L_1\to L_2$ and $\gamma_2:L_2\times L_1\to L_3$ are the restrictions of the map $[\,,]$ to $L_1\times L_1$ and $L_2\times L_1$ respectively. Note that since $[\,,]$ satisfies the identities (20.11) to (20.14), we find that

$$\gamma_1(x,x)=0 \text{ for all } x\in L_1,$$

$$\gamma_i(x+y,z)=\gamma_i(x,z)+\gamma_i(y,z) \text{ for all } i\in\{1,2\}, x,y\in L_1 \text{ and } z\in L_1,$$

$$\gamma_i(x,y+z)=\gamma_i(x,y)+\gamma_i(x,z) \text{ for all } i\in\{1,2\}, x\in L_1 \text{ and } y,z\in L_1,$$

$$-\gamma_2(\gamma_1(y,z),x)-\gamma_2(\gamma_1(z,x),y)=\gamma_2(\gamma_1(x,y),z) \text{ for all } x,y,z\in L_1.$$
$$(20.16)$$

For an element $x\in L_1$, the set $\gamma_2(L_2,x)$ is a subgroup H of L_3. If $|H|=p^k$, we say that x has *breadth* k, and write $b(x)=k$. We define the *breadth* b of γ_2 by $b=\max\{b(x)\,|\,x\in L_1\}$. Clearly, $0\leqslant b\leqslant\min\{s_2,s_3\}$.

There are at most s_2+1 possibilities for b. Assume that b is fixed. We claim that the number of possibilities for γ_2 is at most $p^{bs_1(s_2+s_3-b+1)}$. Let $g_1,g_2,\ldots,g_{r_1}$ be a generating set for L_1 and let $h_1,h_2,\ldots,h_{r_2}$ be a generating set for L_2. We may assume that $r_1 \leqslant s_1$ and $r_2 \leqslant s_2$. We may also assume that $b(g_1) = b$. Now, γ_2 is determined by the $r_1 r_2$ elements $\gamma_2(h_j, g_i) \in L_3$ where $1 \leqslant i \leqslant r_1$ and $1 \leqslant j \leqslant r_2$. For $i \in \{1, 2, \ldots, r_1\}$, let H_i be a subgroup of order p^b containing $\{\gamma_2(h_j, g_i) \mid 1 \leqslant j \leqslant r_2\}$. Such a subgroup exists since γ_2 has breadth b. By Lemma 20.15, there are at most $\left(p^{b(s_3 - b + 1)}\right)^{r_1}$ choices for the subgroups H_i. Once the subgroups H_i are fixed, there are at most p^b choices for each element $\gamma_2(h_j, g_i) \in H_i$ and so at most $p^{b r_1 r_2}$ choices for γ_2. Therefore, using the fact that $r_1 \leqslant s_1$ and $r_2 \leqslant s_2$, we find that the number of choices for γ_2 of breadth b is at most $p^{b s_1 (s_2 + s_3 - b + 1)}$, as claimed.

Suppose now that γ_2 is fixed. We claim that there are at most p to the power

$$s_2(s_1 - 1) + \tfrac{1}{2}(s_2 - b)(s_1 - 1)(s_1 - 2)$$

choices for γ_1. Let $g_1, g_2, \ldots, g_{r_1}$ be the generating set for L_1 chosen above. The map γ_1 is determined by the elements $\gamma(g_i, g_j)$ where $1 \leqslant i < j \leqslant r_1$. There are at most $p^{s_2(r_1 - 1)}$ choices for the $r_1 - 1$ elements $\gamma_1(g_1, g_j)$ where $j \in \{2, 3, \ldots, r_1\}$. Assume that these elements have been chosen. Now, by (20.16), for all $i, j \in \{2, 3, \ldots, r_1\}$,

$$\gamma_2(\gamma_1(g_i, g_j), g_1) = -\gamma_2(\gamma_1(g_j, g_1), g_i) - \gamma_2(\gamma_1(g_1, g_i), g_j).$$

But the right-hand side of this equation is determined by the choices we have already made $\big($note that $-\gamma_2(\gamma_1(g_j, g_1), g_i) = \gamma_2(\gamma_1(g_1, g_j), g_i)\big)$. And so the image of $\gamma_1(g_i, g_j)$ under the homomorphism $x \mapsto \gamma_2(x, g_1)$ is determined. Since g_1 has breadth b, the kernel of this homomorphism has order $p^{s_2 - b}$, and so there are at most $p^{s_2 - b}$ choices for each of the $\tfrac{1}{2}(s_1 - 1)(s_1 - 2)$ elements $\gamma_1(g_i, g_j)$. Hence the number of choices for γ_1 once γ_2 is fixed is at most $p^{s_2(r_1 - 1) + \frac{1}{2}(s_2 - b)(s_1 - 1)(s_1 - 2)}$, and since $r_1 \leqslant s_1$ our claim follows.

Collecting together all our estimates, and using the fact that s_1, s_2, s_3 and b are all bounded above by m, we find that the proposition follows.

Theorem 20.17 *Let p be a fixed prime number, and let m be a positive integer. Then the number of isomorphism classes of groups of order p^m and nilpotency class at most 3 is bounded above by*

$$p^{\frac{2}{27} m^3 + O(m^2)}.$$

Proof: We begin by recalling a standard way of associating a Lie ring with a (usually nilpotent) group (see Gorenstein [36, Section 5.6]). A Lie ring L may be constructed as follows. Let G be a group, and let $G_1, G_2, \ldots$ be its lower central series. For all positive integers i, define $L_i = G_i / G_{i+1}$. Since the subgroups G_i are the terms in the lower central series of G, the groups L_i are abelian. We write the groups L_i additively, and define $L = L_1 \oplus L_2 \oplus \cdots$. To define the Lie multiplication, let i and j be positive integers and suppose that $x \in L_i$ and $y \in L_j$. Then $x = gG_{i+1}$ and $y = hG_{j+1}$ for some $g \in G_i$ and $h \in G_j$. By Proposition 3.6, $[g, h] \in G_{i+j}$. Moreover, it is not difficult to show that $[g, h]G_{i+j+1}$ depends only on the cosets gG_{i+1} and hG_{j+1} rather than on the elements g and h themselves. We define $[x, y] = [g, h]G_{i+j+1} \in L_{i+j}$. We then extend the Lie multiplication to the whole of L in the natural way, by $\mathbb{Z}$-linearity. It is not difficult to check that this definition of Lie multiplication does indeed make L into a Lie ring. Moreover, L is a graded Lie ring that is generated by its first grading.

Let G be a group of order p^m and nilpotency class at most 3. We will bound the number of possibilities for G by first choosing its associated Lie ring L. To this end, let $L = L_1 \oplus L_2 \oplus L_3$ be a graded Lie ring of order p^m that is generated by its first grading. Define integers s_1, s_2 and s_3 by $|L_i| = p^{s_i}$, so that $s_1 + s_2 + s_3 = m$. We aim to bound the number of choices for G once L is fixed.

For $i \in \{1, 2, 3\}$, let r_i be the minimal number of generators of L_i. Clearly, $r_i \leqslant s_i$ for all $i \in \{1, 2, 3\}$. Define $t_i = \sum_{j=1}^{i} r_j$. For all $i \in \{1, 2, 3\}$, choose a minimal generating set $x_{t_{i-1}+1}, x_{t_{i-1}+2}, \ldots, x_{t_i}$ for L_i. For all $j \in \{1, 2, \ldots, t_3\}$, let $a(j)$ be the smallest positive integer such that $p^{a(j)}x_j \in \langle x_{j+1}, x_{j+2}, \ldots, x_{t_3} \rangle$. So $\sum_{j=t_{i-1}+1}^{t_i} a(j) = s_i$, and every element of L_i may be written uniquely in the form $\sum_{j=t_{i-1}+1}^{t_i} e(j)x_j$ where $e(j) \in \{0, 1, \ldots, p^{a(j)} - 1\}$.

Let G be a group whose associated Lie ring is isomorphic to L. For simplicity, we identify the associated Lie ring of G with L. Let

$$G = G_1 > G_2 > G_3 > G_4 = \{1\}$$

be the lower central series of G. Choose elements $g_1, g_2, \ldots, g_{t_3} \in G$ such that $g_j G_{i+1} = x_j$ whenever $t_{i-1} + 1 \leqslant j \leqslant t_i$. Such elements exist since L is the associated Lie ring of G. Because of the way we have chosen the elements x_j of L, and since $L_i = G_i/G_{i+1}$ for $i \in \{1, 2, 3\}$, every element of G may be written uniquely in the form

$$g_1^{e(1)} g_2^{e(2)} \cdots g_{t_3}^{e(t_3)}$$

for some integers $e(j) \in \{0, 1, \ldots, p^{a(j)} - 1\}$. We write this product as $\prod g_u^{e(u)}$. Note that the product lies in G_i if and only if $e(u) = 0$ whenever $u \leqslant t_{i-1}$.

For all $j \in \{1, 2, \ldots, t_3\}$, there exist integers $b(j, u) \in \{0, 1, \ldots, p^{a(u)} - 1\}$ such that the relations

$$g_j^{p^{a(j)}} = \prod g_u^{b(j,u)} \tag{20.17}$$

hold in G. By our choice of the elements x_j, we find $b(j, u) = 0$ whenever $u \leqslant j$. Moreover, for all $i, j \in \{1, 2, \ldots, t_3\}$ such that $i < j$, there exist integers $c(i, j, u) \in \{0, 1, \ldots, p^{a(u)} - 1\}$ such that

$$[g_j, g_i] = \prod g_u^{c(i,j,u)}. \tag{20.18}$$

Since the subgroups G_i form the terms of the lower central series of G, we have that $c(i, j, u) = 0$ whenever one of the following conditions holds:

$$j \geqslant t_2 + 1,$$
$$t_1 + 1 \leqslant i \leqslant t_2 \text{ and } t_1 + 1 \leqslant j \leqslant t_2,$$
$$1 \leqslant i \leqslant t_1, \ t_1 + 1 \leqslant j \leqslant t_2 \text{ and } u \leqslant t_2,$$
$$1 \leqslant i \leqslant t_1, \ 1 \leqslant j \leqslant t_1 \text{ and } u \leqslant t_1.$$

The relations (20.17) and (20.18) form a power-commutator presentation for G. Hence the isomorphism class of G is determined by the Lie ring L together with the integers $a(j)$ for $1 \leqslant j \leqslant t_3$, the integers $b(j, u)$ for $1 \leqslant j \leqslant t_3$ and $j < u \leqslant t_3$, and the integers $c(i, j, u)$ where

$$1 \leqslant i < j \leqslant t_1 \text{ and } t_1 + 1 \leqslant u \leqslant t_3, \text{ or}$$
$$1 \leqslant i \leqslant t_1, \ t_1 + 1 \leqslant j \leqslant t_2 \text{ and } t_2 + 1 \leqslant u \leqslant t_3.$$

We are now in a position to bound the number of choices for the group G. There are at most m^3 possibilities for the integers s_1, s_2 and s_3. By Proposition 20.16, the number of choices for the Lie ring L is at most $(s_2 + 1)p^{M + O(m^2)}$, where

$$M = \max_{0 \leqslant b \leqslant \min\{s_2, s_3\}} \{bs_1(s_2 + s_3 - b) + \tfrac{1}{2}(s_2 - b)s_1^2\}.$$

Once L is fixed, the integers r_i are determined. We choose a generating set $x_1, x_2, \ldots, x_{t_3}$ for L as above. Once this generating set has been chosen the integers a_i are determined. For a fixed integer j, there are at most $p^{a(j+1) + a(j+2) + \cdots + a(t_3)}$ choices for the integers $b(j, u)$. Since $a(j+1) + a(j+2) + \cdots + a(t_3) < m$, there are at most p^m choices for the integers $b(j, u)$ for a fixed integer j and so there are at most p^{m^2} choices for the integers $b(j, u)$ where $1 \leqslant j, u \leqslant t_3$. Since L is the associated Lie ring of G, the Lie multiplication in L determines $[g_i, g_j]G_3$ whenever $1 \leqslant i < j \leqslant t_1$ and determines $[g_i, g_j]$ whenever $1 \leqslant i \leqslant t_1 < j \leqslant t_2$. Hence the integers $c(i, j, u)$ are determined unless $1 \leqslant i < j \leqslant t_1$ and $t_2 + 1 \leqslant u \leqslant t_3$. So the number of choices for the integers $c(i, j, u)$ is at most

$$\prod_{1 \leqslant i < j \leqslant t_1} \prod_{u = t_2 + 1}^{t_3} p^{a(u)} = \prod_{1 \leqslant i < j \leqslant t_1} p^{s_3} \leqslant p^{\frac{1}{2} t_1^2 s_3} \leqslant p^{\frac{1}{2} s_1^2 s_3}.$$

To summarise, we have shown that the number of choices for G is at most $p^{N+O(m^2)}$, where N is the maximum value of

$$bs_1(s_2 + s_3 - b) + \tfrac{1}{2}(s_2 + s_3 - b)s_1^2$$

on the region defined by $s_i \geqslant 0$, $s_1 + s_2 + s_3 = m$ and $0 \leqslant b \leqslant \min\{s_2, s_3\}$. To establish the theorem, it is therefore sufficient to show that $N \leqslant \tfrac{2}{27}m^3$.

Let s_1, s_2 and s_3 be fixed non-negative integers such that $s_1 + s_2 + s_3 = m$. We claim that the function $F(b)$ defined by

$$F(b) = bs_1(s_2 + s_3 - b) + \tfrac{1}{2}(s_2 + s_3 - b)s_1^2 \qquad (20.19)$$

satisfies $F(b) \leqslant \tfrac{2}{27}m^3$ for any real number such that $0 \leqslant b \leqslant s_2 + s_3$. Since $\min\{s_2, s_3\} \leqslant s_2 + s_3$, proving this claim is sufficient to establish that $N \leqslant \tfrac{2}{27}m^3$.

It is clear that $F(s_2 + s_3) = 0$. Moreover,

$$F(0) = \tfrac{1}{2}(s_2 + s_3)s_1^2 = \tfrac{1}{2}(m - s_1)s_1^2 \leqslant \tfrac{2}{27}m^3.$$

Thus it remains to show that $F(b) \leqslant \tfrac{2}{27}m^3$ at an internal maximum. When $s_1 = 0$ this is clear, and so we may assume that $s_1 \neq 0$. Now

$$\frac{\mathrm{d}F}{\mathrm{d}b} = s_1(s_2 + s_3) - 2s_1 b - \tfrac{1}{2}s_1^2.$$

At an internal maximum $\frac{\mathrm{d}F}{\mathrm{d}b} = 0$, and since $s_1 \neq 0$ we may deduce that $b = \tfrac{1}{2}(s_2 + s_3) - \tfrac{1}{4}s_1$ in this situation. When $s_1 > \tfrac{2}{3}m$ (and so $s_2 + s_3 < \tfrac{1}{3}m$), this equality contradicts the fact that $b \geqslant 0$, and so $F(b)$ has no internal maxima when $s_1 > \tfrac{2}{3}m$. So we may suppose that $s_1 = cm$, where c is a real number such that $0 < c \leqslant \tfrac{2}{3}$. Substituting the equalities $b = \tfrac{1}{2}(s_2 + s_3) - \tfrac{1}{4}s_1$, $s_1 = cm$ and $s_2 + s_3 = (1 - c)m$ into (20.19), we find that

$$F(b) = \left(\tfrac{1}{4}c - \tfrac{1}{4}c^2 + \tfrac{1}{16}c^3\right)m^3$$

at an internal maximum. However, it is easy to see that $\tfrac{1}{4}c - \tfrac{1}{4}c^2 + \tfrac{1}{16}c^3 \leqslant \tfrac{2}{27}$ on the interval $0 \leqslant c \leqslant \tfrac{2}{3}$, and so $F(b) \leqslant \tfrac{2}{27}m^3$ at an internal maximum. Thus our claim follows, and the theorem is proved.

21

Survey of other results

In this chapter we survey related topics. The treatment will be brief and proofs will rarely be more than sketched. The intention is merely to draw attention to some points of interest that are connected with the main part of this book.

21.1 Graham Higman's PORC conjecture

In his second paper [46] on enumerating p-groups, Graham Higman introduced the acronym PORC to stand for *polynomial on residue classes*. Specifically, for integers N, k let

$$R_N(k) = \{n \in \mathbb{Z} \mid n \equiv k \ (\mathrm{mod}\, N)\},$$

and then:

Definition. A function f on a set of integers (for example, the set of primes) is said to be PORC if for some integer N, for each residue class $R_N(k)$ there is a polynomial f_k such that, whenever $n \in R_N(k)$ and $f(n)$ is defined, we have $f(n) = f_k(n)$.

We leave it to the reader to check the following fact (recall that the rth cyclotomic polynomial is the polynomial with rational integer coefficients whose roots are the primitive r^{th} roots of 1 in $\mathbb{C}$).

Proposition 21.1 *A function* $f : \mathbb{N} \to \mathbb{C}$ *is* PORC *if and only if the power-series generating function* $1 + \sum f(n)z^n$ *converges for* $|z| < 1$ *to a function* $F(z)$ *which is a rational function whose denominator is a product of cyclotomic polynomials.*

222

The original context for Higman's definition was this:

Conjecture 21.2 (the PORC conjecture) *Let $g_n(p) = f(p^n)$, where p is prime. For fixed n, the function g_n is PORC as a function of p.*

This conjecture is still open. Some evidence has accumulated, however.

(1) For $n = 1, 2, 3, 4$ the function $g_n(p)$ is essentially constant: $g_1(p) = f(p) = 1$, $g_2(p) = f(p^2) = 2$, $g_3(p) = f(p^3) = 5$ and $g_4(p) = f(p^4) = 15$ if $p \geqslant 3$ while $g_4(2) = f(2^4) = 14$. This was proved by Otto Hölder in 1893 (see [47]). The fact that

$$g_5(p) = f(p^5) = 2p + 61 + 2\gcd(p-1, 3) + \gcd(p-1, 4)$$

if $p \geqslant 5$ was proved by G. Bagnera [5] in 1898. Recently the groups of order p^6 have been enumerated by M. F. Newman, E. A. O'Brien and M. R. Vaughan-Lee [78] and those of order p^7 by E. A. O'Brien and M. R. Vaughan-Lee [79]. The reader is referred to these papers for excellent critical reviews of the literature on enumeration of the groups of order p^m for 'small' values of m. In each case the enumeration function turns out to be PORC. Thus g_n is known to be PORC for $n \leqslant 7$.

(2) Higman himself used a small amount of cohomology theory together with theory of algebraic representations of algebraic groups to show in [46] that if $\varphi_n(p)$ denotes the number of isomorphism classes of groups of order p^n and Φ-class 2, then for a fixed value of n, φ_n is PORC as a function of p. In his paper [31] Anton Evseev considerably extends Higman's methods to prove that for a fixed natural number n the number of isomorphism classes of groups of order p^n in which the Frattini subgroup is central is PORC as a function of p.

(3) Marcus du Sautoy has brought the theory of zeta functions of finitely generated torsion-free nilpotent groups to bear on the problem (see [24] and [25]) and has made some progress on certain special classes of groups more general than Higman's groups of Φ-class 2.

Higman has made the following point: even if the PORC conjecture is unattainable it might be possible to prove that there is a PORC function G_n such that $G_n(p) - g_n(p)$ is significantly smaller than $p^{\frac{2}{27}n^3}$.

There are other contexts in which PORC functions arise in group theory. Consider first abelian p-groups G of bounded exponent, say exponent dividing

p^e, where p and e are fixed. If $|G| = p^n$, then we can write G as a direct sum of x_i cyclic summands of order p^i for $i = 1, \ldots, e$ where

$$x_1 + 2x_2 + \cdots + ex_e = n.$$

Two such groups are isomorphic if and only if they have the same number of cyclic summands of order p^i for each i. Thus the number $g(n)$ of such groups of order p^n is the number of solutions to the above equation, that is, the number of partitions of n into parts of size at most e. We have

$$\sum g(n)x^n = \frac{1}{(1-x)(1-x^2)\cdots(1-x^e)}.$$

Expanding the right-hand side in partial fractions we find that it is a function of the form

$$\sum \frac{A_i}{\Psi_i(x)}$$

where the $\Psi_i(x)$ are cyclotomic polynomials. It follows that $g(n)$ is PORC.

Secondly, consider p-groups G with $|G'| = p$ and $p \neq 2$. Then $\Phi(G) \leqslant Z(G)$ and the commutator structure of G may be thought of as a function of two variables $G/\Phi(G) \to C_p$, and hence as an alternating bilinear form on $G/\Phi(G)$ construed as a vector space over $\mathbb{F}_p$. Exploiting this one can calculate that for odd primes p, the number of isomorphism classes of groups of order p^n and exponent dividing p^2 with derived group of order p is independent of p and is the following:

$$
\begin{array}{ll}
\frac{1}{288}n^4 + \frac{7}{144}n^3 + \frac{2}{9}n^2 - \frac{2}{3}n & \text{if } n \equiv 0 \text{ or } 2 \;(\text{mod}\,6), \\[4pt]
\frac{1}{288}n^4 + \frac{7}{144}n^3 + \frac{2}{9}n^2 - \frac{2}{3}n + \frac{1}{9} & \text{if } n \equiv 4 \;(\text{mod}\,6), \\[4pt]
\frac{1}{288}n^4 + \frac{7}{144}n^3 + \frac{2}{9}n^2 - \frac{29}{48}n + \frac{7}{32} & \text{if } n \equiv 3 \text{ or } 5 \;(\text{mod}\,6), \\[4pt]
\frac{1}{288}n^4 + \frac{7}{144}n^3 + \frac{2}{9}n^2 - \frac{29}{48}n + \frac{95}{288} & \text{if } n \equiv 1 \;(\text{mod}\,6).
\end{array}
$$

This result (unpublished, but we have checked it) was announced by Higman in his lecture on Wednesday 6 November 1991 in the series out of which this book has grown. A classification of all groups of order p^n with derived group of order p is given by Simon Blackburn [9].

21.2 Isoclinism classes of p-groups

The notion of isoclinism was introduced by Philip Hall in 1940 to help with the classification of finite p-groups (see [42]). For a group G there is a natural map $\gamma_G : G/Z(G) \times G/Z(G) \to G'$ defined by $\gamma_G : (aZ(G), bZ(G)) \mapsto [a, b]$

where, as usual, $[a, b]$ is the commutator $a^{-1}b^{-1}ab$. Groups G, H are said to be *isoclinic* if they have essentially the same commutator map. To be precise, this means that there exist isomorphisms

$$\phi : G/Z(G) \to H/Z(H) \quad \text{and} \quad \psi : G' \to H'$$

such that

$$\gamma_H \circ (\phi \times \phi) = \psi \circ \gamma_G,$$

in other words, the diagram

$$
\begin{array}{ccc}
G/Z(G) \times G/Z(G) & \overset{\gamma_G}{\longrightarrow} & G' \\
{\scriptstyle \phi \times \phi}\big\downarrow & & \big\downarrow{\scriptstyle \psi} \\
H/Z(H) \times H/Z(H) & \overset{\gamma_H}{\longrightarrow} & H'
\end{array}
$$

commutes. It is easy to see that isoclinism is an equivalence relation on groups. There are two obvious invariants of an isoclinism class $\mathfrak{X}$. They are the groups X, Y such that $G/Z(G) \cong X$ and $G' \cong Y$ for all $G \in \mathfrak{X}$. Then G may be thought of either as a central extension of an abelian group Z by X or as an extension of Y by an abelian group A. Note that X, Y are not independent since $|Y|$ is bounded by a function of $|X|$ and if, for example, X is abelian, then Y must also be abelian and must be a homomorphic image of the exterior square $X \wedge X$. Amongst other things Philip Hall proved that the smallest groups G in an isoclinism class have the property that $Z(G) \leqslant G'$, and he called such groups *stem* groups for the class.

Our interest is in the enumeration function $f_{\mathfrak{X}}(p^m)$ which gives the number of groups of order p^m (up to isomorphism of course) in an isoclinism class $\mathfrak{X}$, when the invariants X, Y of $\mathfrak{X}$ are finite p-groups for some prime number p. Trivially, the abelian groups form a single isoclinism class $\mathfrak{A}$. An immediate consequence of the structure theorem for finite abelian groups, which says that up to isomorphism a finite abelian p-group is uniquely the direct product (or sum) of cyclic groups, is that $f_{\mathfrak{A}}(p^m) = p(m)$, where $p(m)$ is the partition function that enumerates the integer partitions of m (see Theorem 17.2). A famous insight of Ramanujan, as developed by Hardy and Ramanujan jointly, and then by Rademacher (see, for example, [3, Chapter 5] or [4, Chapter 14]; see also Section 17.2 above), yields that

$$p(m) \sim \frac{1}{4m\sqrt{3}} \, e^{K\sqrt{m}} \quad \text{where} \quad K = \sqrt{2\pi/3}.$$

Now if finite p-groups X and Y are given, and $\mathfrak{X}$ is an isoclinism class of which they are the invariants, then we might expect that $f_{\mathfrak{X}}(p^m)$ should, when m is large, depend heavily on the number $p(m - m_0)$ of abelian groups available to serve as centre of a group in $\mathfrak{X}$, where $p^{m_0} = |X|$. That this is so has been made precise and proved by one of us in [8]. There the following theorems are proved.

Theorem 21.3 (Blackburn 1994) *Let $\mathfrak{X}$ be an isoclinism class of groups and let P be one of its stem groups (so that, recall, $Z(P) \leqslant P'$). Suppose that P is a finite p-group for some prime number p, and define non-negative integers a, c by $p^a = |P/P'|$, $p^c = |Z(P)|$. Then there exist positive real numbers k_1, k_2 (depending only on $\mathfrak{X}$) such that*

$$k_1\left(\frac{m}{(\log m)^2}\right)^{(a+c)/2} \leqslant \frac{f_{\mathfrak{X}}(p^m)}{p(m)} \leqslant k_2\, m^{(a+c)/2}.$$

Theorem 21.4 (Blackburn 1994) *If $\mathfrak{X}$ is an isoclinism class of groups then $f_{\mathfrak{X}}(p^m) \leqslant p^{m^2/3} p(m)$. Moreover, given a positive integer m there exists an isoclinism class $\mathfrak{X}$ (depending on m) such that $f_{\mathfrak{X}}(p^m) \geqslant p^{(m^2-9)/8}$.*

The first of these theorems confirms that when m is large the number of groups of order p^m in a fixed isoclinism class is something like $\Phi(m)\mathrm{e}^{K\sqrt{m}}$ where $K = \sqrt{2\,\pi/3}$ and Φ is bounded by a polynomial. Combining the second with the theorems of Higman and Sims (as modified by Newman and Seeley) we see that for large m the groups of order p^m fall into $p^{\frac{2}{27}m^3 + O(m^{5/2})}$ different isoclinism classes. For further insights about the import of the theorems, and for proofs, the reader is referred to the original paper [8].

Hall introduced isoclinism as a tool for the classification of finite p-groups. But the concept makes good sense for groups in general. Isoclinism classes of finite nilpotent groups can easily be understood in terms of isoclinism classes of p-groups. For, if G is a nilpotent group of order $p_1^{\alpha_1} p_2^{\alpha_2} \cdots p_k^{\alpha_k}$, where $p_1, p_2, \ldots, p_k$ are distinct prime numbers, then the isoclinism class of G consists of the nilpotent groups $P_1 \times P_2 \times \cdots \times P_k$, where P_i lies in the isoclinism class of the Sylow p_i-subgroup of G. By way of contrast, some isoclinism classes are essentially 'trivial'. For example, if G is a perfect group with trivial centre and trivial Schur multiplier, such as the Matthieu group M_{11} (see [19], for example), then the isoclinism class of G consists of groups of the form $G \times A$ where A is abelian. It seems a little unlikely that isoclinism would tell us much of interest about enumeration of finite groups in general, but so far as we know this question has not been investigated.

21.3 Groups of square-free order

It was proved by G. Frobenius in 1893 (see [34]) that groups of square-free order are soluble, indeed, that they are Sylow tower groups in the following strong sense:

Definition. A finite group X will be said to be a *Sylow tower group* if a Sylow q-subgroup Q for the largest prime divisor q of X is normal in X and X/Q is a Sylow tower group. To complete this recursive definition of course we need to specify that the trivial group $\{1\}$ is a Sylow tower group.

Note that some authors use the terminology to mean something a little weaker – their definition does not require q to be the largest prime divisor of $|X|$.

Otto Hölder [48] went a little further and showed that groups of square-free order are metacyclic – that is, they have a cyclic normal subgroup with cyclic factor group. And of course, since the orders are coprime, such a group is, in fact, a split extension of one cyclic group by another. He gave a remarkably explicit formula for the number of such groups:

Proposition 21.5 (Hölder 1895) *If n is square-free then*

$$f(n) = \sum_{m|n} \prod_p \frac{p^{c(p)} - 1}{p - 1},$$

where in the product p ranges over prime divisors of n/m and $c(p)$ denotes the number of primes q dividing m such that $q \equiv 1 \pmod{p}$.

It is not hard to see where this formula comes from. Let G be a group of square-free order n and let $H = F(G)$, the Fitting subgroup. Then G is the split extension of H, which is cyclic, by a cyclic group K of order n/m. Since $C_G(H) = H$, the complement K acts faithfully on H. Since Aut H is abelian, the isomorphism class of the extension is determined simply by the subgroup of Aut H induced by K. That subgroup has order n/m and is determined by its p-primary subgroups for the prime divisors p of n/m. If $m = q_1 \cdots q_r$, where $q_1, \ldots, q_r$ are distinct prime numbers, then Aut $H \cong C_{q_1-1} \times \cdots \times C_{q_r-1}$ and the p-primary subgroup of this has rank $c(p)$, where $c(p)$ is the number of i such that $q_i \equiv 1 \pmod{p}$. Therefore the number of its subgroups of order p is $(p^{c(p)} - 1)/(p - 1)$, and the result follows.

It was proved by Annabelle McIver and Peter M. Neumann [67] that $f(n) \leqslant n^{\mu^2 + \mu + 2}$ (where $\mu(n)$ is the maximum power to which any prime divides n – as defined in Chapter 1), a result which is significantly weaker than Pyber's theorem, though significantly easier to prove. For square-free n this tells us that $f(n) \leqslant n^4$. Much better results are known. The simplest to state is

Theorem 21.6 *If n is square-free then $f(n) \leqslant \phi(n)$ where $\phi(n)$ is Euler's function.*

This is proved by Murty and Murty in [71]. It follows of course that $f(n) \leqslant n$ for all square-free n, and $f(n) \leqslant \frac{1}{2}n$ if n is even. More detailed results are to be found in various papers such as the one just cited, [72], [29] and [73]. In particular, in this last paper Murty and Srinivasan prove (a little more than) the following lovely result:

Theorem 21.7 *There exist real numbers $A, B > 0$ such that*

$$f(n) \leqslant O\left(\frac{n}{(\log n)^{A \log\log\log n}}\right) \quad \text{for all square-free } n$$

and

$$f(n) > \frac{n}{(\log n)^{B \log\log\log n}} \quad \text{for infinitely many square-free } n.$$

Compared with theirs, our statement of this theorem is modified in two small ways. First, their upper bound is proved more generally for the enumeration of groups all of whose Sylow subgroups are cyclic. Second, in the fractions on the right of the inequalities they have Euler's function $\phi(n)$ instead of n in the numerator. This makes no difference since $\phi(n) \leqslant n$ for all n and $\phi(n) \geqslant n/(\log\log n + 2\log\log\log n)$ for large square-free n. They use some quite sophisticated methods of analytic number theory. The much weaker result that $f(n)/n \to 0$ as $n \to \infty$ through square-free integers can be proved combinatorially as follows.

Let $n = p_1 p_2 \cdots p_k$ where $p_1 > p_2 > \cdots > p_k$ and the p_i are prime numbers. Write $[1, k]$ for $\{1, 2, \ldots, k\}$. For $i, j \in [1, k]$ define

$$\varepsilon(i, j) = \begin{cases} 1 & \text{if } p_i \equiv 1 \pmod{p_j}, \\ 0 & \text{otherwise}, \end{cases}$$

and

$$b_i = \prod_{j \in [1,k]} p_j^{\varepsilon(i,j)}, \quad a_i = \frac{p_i - 1}{b_i},$$

so that $p_i = a_i b_i + 1$. Also for a subset $I \subseteq [1, k]$ define $I' = [1, k] \setminus I$, define $c_j(I) = \sum_{i \in I} \varepsilon(i, j)$ (so that $c_j(I)$ is the number of members i of I such that $p_i \equiv 1 \pmod{p_j}$) and define

$$f_I = f_I(n) = \prod_{j \in I'} \frac{p_j^{c_j(I)} - 1}{p_j - 1}.$$

In this notation Hölder's theorem states that $f(n) = \sum_{I \subseteq [1,k]} f_I$.

Lemma 21.8 *With the notation and the assumptions of the above paragraph,*

$$f_I \leqslant \frac{n}{\prod_{i \in I} a_i \prod_{i \in I} p_i^{c_i(I)} \prod_{j \in I'} p_j (p_j - 1)}.$$

Proof: We have

$$f_I = \prod_{j \in I'} \frac{p_j^{c_j(I)} - 1}{p_j - 1}$$

$$\leqslant \prod_{j \in I'} \frac{p_j^{c_j(I)}}{p_j - 1}$$

$$= \frac{n}{\prod_{i \in I} p_i} \times \prod_{j \in I'} \frac{p_j^{c_j(I)}}{p_j (p_j - 1)}$$

$$\leqslant \frac{n}{\prod_{i \in I} a_i b_i} \times \prod_{j \in I'} \frac{p_j^{c_j(I)}}{p_j (p_j - 1)}.$$

Now

$$\prod_{i \in I} b_i = \prod_{i \in I} \prod_{j \in [1,k]} p_j^{\varepsilon(i,j)} = \prod_{j \in [1,k]} \prod_{i \in I} p_j^{\varepsilon(i,j)}$$

$$= \prod_{j \in [1,k]} p_j^{c_j(I)} = \prod_{i \in I} p_i^{c_i(I)} \times \prod_{j \in I'} p_j^{c_j(I)}.$$

Therefore

$$f_I \leqslant \frac{n}{\prod_{i \in I} a_i \prod_{i \in I} p_i^{c_i(I)} \prod_{j \in I'} p_j (p_j - 1)},$$

as the lemma states.

Proposition 21.9 *Let k be a fixed natural number. Then $f(n)/n \to 0$ as $n \to \infty$ through square-free numbers with at most k prime factors.*

Proof: We use the notation introduced in the paragraph preceding the foregoing lemma. Let $I \subseteq [1, k]$ and suppose that $f_I > n/A$ where A is some (large) positive real number. By the lemma,

$$\prod_{i \in I} a_i \prod_{i \in I} p_i^{c_i(I)} \prod_{j \in I'} p_j(p_j - 1) < A.$$

Therefore $\prod_{j \in I'} p_j < A$ and $a_i < A$ for all $i \in I$. We bound p_i for $i \in I$ as follows. Given that $i \in I$,

$$b_i = \prod_{j \in [1,k]} p_j^{\varepsilon(i,j)} = \prod_{j \in I} p_j^{\varepsilon(i,j)} \prod_{j \in I'} p_j^{\varepsilon(i,j)} \leqslant \prod_{j \in I} p_j^{c_j(I)} \prod_{j \in I'} p_j < A,$$

and so $p_i = a_i b_i + 1 < A^2 + 1$. It follows that $n < (A^2 + 1)^{|I|} A < (A^2 + 1)^{k+1}$ and, given that k is fixed, this shows that n is bounded as a function of A.

Now if $f(n) > \epsilon n$ for some $\epsilon > 0$ then, since Hölder's theorem expresses $f(n)$ as the sum of 2^k summands f_I (some of which are 0 of course), there must exist $I \subseteq [1, k]$ such that $f_I > \epsilon n/2^k$. Taking $A = 2^k/\epsilon$ we see that n is bounded. This proves the proposition.

For use in the next two results we need the following calculation:

Lemma 21.10 $\displaystyle \prod_{\substack{p \geqslant 3, \\ p \text{ prime}}} \left(1 + \frac{2}{p(p-1)}\right) < 2.$

Proof: The product over the odd prime numbers may be compared with $\prod(1 + 1/r(2r+1))$ and this converges since the sum $\sum 1/r(2r+1)$ converges. Let

$$P = \prod_{\substack{p \geqslant 3, \\ p \text{ prime}}} \left(1 + \frac{2}{p(p-1)}\right) \quad \text{and} \quad Q = \prod_{r \geqslant 1} \left(1 + \frac{1}{r(2r+1)}\right).$$

Then $P < Q$ and, since $\log(1 + x) < x$ if $0 < x < 1$,

$$\log Q = \sum_{r \geqslant 1} \log \left(1 + \frac{1}{r(2r+1)}\right) < \sum_{r \geqslant 1} \frac{1}{r(2r+1)}.$$

Giving the first summand special treatment we see that

$$\sum_{r \geqslant 1} \frac{1}{r(2r+1)} < \frac{1}{3} + \frac{1}{2}\left(\sum_{r \geqslant 2} \frac{1}{r^2}\right) = \frac{\pi^2}{12} - \frac{1}{6} < \frac{2}{3}.$$

Therefore $Q < e^{2/3} < 2$ and so $P < 2$ as the lemma states.

Proposition 21.11 *Let n be an odd square-free number with k prime factors. Then* $f(n) \leqslant n/2^{k-1}$.

Proof: Again we use the notation introduced in the paragraph preceding Lemma 21.8. Since n is odd, a_i must be even for all $i \in [1, k]$ and so $a_i \geqslant 2$. It follows from Lemma 21.8 that

$$f_I \leqslant \frac{n}{2^{|I|} \prod\limits_{j \in I'} p_j(p_j - 1)} \leqslant \frac{n}{2^k} \prod_{j \in I'} \frac{2}{p_j(p_j - 1)}.$$

Thus

$$f(n) = \sum_{I \subseteq [1,k]} f_I \leqslant \frac{n}{2^k} \sum_{I \subseteq [1,k]} \prod_{j \in I'} \frac{2}{p_j(p_j - 1)} = \frac{n}{2^k} \prod_{j \in [1,k]} \left(1 + \frac{2}{p_j(p_j - 1)}\right).$$

The result now follows from Lemma 21.10.

This simple argument does not work for even n, but a slightly more sophisticated approach proves the following.

Proposition 21.12 *Let n be a square-free number with k prime factors. If k is sufficiently large then* $f(n) \leqslant \dfrac{\log k}{k} n$.

Proof: Note that, although it is not a matter of great importance, here, as elsewhere in the book, we are using $\log k$ to mean $\log_2 k$. We continue with the same notation and seek to prove that if k is large enough then for any subset I of $[1, k]$ we have

$$(\star) \qquad f_I \leqslant n \frac{\log k}{4k} \prod_{j \in I'} \frac{2}{p_j(p_j - 1)},$$

for then Lemma 21.10 adjusted to include a factor for $p = 2$ will deliver the result that

$$f(n) < n \frac{\log k}{4k} \sum_{I \subseteq J} \prod_{j \in I'} \frac{2}{p_j(p_j - 1)} = n \frac{\log k}{4k} \prod_{j \in [1,k]} \left(1 + \frac{2}{p_j(p_j - 1)}\right) < \frac{\log k}{k} n.$$

Now by Lemma 21.8

$$f_I \leqslant \frac{n}{2^{|I'|} \prod\limits_{i \in I} a_i \prod\limits_{i \in I} p_i^{c_i(I)} \prod\limits_{j \in I'} \frac{1}{2} p_j(p_j - 1)}.$$

Define

$$I_0 = \{i \in I \mid c_i(I) \neq 0\}, \quad I_1 = I \setminus I_0, \quad J = [1, k] \setminus I_1.$$

If $2^{|J|} \geqslant 4k/\log k$ then $2^{|I'|} \prod_{i \in I} p_i^{c_i(I)} \geqslant 4k/\log k$ and the desired inequality $(\star)$ certainly holds.

We may now suppose that $2^{|J|} < 4k/\log k$. In this case $|I_1| = k - |J| > k - (2 + \log k - \log\log k)$, so, given that k is large ($k \geqslant 16$ certainly suffices), $|I_1| > \frac{1}{2}k$. For $i \in [1, k]$ define $J_i = \{j \in [1, k] \mid \varepsilon(i, j) \neq 0\}$, so that $b_i = \prod_{j \in J_i} p_j$. We show that if $i \in I_1$ then $J_i \subseteq J$. For, let $i \in I_1$ and $j \in J_i$: if $j \in I'$ then certainly $j \in J$; otherwise $j \in I$ and then, since $\varepsilon(i, j) \neq 0$, $c_j(I) \neq 0$ and so $j \in I_0$. Thus $J_i \subseteq I' \cup I_0 = J$. Now J has $2^{|J|}$ subsets and so there must exist distinct indices $i_1, i_2, \ldots, i_r \in I_1$ such that $J_{i_1} = J_{i_2} = \cdots = J_{i_r}$ and $r \geqslant |I_1|/2^{|J|} > \frac{1}{8}\log k$. Then, since $b_{i_1} = b_{i_2} = \cdots = b_{i_r}$, the numbers a_{i_1}, a_{i_2}, $\ldots, a_{i_r}$ must all be different. It follows that $\prod_{i \in I} a_i \geqslant r!$. It is easy to see that $r! > 2^{8r}$ when $r \geqslant 2^{32}$ and so $r! > k$ when $\log k \geqslant 2^{35}$. Thus, given that k is large, $\prod_{i \in I} a_i > k > 4k/\log k$, so again the inequality $(\star)$ holds. This proves the proposition.

Putting Propositions 21.9 and 21.12 together we have the promised fact (which, we stress, is much weaker than Theorem 21.7)

Theorem 21.13 *For square-free* n, $\dfrac{f(n)}{n} \to 0$ *as* $n \to \infty$.

Our proof does not give much insight into what a good upper bound for $f(n)/n$ might be for square-free n: it would be useful to have a combinatorial proof of the theorem of M. Ram Murty and S. Srinivasan in case it could be adapted to other situations—in particular to an analysis of $f(n)$ for cube-free n (see below). Their theorem has as a consequence that there does not exist $\gamma < 1$ such that $f(n) < O(n^\gamma)$ for all square-free n. This was proved analytically by Erdős, Murty and Murty in [29], who show that if $n = q_1 q_2 \cdots q_k$, where $q_1, q_2, \ldots, q_k$ are the first k prime numbers, then $\log f(n)/\log n \to 1$ as $k \to \infty$. The following simple heuristic argument, although not a proof, should give some insight into why the result is true. Given $\gamma < 1$ choose r such that $(r-1)/(r+1) > \frac{1}{2}(\gamma+1)$. Let p_0 be a prime number $> r$. Choose $k_1, k_2, \ldots, k_r$ such that each of $k_i p_0 + 1$ is prime for $1 \leqslant i \leqslant r$. This is possible by Dirichlet's theorem that the arithmetic progression $(kp_0 + 1)_{k=1,2,\ldots}$ contains infinitely many primes. There is then no obvious reason why there should not exist arbitrarily large prime numbers p such that $k_i p + 1$ is prime for $1 \leqslant i \leqslant r$. So let us assume this to be true and let p be a very large prime with this property. Define

$$m = (k_1 p + 1)(k_2 p + 1) \cdots (k_r p + 1) \quad \text{and} \quad n = pm.$$

Consider groups G constructed as semidirect product of a cyclic normal subgroup H of order m by a group of order p. Now $\operatorname{Aut} H$ contains an elementary abelian subgroup of order p^r, and this has more than p^{r-1} subgroups of order p. It follows that $f(n) > p^{r-1}$. On the other hand, $n < K p^{r+1}$, where $K = \prod(k_i + 1)$. Therefore $f(n) > (n/K)^{(r-1)/(r+1)} > B n^\gamma n^{(1-\gamma)/2}$, where $B = K^{-(r-1)/(r+1)} > 0$. So for any $A > 0$, if n is sufficiently large, that is if p is sufficiently large, we have $f(n) > An^\gamma$.

Here is an example: if $p = 5231$, then $2p+1 = 10\,463$, $6p+1 = 31\,387$, $8p+1 = 41\,849$, $12p+1 = 62\,773$, and all these are prime. Let

$$n = p(2p+1)(6p+1)(8p+1)(12p+1)$$
$$= 4\,512\,827\,385\,263\,473\,497\,647 < 5 \times 10^{21}.$$

Then

$$f(n) = 1 + (p^4 - 1)/(p - 1) = 143\,165\,109\,985 > 1.42 \times 10^{11}.$$

Thus $f(n)^2 > 2 \times 10^{22} > 4n$, so $f(n) > 2n^{1/2}$. Of course there are likely to be much smaller square-free values of n for which $f(n) > 2n^{1/2}$, but the example still has value as illustrating the method.

21.4 Groups of cube-free order

The result of McIver and Neumann quoted on p. 227 above tells us that if n is cube-free then $f(n) < n^8$. Again, this is likely not to be a realistic upper bound and we conjecture (see below) that $f(n) = o(n^2)$ for cube-free n. The situation is different from the square-free case insofar as groups of cube-free order need not be soluble, and, even if soluble, need not be Sylow tower groups in our (strong) sense. Nevertheless, their structure is very restricted.

Proposition 21.14 *Let G be a group of order n, where n is cube-free, and let $S(G)$ be its soluble radical (the largest soluble normal subgroup) and $O(G)$ the largest normal subgroup of odd order. Then*

(1) *$O(G)$ is a Sylow tower group.*
(2) *Either $S(G)/O(G) \cong \operatorname{Alt}(4)$ or $S(G)$ is a Sylow tower group.*
(3) *If G is not soluble then $G/S(G) \cong \operatorname{PSL}(2, p)$ for some prime number $p \equiv \pm 3 \pmod 8$.*

Proof: For (1) we need to show that if n is odd then G is a Sylow tower group. Let p be the smallest prime divisor of n and let P be a Sylow p-subgroup. If P is cyclic of order p^a then $|\operatorname{Aut} P| = p^{a-1}(p-1)$, and if P

is elementary abelian of order p^2 then $|\operatorname{Aut} P| = p(p-1)^2(p+1)$. In either case, $|\operatorname{Aut} P|$ is not divisible by any prime larger than p, and therefore P lies in the centre of its normaliser $N_G(P)$. By Burnside's transfer theorem G has a normal p-complement H. As inductive assumption we may assume that H is a Sylow tower group and it follows that so is G.

For (2) suppose that G is soluble and define $X = S(G)/O(G)$. If X is a 2-group then G is a Sylow tower group by (1). So suppose that X is not a 2-group. Let $Y = F(X)$, the Fitting subgroup of X. Since X has no non-trivial normal subgroup of odd order, Y is a 2-group. We know that Y is abelian (it has order $\leqslant 4$) and that $C_X(Y) = Y$. It follows that $X/Y \leqslant \operatorname{Aut} Y$. Since $|X|$ is not divisible by 8 and is not a power of 2, the only possibility is that $X \cong \operatorname{Alt}(4)$.

For (3) suppose that G is not soluble and let $X = G/S(G)$. Let Y be a minimal non-trivial normal subgroup of X. We know then that $Y = Y_1 \times \cdots \times Y_k$ where $Y_1, \ldots, Y_k$ are isomorphic simple groups. By (1) they must all be of even order; and indeed, their order must be divisible by 4 since a group of twice odd order has a normal subgroup of index 2. Since 4 is the highest power of 2 that can divide $|X|$, we must have $k = 1$ so that Y is simple. The simple groups with Sylow 2-subgroup of order 4 were classified by Gorenstein and Walter (see [37]). Only groups $\operatorname{PSL}(2, q)$ where q is a prime power and $q \equiv \pm 3 \pmod 8$ occur. Suppose that $q = p^e$. Then $e \leqslant 2$ since q divides the group order, which is cube-free. And if p is odd and $e = 2$ then $q \equiv 1 \pmod 8$. Thus we must have that either $q = p = $ a prime number $\equiv \pm 3 \pmod 8$ or $q = 4$. But $\operatorname{PSL}(2, 4) \cong \operatorname{PSL}(2, 5)$. Finally, since $C_X(Y) \lhd X$, and Y must be the only minimal non-trivial normal subgroup of X, we have $Y \lhd X \leqslant \operatorname{Aut} Y = \operatorname{PGL}(2, p)$. But $|\operatorname{PGL}(2, p) : \operatorname{PSL}(2, p)| = 2$ so since $|X|$ is cube-free, $X \cong \operatorname{PSL}(2, p)$.

Proposition 21.15 *Let G be a group of order n, where n is cube-free, let $S(G)$ be its soluble radical, $R(G)$ its soluble residual (the smallest normal subgroup with soluble factor group) and, for a prime number p, let $O_p(G)$ be its largest normal p-subgroup. Then*

(1) *$G/C_G(O_p(G))$ has an abelian subgroup of index $\leqslant 2$.*
(2) *If n is odd or if $G/O(G) \cong \operatorname{Alt}(4)$ then $G' \leqslant F(G)$.*
(3) *$G = R(G) \times S(G)$.*

Proof: For (1) let $M = O_p(G)$ for some prime number p and let $A = G/C_G(M)$. Then $M \cong C_p$, $M \cong C_{p^2}$ or $M \cong C_p \times C_p$, so $\operatorname{Aut} M$ is either cyclic or isomorphic to $\operatorname{GL}(2, p)$. Now $A \leqslant \operatorname{Aut} M$ and p does not divide $|A|$. Subgroups of $\operatorname{GL}(2, p)$ whose order is odd and not divisible by p are abelian. Therefore if A has odd order then it is abelian. If $|A| = 2m$ where m is odd then A has a normal subgroup of order m and this is abelian. Suppose that $|A| = 4m$, where m is odd. Consider first the possibility that the Sylow 2-subgroups of A are cyclic, generated by an element a, say. In this case A has a normal subgroup B of order m and, since m is odd, this is abelian. If B is cyclic and b is a generator then, construed as a 2×2 matrix, b has at most two eigenvalues (in $\mathbb{F}_p$ or in $\mathbb{F}_{p^2}$) and a must permute its eigenspaces. Therefore a^2 centralises b, and so $\langle a^2, b \rangle$ is cyclic and it is a subgroup of index 2 in A. If B is non-cyclic then it must act reducibly on M and has precisely two one-dimensional invariant subspaces: again, since $\langle a \rangle$ normalises B, a^2 must fix both of these and therefore centralises B, so that $\langle a^2, B \rangle$ is an abelian subgroup of index 2 in A. The second possibility is that a Sylow 2-subgroup T of A is elementary abelian of order 4. In this case $T \cap \mathrm{SL}(2, p)$ must be non-trivial (because the image of T under the determinant map is cyclic). Let $a \in T \cap \mathrm{SL}(2, p)$, $a \neq I_2$ (we use I_2 here to denote the 2×2 identity matrix). Since the eigenvalues of a are ± 1 and $\det a = 1$, the only possibility is $a = -I_2$. Therefore a is central in A and $A/\langle a \rangle$, being of order $2m$, has a normal subgroup $C/\langle a \rangle$ of order m. Then, $|C| = 2m$ and $C = \langle a \rangle \times B$, where B is a subgroup of order m. Being of odd order B must be abelian, and therefore C is an abelian subgroup of index 2 in A. This proves (1).

Now $F(G) = \prod O_p(G)$ and it follows immediately that if n is odd then $G/F(G)$ is abelian, so $G' \leqslant F(G)$. Suppose that $G/O(G) \cong \mathrm{Alt}(4)$. Then G has no subgroups of index 2 and therefore again $G/C_G(O_p(G))$ is abelian for every prime p, so $G' \leqslant F(G)$. This proves (2).

Clause (3) is trivial if $R(G) = \{1\}$. Suppose therefore that $R(G) \neq \{1\}$ and let $K = S(R(G))$. By part (3) of the previous proposition, $R(G)/K \cong \mathrm{PSL}(2, p)$ for some prime number $p \equiv \pm 3 \pmod 8$ (and of course $p > 3$ since $R(G)/K$ is simple). Define $A = K/K'$ and consider the action of $R(G)/K$ on A. If q is a prime dividing $|A|$ and Q is its q-primary constituent then, since $\mathrm{PSL}(2, p)$ contains a subgroup isomorphic to $\mathrm{Alt}(4)$ which cannot act faithfully on a group of order q or q^2, the action of $R(G)/K$ on Q must be trivial. Therefore the action of $R(G)/K$ on A is trivial. Thus $K/K' \leqslant Z(R(G)/K')$. But the Schur multiplier of $\mathrm{PSL}(2, p)$ has order 2 and it follows (since K has odd order) that $R(G)/K' \cong \mathrm{PSL}(2, p) \times K/K'$. Since $R(G)' = R(G)$ we must have $K/K' = \{1\}$ and so $K = \{1\}$. Then $R(G) \cong \mathrm{PSL}(2, p)$, and since $R(G) \cap S(G) = \{1\}$ we have $G = R(G) \times S(G)$ by Proposition 21.14 (3). This completes the proof.

No formula quite as simple as Hölder's can exist for $f(n)$ when n is cube-free. Nevertheless, since up to conjugacy the number of faithful actions of an abelian group K of odd order n/m on an abelian group H of odd cube-free order m is not very large, and since $O(G)$ is a large part of G for a group G of odd cube-free order n, it should be possible to get good estimates. In particular, it should be possible to settle the following conjecture by a modification of the method that we used to prove Theorem 21.13.

Conjecture 21.16 *If n is cube-free then $f(n) < n^2$. Also, $f(n)/n^2 \to 0$ as $n \to \infty$ through cube-free numbers.*

An algorithm for constructing the groups of cube-free order computationally has been created by Heiko Dietrich and Bettina Eick [22] and exploited by them to tabulate all these groups up to order 50 000.

21.5 Groups of arithmetically small orders

By an *arithmetically small* number we mean a natural number n for which $\lambda(n)$ is small. If $\lambda(n) = 1$ then n is prime and $f(n) = 1$. If $\lambda(n) = 2$ then, as is well known, $f(n) \leqslant 2$. More precisely,

$$f(pq) = \begin{cases} 2 & \text{if } p = q, \\ 2 & \text{if } p < q \text{ and } q \equiv 1 \pmod{p}, \\ 1 & \text{if } p < q \text{ and } q \not\equiv 1 \pmod{p}. \end{cases}$$

The case $\lambda(n) = 3$ was treated by Otto Hölder in a classic paper [47] published in 1893. As is now well known and easy to derive, there are three abelian and two non-abelian groups of order p^3, so $f(p^3) = 5$.

For $n = pq^2$ there are a number of cases. Let G be a group of order pq^2 where $p \neq q$ and let P, Q be a Sylow p-subgroup and a Sylow q-subgroup respectively. Consider first the case where $q \equiv 1 \pmod{p}$. Then $Q \lhd G$ and G is a semidirect product of Q by P. The semidirect product is determined by an action of P on Q and the isomorphism class of G depends only on the conjugacy class of the image of P in $\mathrm{Aut}\, Q$. If Q is cyclic then there is just one subgroup of $\mathrm{Aut}\, Q$ of order P and so up to isomorphism there are just two possibilities for G, namely the cyclic group C_n and one non-abelian extension of Q by P. If $Q \cong C_q \times C_q$ then $\mathrm{Aut}\, Q \cong \mathrm{GL}(2, q)$ and the number of conjugacy classes of subgroups of order p is 2 if $p = 2$ and $\frac{1}{2}(p+3)$ if p is odd. To see this let ρ be a primitive p^{th} root of 1 modulo q. Any subgroup of order p is conjugate to the subgroup P_k generated by a diagonal matrix $\mathrm{Diag}\,(\rho, \rho^k)$ with eigenvalues ρ, ρ^k for some k. If $k \not\equiv 0 \pmod{p}$ then there exists k' such that $kk' \equiv 1 \pmod{p}$. Then P_k contains $\mathrm{Diag}\,(\rho^{k'}, \rho)$, which is conjugate to $\mathrm{Diag}(\rho, \rho^{k'})$, so that P_k and $P_{k'}$ are conjugate. These are the only conjugacies between the groups P_k, however, and so P_0, P_1, P_{-1} and groups P_k, one for each pair $\{k, k'\}$ of elements of $\mathbb{Z}_p$ such that $kk' = 1$ and $k \neq k'$ in $\mathbb{Z}_p$, form a complete set of representatives of the conjugacy classes of subgroups of order p in $\mathrm{GL}(2, p)$. Note that $P_1 = P_{-1}$ if $p = 2$. Thus we have two non-abelian groups of order $p q^2$ with elementary abelian Sylow q-subgroup if $p = 2$ and $3 + \frac{1}{2}(p - 3)$, that is $\frac{1}{2}(p + 3)$, such groups if p is odd. Adding in the abelian group $C_{pq} \times C_q$ and putting together the numbers of groups with cyclic and non-cyclic Sylow q-subgroups we find that if $q \equiv 1 \pmod{p}$ then $f(p q^2) = 5$ if $p = 2$ while $f(p q^2) = \frac{1}{2}(p + 9)$ if p is odd.

The second case is that in which $p \equiv 1 \pmod{q^2}$. In this case $P \lhd G$ and so G is a semidirect product of C_p by a group of order q^2. If Q is elementary abelian then there are just two possibilities – either it acts trivially on P or it acts non-trivially. If $Q \cong C_{q^2}$ then there are three possibilities corresponding to whether the group of automorphisms induced by Q on P has order 1, q or q^2. Thus in this case $f(n) = 5$.

Thirdly, it is possible that $p \not\equiv 1 \pmod{q^2}$ but $p \equiv 1 \pmod{q}$. If $p = 3$ then $q = 2$ and $f(n) = f(12) = 5$. Otherwise, if $p > 3$ this is very similar to the second case except that if Q is cyclic then it cannot act faithfully on P, so we find that $f(n) = 4$.

The fourth case is that in which $q \equiv -1 \pmod{p}$. For $p = 2$ this comes under the first case and for $q = 2$ (in which case $p = 3$) it comes under the third, so suppose now that both p and q are odd. Then $Q \lhd G$ and we treat G as a split extension of Q by P. If Q is cyclic then $|\mathrm{Aut}\, Q| = q(q - 1)$ which is not divisible by p and so the only group that arises is C_n. If Q is elementary abelian then $\mathrm{Aut}\, Q$, which is $\mathrm{GL}(2, q)$, has one conjugacy class of subgroups of order p and so there are two possibilities for G, the abelian group $C_{pq} \times C_q$ and a Frobenius group. Thus in this case $f(n) = 3$.

Finally, there is the possibility that p and q are arithmetically independent in the sense that none of the congruences $q \equiv \pm 1 \pmod{p}$, $p \equiv 1 \pmod{q}$ hold. In this case G must be abelian and so $f(n) = 2$. Thus

Proposition 21.17 (Hölder 1893) *If p, q are distinct primes then*

$$
f(p q^2) = \begin{cases}
5 & \text{if } p = 2 \text{ or if } p = 3, q = 2, \\
\frac{1}{2}(p + 9) & \text{if } p \text{ is odd and } q \equiv 1 \pmod{p}, \\
5 & \text{if } p \equiv 1 \pmod{q^2}, \\
4 & \text{if } p > 3, \ p \equiv 1 \pmod{q} \text{ but } p \not\equiv 1 \pmod{q^2}, \\
3 & \text{if } p > 2, \ q > 3 \text{ and } q \equiv -1 \pmod{p}, \\
2 & \text{in all other cases.}
\end{cases}
$$

Hölder analyses the case $n = pqr$ where $p > q > r$ in a similar way. For us it is easier to let $H = F(G)$, the Fitting subgroup of G and use the facts first that $C_G(H) = H$, second that G is a split extension of H by a cyclic group isomorphic to G/H and third that if P, Q, R are a Sylow p-subgroup, a Sylow q-subgroup and a Sylow r-subgroup respectively, then $R \leqslant H$. Thus there are four possibilities: that $H = R$, that $H = Q \times R$, that $H = P \times R$ and that $H = G$. This leads easily to the following result:

Proposition 21.18 (Hölder 1893) *If p, q, r are prime and $p < q < r$ then*

$$
f(pqr) = \begin{cases}
p+4 & \text{if } r \equiv 1 \pmod{p}, \ r \equiv 1 \pmod{q} \text{ and } q \equiv 1 \pmod{p}, \\
p+2 & \text{if } r \equiv 1 \pmod{p}, \ r \not\equiv 1 \pmod{q} \text{ and } q \equiv 1 \pmod{p}, \\
4 & \text{if } r \equiv 1 \pmod{p}, \ r \equiv 1 \pmod{q} \text{ and } q \not\equiv 1 \pmod{p}, \\
2 & \text{if } r \equiv 1 \pmod{p}, \ r \not\equiv 1 \pmod{q} \text{ and } q \not\equiv 1 \pmod{p}, \\
3 & \text{if } r \not\equiv 1 \pmod{p}, \ r \equiv 1 \pmod{q} \text{ and } q \equiv 1 \pmod{p}, \\
2 & \text{if } r \not\equiv 1 \pmod{p}, \ r \not\equiv 1 \pmod{q} \text{ and } q \equiv 1 \pmod{p}, \\
2 & \text{if } r \not\equiv 1 \pmod{p}, \ r \equiv 1 \pmod{q} \text{ and } q \not\equiv 1 \pmod{p}, \\
1 & \text{if } r \not\equiv 1 \pmod{p}, \ r \not\equiv 1 \pmod{q} \text{ and } q \not\equiv 1 \pmod{p}.
\end{cases}
$$

Note that these eight cases cover all possibilities.

21.6 Surjectivity of the enumeration function

An entertaining but mildly eccentric question which has been raised by a number of people at different times (for example, there are faint intimations of it in some short papers by G. A. Miller early in the 1930s, and it was asked of the second author by Des MacHale orally some time in the late 1970s and again in writing in January 1991) is whether the enumeration function f is surjective. Thus the question is whether for every $m \geqslant 1$ there exists n such that $f(n) = m$.

Definition. Powers p^α, q^β of distinct prime numbers p, q will be said to be *arithmetically independent* if $p^\mu \not\equiv 1 \pmod{q}$ for $1 \leqslant \mu \leqslant \alpha$ and $q^\nu \not\equiv 1 \pmod{p}$ for $1 \leqslant \nu \leqslant \beta$. Let

$$
n_1 = p_1^{\alpha_1} \cdots p_r^{\alpha_r}, \quad n_2 = q_1^{\beta_1} \cdots q_s^{\beta_s},
$$

where $p_1, \ldots, p_r$ are distinct prime numbers, as are $q_1, \ldots, q_s$. Then n_1, n_2 will be said to be *arithmetically independent* if $\gcd(n_1, n_2) = 1$ and $p_i^{\alpha_i}$, $q_j^{\beta_j}$ are arithmetically independent for all i, j.

Lemma 21.19 *Let G be a group of order n, where $n = n_1 n_2$ and n_1, n_2 are arithmetically independent. Then $G = G_1 \times G_2$ where $|G_1| = n_1$ and $|G_2| = n_2$.*

Proof: We may certainly suppose that $n_1 > 1$ and $n_2 > 1$. Write

$$n_1 = p_1^{\alpha_1} \cdots p_r^{\alpha_r}, \quad n_2 = q_1^{\beta_1} \cdots q_s^{\beta_s},$$

where $p_1, \ldots, p_r, q_1, \ldots, q_s$ are distinct prime numbers. Consider first the case where $s = 1$ and let $q = q_1$. Since for $1 \leqslant i \leqslant r$ and $1 \leqslant \beta \leqslant \beta_1$ the number $q^\beta - 1$ is not divisible by p_i, by the Frobenius Transfer Theorem (see, for example, [36, Section 7.4, Theorem 4.5]) G is q-nilpotent, that is, it has a normal subgroup G_1 of order n_1. Let Q be a Sylow q-subgroup. For $1 \leqslant i \leqslant r$ the number of Sylow p_i-subgroups of G_1 is not divisible by q and therefore there is at least one such subgroup P_i that is normalised by Q. Now since $|P_i|$ and $|Q|$ are arithmetically independent by Sylow's theorem both P_i and Q are normal in $P_i Q$ and therefore Q centralises P_i. It follows that Q centralises G_1 and so $G = G_1 \times Q$. Thus the desired result is true when $s = 1$.

Now suppose that $s > 1$. Since n_1, n_2 are arithmetically independent they must both be odd. By the Feit–Thompson Theorem G is soluble. Let $P_1, \ldots, P_r$, $Q_1, \ldots, Q_s$ be a Sylow system (see p. 48) of G, where $|P_i| = p_i^{\alpha_i}$ and $|Q_j| = q_j^{\beta_j}$. Define $G_1 = P_1 \cdots P_r$ and $G_2 = Q_1 \cdots Q_s$. Then certainly $|G_1| = n_1$ and $|G_2| = n_2$ and what is left to prove is that G_1, G_2 centralise each other. Define $M_j = G_1 Q_j$, so that $M_j \leqslant G$ and $|M_j| = n_1 q_j^{\beta_j}$. Applying what has already been proved in the case $s = 1$ to M_j we see that Q_j centralises G_1. Since this is true for $1 \leqslant j \leqslant s$ we see that G_2 centralises G_1, as required. Thus $G = G_1 \times G_2$.

Corollary 21.20 *If n_1, n_2 are arithmetically independent then $f(n_1 n_2) = f(n_1) f(n_2)$.*

We focus for a while on values $f(n)$ for square-free n. The following ideas were developed independently by R. Keith Dennis and two of the authors (SRB and PMN) early in the 1990s—both unpublished. Let Γ be a rooted directed tree with vertex set $[1, k]$ (which, recall, is the set $\{1, 2, \ldots, k\}$), with root 1, and such that if $(i, j) \in \Gamma$ (that is, (i, j) is an edge) then $i < j$. Consider prime numbers $p_1, p_2, \ldots, p_k$ satisfying the conditions

$$C_\Gamma : \begin{cases} p_1 > p_2 > \cdots > p_k > 2, \\ p_i \equiv 1 \pmod{p_j} \text{ if } (i, j) \in \Gamma, \\ p_i \not\equiv 1 \pmod{p_j} \text{ if } (i, j) \notin \Gamma. \end{cases}$$

We make two points about this.

Lemma 21.21 *Let Γ be a rooted directed tree with vertex set $[1, k]$ as above and let $n_1, X \in \mathbb{N}$ with n_1 odd and $X > n_1$. Then there exist k prime numbers satisfying conditions C_Γ and with $p_k > X$ and $p_i \equiv 2 \pmod{n_1}$ for all i.*

Proof: For notational convenience define $\varepsilon(i, j)$ to be 1 or 0 according as (i, j) is an edge of Γ or not. By Dirichlet's theorem on primes in arithmetic progressions, since n_1 is odd there exists a prime p_k such that $p_k > X$ and $p_k \equiv 2 \pmod{n_1}$. Suppose that $1 \leqslant r < k$ and primes $p_{r+1}, \ldots, p_k$, all congruent to $2 \pmod{n_1}$, have been chosen to satisfy those conditions of C_Γ which refer to them. Define

$$x = \prod_{r < j \leqslant k} p_j^{\varepsilon(r,j)}, \quad y = n_1 \times \prod_{r < j \leqslant k} p_j^{1-\varepsilon(r,j)}.$$

Then x and y are coprime and so the Chinese Remainder Theorem may be used to find $z \in \mathbb{N}$ such that $z \equiv 1 \pmod{x}$ and $z \equiv 2 \pmod{y}$. Then since n_1 is odd xy and z must be coprime and by Dirichlet's theorem there exists a prime number $p \equiv z \pmod{xy}$. Then of course $p \equiv 1 \pmod{x}$ and $p \equiv 2 \pmod{y}$, and so we take $p_r = p$. Ultimately this process gives the desired primes $p_1, p_2, \ldots, p_k$.

The second point we wish to make about the rooted directed trees Γ introduced above is that if $n = p_1 \cdots p_k$ where the prime numbers $p_1, \ldots, p_k$ satisfy the conditions C_Γ then, because each vertex is the terminal vertex of at most one edge, the number of groups of order n depends on Γ but not on n, that is to say, not on the particular choice of the set of primes satisfying C_Γ. To prove and exploit this it is convenient to have some *ad hoc* terminology:

Definition. Let Γ be a rooted directed tree with vertex set $[1, k]$. A subgraph Δ of Γ that contains all the vertices and has no directed paths of length 2 will be called a *log-pile* obtained by cutting up the tree Γ. Define $\lambda(\Gamma)$ to be the number of log-piles obtained by cutting up Γ.

Proposition 21.22 *Let Γ be a rooted directed tree with vertex set $[1, k]$, and let $n = p_1 p_2 \cdots p_k$ where $p_1, p_2, \ldots, p_k$ are primes satisfying the conditions C_Γ. Then*

$$f(n) = \lambda(\Gamma).$$

Proof: Let G be a group of order n and for $i \in [1, k]$ let P_i be a Sylow p_i-subgroup of G. We define a graph $\Delta(G)$ with vertex set $[1, k]$ by specifying that (i, j) is an edge if $i < j$ and P_i is normalised but not centralised by a conjugate of P_j. For this to be possible it is necessary that $p_i \equiv 1 \pmod{p_j}$ and so $\Delta(G) \subseteq \Gamma$. In fact, $\Delta(G)$ is a log-pile. For, if $(i, j) \in \Delta(G)$ and P_i is a Sylow p_i-subgroup normalised but not centralised by a Sylow p_j-subgroup P_j, then we know that P_i must lie in G'. Since G is the split extension of its Fitting subgroup H by a cyclic subgroup K of coprime order, $G' \leqslant H$ and so $P_i \leqslant H$. Since H is cyclic, $P_j \not\leqslant H$ and so $P_j \not\leqslant G'$, whence P_j lies in the centre of its normaliser. Thus if (i, j) is an edge of $\Delta(G)$ then there are no directed edges originating in j, that is $\Delta(G)$ is a log-pile.

Conversely, given a log-pile $\Delta \subseteq \Gamma$ there is a group G, unique up to isomorphism, such that $\Delta(G) = \Delta$. For, define I to be the set of those elements of $[1, k]$ which are either isolated (belong to no edge) or initial vertices of edges of Δ, and define J to be the set of terminal vertices of edges of Δ. Clearly $I \cap J = \emptyset$ and the condition that Δ has no paths of length 2 ensures that $I \cup J = [1, k]$. Define $a = \prod_{i \in I} p_i$ and $b = \prod_{j \in J} p_j$. Up to equivalence there is a unique action of the cyclic group C_b on the cyclic group C_a in which the Sylow p_j-subgroup of C_b acts non-trivially on the Sylow p_i-subgroup of C_a if and only if $(i, j) \in \Delta$. If G is the corresponding semidirect product of C_a by C_b then $\Delta(G) = \Delta$.

It follows immediately that $f(n) = \lambda(\Gamma)$, as stated.

Now log-pile numbers can easily be computed recursively. For Γ with k vertices as above define $\mu(\Gamma)$ to be the number of log-piles obtained by cutting up Γ in such a way that the root becomes isolated. Now define

$$S = \{(k, a, b) \mid \exists \Gamma, \text{a rooted tree on } [1, k] \mid a = \lambda(\Gamma), b = \mu(\Gamma)\}.$$

Proposition 21.23 *The set S can be generated recursively by the rules*

(1) $(1, 1, 1) \in S$;
(2) $(k, a, b) \in S \Rightarrow (k+1, a+b, a) \in S$;
(3) $(k_1, a_1, b_1), (k_2, a_2, b_2) \in S \Rightarrow (k_1 + k_2 - 1, a_1 a_2, b_1 b_2) \in S$.

Rule (1) is clear. Rule (2) is the fact that a rooted tree of size $k+1$ can be created from one of size k by adding a new root below the old one. And rule (3) comes from creating a new tree Γ from two rooted trees Γ_1 and Γ_2 by amalgamating their roots. Conversely, if Γ is any non-trivial rooted tree then either there is a unique edge upwards from the root, in which case Γ is obtained from a rooted tree with one fewer vertex by adjoining a new root, or, if there are several edges emanating from the root then Γ is obtainable (usually in many ways) by amalgamation.

Now define $A = \{m \in \mathbb{N} \mid \exists n \in \mathbb{N} : n \text{ is } \underline{\text{odd}} \text{ and } f(n) = m\}$. It seems very likely that $A = \mathbb{N}$. As an immediate consequence of Corollary 21.20, Lemma 21.21 and Proposition 21.22, we note the following:

Proposition 21.24 *If $m \in A$ and a is a log-pile number then $am \in A$.*

In view of this, it is of particular interest to know which prime numbers are log-pile numbers. Let E be the set of natural numbers which do not occur as log-pile numbers. R. Keith Dennis has communicated to us unpublished computations (announced in a lecture he gave at Groups St Andrews in Galway, August 1993) which strongly suggest that E is finite, that it contains just 508 members, 233 of which are prime, and in fact

$$E = \{7, 11, 19, 29, 31, 47, 49, 53, 67, 71, 73, 79, 87, 91, \ldots, 55\,487\}.$$

In fact, his list of 508 members of E is complete up to 50 000 000, and for every $m \in E$ he has found square-free integers n such that $f(n) = m$. For example, by Propositions 21.17 and 21.18 used together with Dirichlet's theorem on primes in arithmetic progressions, every number of the form $\frac{1}{2}(p+9)$, $p+2$ or $p+4$ where p is prime occurs infinitely often as a value of $f(n)$ on square-free integers n, and this already deals with the numbers in E shown above except for 29, 67, 55 487. Thus he has pretty convincing evidence that the enumeration function f is surjective even when restricted just to odd square-free integers.

The machinery described above can be developed to deal with groups of cube-free order. To do this we consider *weighted* rooted trees on the vertex set $[1, k]$. Such an object is a directed tree Γ in which, as for ordinary rooted trees, if (i, j) is an edge then $i < j$, but in which each vertex i has a weight $\alpha(i) \in \{1, 2\}$, corresponding to the fact that there are now two possibilities for the structure of the Sylow p_i-subgroups. Given Γ we consider primes $p_1, \ldots, p_k$ satisfying:

$$C_\Gamma : \begin{cases} p_1 > p_2 > \cdots > p_k > 2; \\ (i, j) \in \Gamma, \ \alpha(i) = 1 \ \Rightarrow \ p_i \equiv 1 \pmod{p_j}, \ p_i \not\equiv 1 \pmod{p_j^2}; \\ (i, j) \in \Gamma, \ \alpha(i) = 2 \ \Rightarrow \ p_i \equiv -1 \pmod{p_j}, \ p_i \not\equiv -1 \pmod{p_j^2}; \\ (i, j) \notin \Gamma \ \Rightarrow \ \mathrm{ord}(p_i) \pmod{p_j} > 2. \end{cases}$$

The analogue of Lemma 21.21 may be proved in a similar way. Then we consider the cube-free integer

$$n = p_1^{\alpha(1)} \, p_2^{\alpha(2)} \cdots p_k^{\alpha(k)},$$

and groups of order n.

To count these we first assign weights also to the edges of Γ: if (i, j) is an edge then define $w(i, j)$ to be the number of inequivalent ways that groups of order $p_j^{\alpha(j)}$ can act non-trivially on groups of order $p_i^{\alpha(i)}$, that is, the number of non-isomorphic semidirect products (other than the direct product) of a group of order $p_i^{\alpha(i)}$ by a group of order $p_j^{\alpha(j)}$. Given that the arithmetical conditions C_Γ hold, this means in fact that $w(i, j) = \alpha(j)$. Now as in the square-free case we define a log-pile derived from Γ to be a graph Δ with the same vertex set $[1, k]$, and whose edge set is a subset of the edge set of Γ with the restriction that Δ has no paths (directed of course) of length 2. We assign a weight to the log-pile Δ as follows: define $I(\Delta)$ to be the set of isolated vertices of Δ and $E(\Delta)$ to be the edge set of Δ; then define

$$w(\Delta) = \prod_{i \in I(\Delta)} \alpha(i) \times \prod_{(i,j) \in E(\Delta)} w(i, j).$$

Now the weighted log-pile number of Γ is defined by

$$\lambda_w(\Gamma) = \sum_\Delta w(\Delta),$$

where the sum is over all log-piles obtained by cutting up Γ.

Let n be as above, let G be a group of order n and for $1 \leqslant i \leqslant k$ let P_i be a Sylow p_i-subgroup of G. As in the square-free case we define a graph $\Delta(G)$ with vertex set $[1, k]$ by specifying that (i, j) is an edge if and only if P_i is normalised but not centralised by some conjugate of P_j. In general, given that n is odd, this would imply that one of the following holds:

(a) $\alpha(i) = 1$ and $p_i \equiv 1 \pmod{p_j}$;

(b) $\alpha(i) = 2$, P_i is cyclic and $p_i \equiv 1 \pmod{p_j}$;

(c) $\alpha(i) = 2$, P_i is elementary abelian and $p_i \equiv -1 \pmod{p_j}$;

(d) $\alpha(i) = 2$, P_i is elementary abelian and $p_i \equiv 1 \pmod{p_j}$.

The third clause of C_Γ ensures, however, that (b) and (d) do not arise. Proposition 21.15 (2) implies that $\Delta(G)$ has no paths of length 2 because if (i, j) is an edge then a Sylow p_i-subgroup P_i must be contained in G' and a Sylow p_j-subgroup cannot be contained in G' (since it does not centralise P_i) and so j cannot be the initial vertex of any edge. Therefore, $\Delta(G)$ is a log-pile obtained by cutting up Γ. Now we assign labels $\beta(i, j) \in \{0, 1\}$ to the edges (i, j) of the log-pile $\Delta(G)$ as follows:

$$\beta(i, j) = \begin{cases} 0 & \text{if } P_j \text{ is cyclic,} \\ 1 & \text{if } P_j \cong C_{p_j} \times C_{p_j}. \end{cases}$$

Thus groups of order n give rise to edge-labelled log-piles obtained from Γ. These labellings have the property that if (i, j) is an edge and $\alpha(j) = 1$ then $\beta(i, j) = 0$. Conversely, given an edge-labelled log-pile Δ derived from Γ and satisfying this condition, there are precisely $\prod_{i \in I(\Delta)} \alpha(i)$ groups G of order n with $\Delta(G) = \Delta$, where $I(\Delta)$ is the set of isolated vertices of Δ as above. This depends upon the fact that when (i, j) is an edge the possibilities that $p_i \equiv \pm 1 \pmod{p_j^2}$ have been excluded, so that even if the Sylow p_j-subgroup is cyclic of order p_j^2 it can act non-trivially on a Sylow p_i-subgroup in only one way. Now for a given log-pile Δ obtained from Γ the number of groups G such that $\Delta(G) = \Delta$ is $w(\Delta)$: the factor $\prod_{i \in I(\Delta)} \alpha(i)$ in $w(\Delta)$ accounts for the number of possibilities for the centre of G (which is the direct product of the Sylow subgroups on which all other Sylow subgroups act trivially), and the factor $\prod_{(i,j) \in E(\Delta)} w(i, j)$ accounts for the number of edge-labellings, that is, of actions on the non-central Sylow subgroups. Thus $f(n) = \lambda_w(\Gamma)$.

Now let Γ be a weighted rooted directed graph with vertex set $[1, k]$. To go with $\lambda_w(\Gamma)$ we define $\mu_w(\Gamma) := \sum_{\Delta_1} w(\Delta_1)$ where the sum is over all log-piles Δ_1 derived from Γ in which the root (vertex 1) is isolated (these correspond to groups in which the Sylow p_1-subgroup P_1 splits off as a direct factor), and $\tau(\Gamma)$ to be $\alpha(1)$. Then define

$$S_1 = \{(k, a, b, t) \mid \exists \Gamma, \text{a tree on } [1, k] : a = \lambda_w(\Gamma), b = \mu_w(\Gamma) \text{ and } t = \tau(\Gamma)\}.$$

Now S_1 can be constructed from the two quadruples $(1, 1, 1, 1)$ and $(1, 2, 2, 2)$ using a richer set of rules than are available in the square-free case. For,

(1) $(k, a, b, 1) \in S_1 \Rightarrow (k+1, a+b, a, 1) \in S_1$;
(2) $(k, a, b, 1) \in S_1 \Rightarrow (k+1, 2a+b, 2a, 2) \in S_1$;
(3) $(k, a, b, 2) \in S_1 \Rightarrow (k+1, a+b, a, 1) \in S_1$;
(4) $(k, a, b, 2) \in S_1 \Rightarrow (k+1, 2a+b, 2a, 2) \in S_1$;
(5) $(k_1, a_1, b_1, 1), (k_2, a_2, b_2, 1) \in S_1 \Rightarrow (k_1 + k_2 - 1, a_1 a_2, b_1 b_2, 1) \in S_1$;
(6) $(k_1, a_1, b_1, 2), (k_2, a_2, b_2, 2) \in S_1 \Rightarrow (k_1 + k_2 - 1, a^*, \frac{1}{2} b_1 b_2, 2) \in S_1$,
    where $a^* = a_1 a_2 - \frac{1}{2}(a_1 b_2 + a_2 b_1 - b_1 b_2)$.

Note that if $\tau(\Gamma) = 2$ then $\mu_w(\Gamma)$ is automatically even. Also, quadruples $(k, a, b, t) \in S_1$ in which a and b are both even are of no great interest since any quadruples derived from them will have the same property and our ambition should be to find graphs with prime log-pile numbers. As a consequence, the quadruples derived from $(1, 2, 2, 2)$ are hardly worth pursuing.

The rules for generating weighted log-pile numbers are of course sufficient to generate all the ordinary log-pile numbers. They also generate most (perhaps all—we have not yet checked) of the members of the exceptional set E described above.

In this treatment of weighted graphs and labelled log-piles related to cube-free integers we have, up to now, restricted attention to one of four interesting sets of arithmetic conditions on the primes p_i. Let Γ be a weighted rooted directed graph with vertex set $[1, k]$. We have considered primes $p_1, \ldots, p_k$ satisfying

$$C: \quad p_1 > p_2 > \cdots > p_k > 2 \quad \text{and} \quad \text{ord}(p_i) \ (\text{mod } p_j) > 2 \text{ if } (i, j) \notin \Gamma$$

together with the first of the following four pairs of additional conditions:

$$C_\Gamma(1): \begin{cases} (i, j) \in \Gamma, \ \alpha(i) = 1 \ \Rightarrow \ p_i \equiv 1 \ (\text{mod } p_j), \ p_i \not\equiv 1 \ (\text{mod } p_j^2), \\ (i, j) \in \Gamma, \ \alpha(i) = 2 \ \Rightarrow \ p_i \equiv -1 \ (\text{mod } p_j), \ p_i \not\equiv -1 \ (\text{mod } p_j^2); \end{cases}$$

$$C_\Gamma(2): \begin{cases} (i, j) \in \Gamma, \ \alpha(i) = 1 \ \Rightarrow \ p_i \equiv 1 \ (\text{mod } p_j^2), \\ (i, j) \in \Gamma, \ \alpha(i) = 2 \ \Rightarrow \ p_i \equiv -1 \ (\text{mod } p_j), \ p_i \not\equiv -1 \ (\text{mod } p_j^2); \end{cases}$$

$$C_\Gamma(3): \begin{cases} (i, j) \in \Gamma, \ \alpha(i) = 1 \ \Rightarrow \ p_i \equiv 1 \ (\text{mod } p_j), \ p_i \not\equiv 1 \ (\text{mod } p_j^2), \\ (i, j) \in \Gamma, \ \alpha(i) = 2 \ \Rightarrow \ p_i \equiv -1 \ (\text{mod } p_j^2); \end{cases}$$

$$C_\Gamma(4): \begin{cases} (i, j) \in \Gamma, \ \alpha(i) = 1 \ \Rightarrow \ p_i \equiv 1 \ (\text{mod } p_j^2), \\ (i, j) \in \Gamma, \ \alpha(i) = 2 \ \Rightarrow \ p_i \equiv -1 \ (\text{mod } p_j^2). \end{cases}$$

To count the groups of order n (where $n = p_1^{\alpha(1)} \cdots p_k^{\alpha(k)}$ as before) we need to use appropriate edge-labellings. Consider groups G where the Sylow p_j-subgroup is of order p_j^2 and normalises but does not centralise the Sylow p_i-subgroup. If $\alpha(i) = 1$ and $p_i \equiv 1 \ (\text{mod } p_j^2)$ then we assign a label $\beta(i, j)$ to the edge (i, j) in the log-pile $\Delta(G)$ as follows:

$$\beta(i, j) = \begin{cases} 0 & \text{if } P_j \cong C_{p_j^2} \text{ and acts faithfully on } P_i, \\ 1 & \text{if } P_j \cong C_{p_j^2} \text{ and induces } C_{p_j} \text{ on } P_i, \\ 2 & \text{if } P_j \cong C_{p_j} \times C_{p_j}. \end{cases}$$

Analogous adjustments are needed if $\alpha(i) = \alpha(j) = 2$ and $p_i \equiv -1 \pmod{p_j^2}$. Thus we should now assign weights to the edges (i, j) of Γ as follows:

$$w(i, j) = \begin{cases} 1 & \text{if } \alpha(j) = 1; \\ 2 & \text{if } \alpha(j) = 2 \text{ in case } C_\Gamma(1); \\ 3 & \text{if } \alpha(i) = 1, \alpha(j) = 2 \text{ in case } C_\Gamma(2); \\ 2 & \text{if } \alpha(i) = \alpha(j) = 2 \text{ in case } C_\Gamma(2); \\ 2 & \text{if } \alpha(i) = 1, \alpha(j) = 2 \text{ in case } C_\Gamma(3); \\ 3 & \text{if } \alpha(i) = \alpha(j) = 2 \text{ in case } C_\Gamma(3); \\ 3 & \text{if } \alpha(j) = 2 \text{ in case } C_\Gamma(4). \end{cases}$$

The weight of a log-pile obtained from Γ is now defined exactly as before, and so are the weighted log-pile numbers $\lambda_w(\Gamma)$, $\mu_w(\Gamma)$.

Case $C_\Gamma(1)$ gives rise to the set S_1 of quadruples that has already been discussed. The other three give rise to sets S_2, S_3, S_4 of quadruples of generalised log-pile parameters all of which may be generated from the initial quadruples $(1, 1, 1, 1)$ and $(1, 2, 2, 2)$ recursively. Rules (1), (2), (5) and (6) are the same in all cases, but for S_2 rule (3) is replaced by

$$(3')\quad (k, a, b, 2) \in S_2 \Rightarrow (k+1, a+\tfrac{3}{2}b, a, 1) \in S_2;$$

for S_3 rule (4) is replaced by

$$(4')\quad (k, a, b, 2) \in S_3 \Rightarrow (k+1, 2a+\tfrac{3}{2}b, 2a, 2) \in S_3;$$

while for S_4 both rules (3) and (4) are replaced by the S_4 versions of (3') and (4') respectively.

The point is that we have now described five systems of recursive rules for generating useful values of $f(n)$. If these do not provide sufficient tools for proving that f is a surjective function then the problem is a hard one.

21.7 Densities of certain sets of group orders

Questions like the following one have frequently occurred informally: what can be said about the set T_1 of integers n for which $f(n) = 1$? It contains all the prime numbers. Of course it also contains many composite numbers, such as $15, 33, 35, \ldots$ The following is an immediate consequence of the fact that there are at least two groups of order p^k whenever p is prime and $k \geqslant 2$ together with Hölder's theorem stated as Proposition 21.5 above.

Observation 21.25 $f(n) = 1$ *if and only if* $\gcd(n, \phi(n)) = 1$, *where* $\phi(n)$ *is Euler's function enumerating the number of* $m \in \mathbb{N}$, $1 \leqslant m \leqslant n$, *that are coprime with n.*

For large real x define $t_1(x)$ to be the number of $n \in T_1$ such that $n \leqslant x$. In [28] Paul Erdős, working with the condition $\gcd(n, \phi(n)) = 1$, proved the following theorem:

Theorem 21.26 (Erdős 1948) $t_1(x) \sim \dfrac{e^{-\gamma} x}{\log \log \log x}$, *where* γ *is Euler's constant* $\lim_{n \to \infty} (\sum_{r=1}^{n} 1/r - \log n)$.

Michael E. Mays considers similar problems in [66], asking for the density of the sets of integers n such that all groups of order n are cyclic, abelian, nilpotent, supersoluble or soluble respectively. Of course all groups of order n are cyclic if and only if $f(n) = 1$. Define

$T_2 = $ the set of all n such that all groups of order n are abelian

and

$T_3 = $ the set of all n such that all groups of order n are nilpotent.

More generally, for a function $c : \{\text{primes}\} \to \mathbb{N} \cup \{\infty\}$ define $\mathfrak{FN}(c)$ to be the class of finite nilpotent groups in which the Sylow p-subgroup has nilpotency class at most $c(p)$. Then define

$T_4 = $ the set of all n such that all groups of order n lie in $\mathfrak{FN}(c)$.

The sets T_2 and T_3 are the extreme cases of T_4 in which $c(p) = 1$ for all p and $c(p) = \infty$ for all p respectively. Now recall that a function $\psi : \mathbb{N} \to \mathbb{N}$ is said to be *multiplicative* if $\psi(m_1 m_2) = \psi(m_1)\psi(m_2)$ whenever m_1, m_2 are coprime. A multiplicative function is determined by the values it takes on prime powers p^a. Let ϕ_2 be the multiplicative function such that

$$\phi_2(p^a) = \begin{cases} (p-1) & \text{if } a = 1, \\ (p-1)(p^2-1) & \text{if } a = 2, \\ p(p-1)(p^2-1) \cdots (p^a-1) & \text{if } a \geqslant 3, \end{cases}$$

and let ϕ_3 be the multiplicative function such that

$$\phi_3(p^a) = (p-1)(p^2-1) \cdots (p^a-1).$$

More generally, given $c : \{\text{primes}\} \to \mathbb{N} \cup \{\infty\}$ let ϕ_4 be the multiplicative function such that

$$\phi_4(p^a) = \begin{cases} (p-1)(p^2-1) \cdots (p^a-1) & \text{if } a \leqslant c(p)+1, \\ p(p-1)(p^2-1) \cdots (p^a-1) & \text{if } a > c(p)+1, \end{cases}$$

where of course $\infty + 1$ is deemed to be ∞.

Observation 21.27

(1) $n \in T_2$ *if and only if* $\gcd(n, \phi_2(n)) = 1$.
(2) $n \in T_3$ *if and only if* $\gcd(n, \phi_3(n)) = 1$.
(3) $n \in T_4$ *if and only if* $\gcd(n, \phi_4(n)) = 1$.

Proof: As we have already observed, (1) and (2) are special cases of (3), so we focus on the latter. Note that $\gcd(n, \phi_4(n)) = 1$ if and only if first, any two prime powers p^α, q^β dividing n are arithmetically independent in the sense defined on p. 238 and second, if p^α divides n then $\alpha \leqslant c(p)+1$. Thus if G is a group of order n and $\gcd(n, \phi_4(n)) = 1$, then by the Frobenius Transfer Theorem (see, for example, [36, Chapter 7, Theorem 4.5]) G has a normal p-complement for every prime p and so G is the direct product of its Sylow subgroups, that is, it is nilpotent. (Compare with Lemma 21.19 but note that this special case does not require the Feit–Thompson Theorem.) Furthermore, the Sylow p-group, being of order at most $p^{c(p)+1}$, has class $\leqslant c(p)$. Conversely, if $\gcd(n, \phi_4(n)) \neq 1$ then either there is a prime p such that $p^{c(p)+2}$ divides n, in which case there is a group of order n of the form $P \times C_{n/p^{c(p)+2}}$ where P is a group of nilpotency class $c(p)+1$ and order $p^{c(p)+2}$ (a p-group of maximal class), or there are prime powers p^α, q^β dividing n that are not arithmetically independent, say q divides $p^\alpha - 1$. In this case there is a non-nilpotent extension H of an elementary abelian group of order p^α by C_q, and so there is a non-nilpotent group $H \times C_{n/p^\alpha q}$ of order n. Thus if $\gcd(n, \phi_4(n)) \neq 1$ then not all groups of order n lie in $\mathfrak{FN}(c)$.

In her paper [85] Eira Scourfield extends Erdős' result, Theorem 21.26, to a very large class of multiplicative functions ψ. Let $s_\psi(x)$ denote the number of positive integers $n \leqslant x$ for which $\gcd(n, \psi(n)) = 1$. She shows that, if for each $a \geqslant 1$, $\psi(p^a) = W_a(p)$ for all prime numbers p, where $W_a(x)$ is a polynomial, $W_1(x)$ is non-constant, and $W_1(0) \neq 0$, then there exists $C \in (0, \infty)$ and there exists $\lambda \in (0, 1]$ such that $s_\psi(x) \sim C x / (\log\log\log x)^\lambda$. She shows also that, in particular, if $\psi(p) = W_1(p) = p - 1$ then $C = e^{-\gamma}$ and $\lambda = 1$. (This is a special case of [85, Corollary 2].) The following theorem follows immediately:

Theorem 21.28 *Define $t_2(x)$ to be the number of $n \leqslant x$ such that all groups of order n are abelian, $t_3(x)$ to be the number of $n \leqslant x$ such that all groups of order n are nilpotent and $t_4(x)$ to be the number of $n \leqslant x$ such that all groups of order n lie in $\mathfrak{F}\mathfrak{N}(c)$, where, as above, $c : \{\text{primes}\} \to \mathbb{N} \cup \{\infty\}$. Then*

$$t_i(x) \sim \frac{e^{-\gamma} x}{\log \log \log x} \quad \text{for } i = 2, 3, 4.$$

Recall that a group is said to be supersoluble if and only if all its chief factors are of prime order. Let

$T_5 = $ the set of all n such that all groups of order n are supersoluble,

$T_6 = $ the set of all n such that all groups of order n are soluble,

and, as before, let $t_i(x)$ be the number of natural numbers $n \leqslant x$ such that $n \in T_i$. In [66] Mays observes that when x is large enough

$$0.607 \, x < t_5(x) < 0.978 \, x \quad \text{and} \quad 0.869 \, x < t_6(x) < 0.978 \, x.$$

For $t_5(x)$ and the lower bound his argument is, in effect, that every group of square-free order is supersoluble. As is well known, the set of square-free numbers has asymptotic density $6/\pi^2$, and $6/\pi^2 > 0.607$. His lower bound for $t_6(x)$ comes from the Feit–Thompson theorem together with a famous theorem of Burnside, which tells us that if a positive integer n is not divisible by 12, 16 or 56 then the groups of order n are soluble. The inclusion–exclusion principle then yields that

$$t_6(x) \geqslant \left(1 - \frac{1}{12} - \frac{1}{16} - \frac{1}{56} + \frac{1}{48} + \frac{1}{112} + \frac{1}{168} - \frac{1}{336} \right) x$$

$$= \frac{73}{84} x > 0.869 \, x.$$

The upper bound for $t_5(x)$ comes from that for $t_6(x)$. For the upper bound $t_6(x) < 0.978 \, x$ he observes that the multiples of $|\mathrm{PSL}(2, p)|$ for $p = 5, 7, 13$, that is the multiples of 60, of 168 and of 1092, all lie in the complement of T_6. In fact this only gives that $t_6(x) < 0.9783 \, x$:

$$t_6(x) < \left(1 - \frac{1}{60} - \frac{1}{168} - \frac{1}{1092} + \frac{1}{840} + \frac{1}{5460} + \frac{1}{2184} - \frac{1}{10\,920} \right) x$$

$$= \frac{5341}{5460} x,$$

and $0.9782 < \frac{5341}{5460} < 0.9783$. But if we use $\mathrm{PSL}(2, 17)$ as well as $\mathrm{PSL}(2, 5)$, $\mathrm{PSL}(2, 7)$ and $\mathrm{PSL}(2, 13)$ then we have enough to prove that $t_6(x) < 0.978 \, x$.

We propose to sketch a proof of the following rather stronger theorem:

Theorem 21.29 *There exist* $c_5, c_6 \in (0, 1)$, *in fact* $c_5 = 0.86\ldots$ *and* $c_6 = 0.97\ldots$, *such that* $t_5(x) \sim c_5 x$ *and* $t_6(x) \sim c_6 x$ *as* $x \to \infty$.

We begin the proof with a general lemma (see Theorem 0.1 of [43]—we are grateful to Dr Eira Schofield for drawing this reference to our attention).

Lemma 21.30 *Let R be a set of positive integers and let*

$$S = \{n \in \mathbb{N} \mid \exists r \in R : r \ \text{divides} \ n\} .$$

For $x > 0$ define $s(x) = |\{n \in S \mid n \leqslant x\}|$. If $\sum_{r \in R} r^{-1}$ converges then $\lim_{x \to \infty} s(x)/x$ exists.

Proof: For $x > 0$ define $m(x) = \mathrm{lcm}\{r \in R \mid r \leqslant x\}$ and then

$$\lambda(x) = \frac{\left|\{n \in [1, m(x)] \mid \exists r \in R : r \leqslant x \ \text{and} \ r \ \text{divides} \ n\}\right|}{m(x)} ,$$

where $[1, m]$ denotes the set $\{1, 2, 3, \ldots, m\}$. If $x \leqslant y$ then $m(x)$ divides $m(y)$, and, since $\{n \leqslant m(y) \mid \exists r \in R : r \leqslant x \ \text{and} \ r \ \text{divides} \ n\}$ consists of $m(y)/m(x)$ disjoint shifts of $\{n \leqslant m(x) \mid \exists r \in R : r \leqslant x \ \text{and} \ r \ \text{divides} \ n\}$, we see that $\lambda(x) \leqslant \lambda(y)$. Since λ is a non-decreasing function that is bounded above by 1, there exists $\delta \leqslant 1$ such that $\lambda(x) \to \delta$ as $x \to \infty$.

Now suppose that $\sum_{r \in R} r^{-1}$ converges. Let $\varepsilon > 0$. Choose x_0 so that $\sum_{r \in R, \ r > x_0} r^{-1} < \frac{1}{2}\varepsilon$ and $\delta - \lambda(x_0) < \frac{1}{2}\varepsilon$, and define $M = m(x_0)$, $\lambda_0 = \lambda(x_0)$. Clearly

$$s(x) \geqslant M\lambda_0 \lfloor x/M \rfloor .$$

But also

$$s(x) < M\lambda_0 \lfloor x/M \rfloor + M + \left(\sum_{r \in R, \ r > x_0} \frac{1}{r} \right) x$$

because $M\lambda_0 \lfloor x/M \rfloor$ counts the number of integers in $[1, M\lfloor x/M \rfloor]$ that are divisible by some member r of R with $r \leqslant x_0$, M is a crude upper bound for the number of members of S that lie between $M\lfloor x/M \rfloor$ and x, and the last term is an upper bound for the members of S that are at most x and are divisible by some member r of R such that $r > x_0$. Since $x - M < M \lfloor x/M \rfloor \leqslant x$ we have

$$\lambda_0(x - M) < s(x) < \lambda_0 x + M + \tfrac{1}{2}\varepsilon x$$

and so

$$\lambda_0 - \lambda_0 \frac{M}{x} < \frac{s(x)}{x} < \lambda_0 + \frac{M}{x} + \tfrac{1}{2}\varepsilon.$$

Now if $x > 2M/\varepsilon$, so that $M/x < \tfrac{1}{2}\varepsilon$, then since $\delta - \tfrac{1}{2}\varepsilon < \lambda_0 \leqslant \delta \leqslant 1$,

$$\delta - \varepsilon < \frac{s(x)}{x} < \delta + \varepsilon.$$

Thus $s(x)/x \to \delta$ as $x \to \infty$, and the lemma is proved.

Corollary 21.31 *With the notation and assumptions of the lemma, if*

$$T = \{n \in \mathbb{N} \mid \forall r \in R \colon r \text{ does not divide } n\}$$

and $t(x) = |\{n \in T \mid n \leqslant x\}|$, *then* $\lim_{x \to \infty} t(x)/x$ *exists.*

For, $T = \mathbb{N} \setminus S$ and so $s(x) + t(x) = \lfloor x \rfloor = x - \xi$ where $0 \leqslant \xi < 1$. Given that $s(x)/x \to \delta$ as $x \to \infty$, it follows that $t(x)/x \to 1 - \delta$ as $x \to \infty$.

For the proof of the theorem we treat supersolubility first. A group fails to be supersoluble if and only if it has a section (a quotient group of a subgroup) which is a minimal non-supersoluble group, that is, a non-supersoluble group all of whose proper subgroups and proper quotient groups are supersoluble. Our next lemma classifies these.

Lemma 21.32 *If G is a minimal non-supersoluble group then G is a semi-direct product $A.B$ where A is an elementary abelian normal subgroup of prime-power order p^a, where $a \geqslant 2$, and B is a group acting faithfully and irreducibly on A and satisfying one of the following conditions:*

(I) *$B \cong C_{q^b}$ where q is prime, a is minimal subject to $p^a \equiv 1 \pmod{q^b}$ and $p \equiv 1 \pmod{q^{b-1}}$.*

(II) *B is a non-abelian q-group of order q^b, where q is prime, $p \equiv 1 \pmod{q}$ and $a = q$.*

(III) *$B \cong \langle x, y \mid x^q = y^{r^c} = 1, \; y^{-1}xy = x^\rho \rangle$, where q, r are prime, $q \equiv 1 \pmod{r}$, $p \equiv 1 \pmod{(q r^{c-1})}$, ρ is a primitive r^{th} root of 1 modulo q and $a = r$.*

Here is a sketch of a proof. Let G be a minimal non-supersoluble group. Suppose first, seeking a contradiction, that G is non-soluble. Then it has a non-abelian simple composition factor and this being non-supersoluble must be equal to G: thus G is simple. Let p be the smallest prime number dividing $|G|$. If Q is any non-trivial p-subgroup of G then $N_G(Q)$, being a proper subgroup of G, is supersoluble. Therefore all its chief factors are cyclic, and since p is the smallest prime dividing $|G|$ while the order of the automorphism group induced on a chief factor of order p must divide $p-1$, all chief factors of $N_G(Q)$ contained in Q must be central. It follows that $N_G(Q)/C_G(Q)$ is a p-group. Now by the Frobenius transfer theorem (see, for example, [36, Chapter 7, Theorem 4.5]) G has a normal p-complement contrary to simplicity. Thus G must be soluble.

Let A be a minimal non-trivial normal subgroup of G. Then $|A| = p^a$ for some prime power p^a, and A is elementary abelian. Since G/A is supersoluble while G is not we must have $a \geqslant 2$. Also, A is the unique minimal non-trivial normal subgroup in G because if X were another then $A \cap X$ would be trivial and A (with the action of $G/C_G(A)$ on it) would be a non-cyclic chief factor of G/X contrary to the fact that G/X should be supersoluble. Let $K = C_G(A)$, so that $A \leqslant K \trianglelefteq G$ and G/K acts as an irreducible group of automorphisms of A, hence may be identified with an irreducible subgroup of $\mathrm{GL}(a, p)$. Also G/K is supersoluble since $K \neq \{1\}$, and it has no non-trivial normal p-subgroup since it acts faithfully and irreducibly on A.

If G/K is abelian then, since it acts irreducibly on A, it is cyclic of order not dividing $p-1$. There is therefore a prime power q^b dividing $|G/K|$ but not $p-1$. Let x be an element of G such that xK has order q^b. Replacing x by a suitable power we may assume that the order of x is q^c for some $c \geqslant b$. Now the subgroup $\langle A, x \rangle$ is not supersoluble and therefore $G = \langle A, x \rangle$. Since $\langle x^{q^b} \rangle$ centralises both A and x it is normal in G, hence must be trivial since it does not contain the unique minimal normal subgroup A. Thus in fact $c = b$, and if $B = \langle x \rangle$ then $G = A.B$ and G is of type (I).

Suppose now that G/K is nilpotent but non-abelian. Let q be a prime such that the Sylow q-subgroup of G/K is non-abelian and let B be a Sylow q-subgroup of G. Note that $q \neq p$ since G/K has no non-trivial normal p-subgroup, so by Maschke's theorem, A is completely reducible as $\mathbb{F}_p B$-module. Since $B/(B \cap K)$ is non-abelian the irreducible summands cannot all have dimension 1. Therefore $A.B$ is not supersoluble, so $G = A.B$. Arguing exactly as in the previous paragraph we find that $B \cap K = \{1\}$, so $K = A$ and B acts faithfully on A. Let $|B| = q^b$ and let C be a subgroup of index q in B. Since $A.C$ is supersoluble, by Maschke's theorem and Clifford's theorem A is a direct product of minimal normal subgroups of $A.C$, each of these has order p and they are permuted transitively by the action of B, that is, of B/C. Therefore $a = q$ and C is abelian. Also $p \equiv 1 \pmod{q^c}$, where q^c is the exponent of C and in particular, $p \equiv 1 \pmod{q}$. Thus G is of type (II).

Suppose now that G/K is not nilpotent. Let M be a maximal normal subgroup of G containing K. Then $G/M \cong C_r$ for some prime number r. Now M is supersoluble and so applying Clifford's theorem we see that A splits as the direct product of cyclic subgroups of order p, which are permuted transitively by conjugation in G, and each of which is normal in M. Therefore $a = r$ and M/K is abelian of exponent dividing $p - 1$. Consider M/K as a $\mathbb{Z}_{p-1}(G/M)$-module. Let $q_1, \ldots, q_k$ be the primes dividing $p - 1$, let $m = q_1 \cdots q_k$ and define $M_i = \{x \in M \mid x^{m^i} \in K\}$. Clearly $M_i \trianglelefteq G$ for all relevant i. Since G/K is non-nilpotent there is a non-central chief factor of G/K in M/K and therefore in M_{i+1}/M_i for some $i \geq 0$. Since M/K is abelian the map $xK \mapsto x^{m^i}K$ is a module homomorphism $M_{i+1}/K \to M_1/K$ and its kernel is M_i/K. Therefore M_1/K contains a non-central chief factor of G/M and it follows that there is a minimal non-trivial normal subgroup Y/K of G/K that is contained in M/K, is not central in G/K and is isomorphic to C_q for some prime $q \equiv 1 \pmod{r}$. Note that q divides $p - 1$. Since G/K is supersoluble Y/K is cyclic of order q. Let Q be a Sylow q-subgroup of Y, so that $QK = Y$ and $Q/(Q \cap K) \cong C_q$. Since $Y \trianglelefteq G$ the Frattini Argument tells us that $G = N_G(Q)Y = N_G(Q)K$. Choose $y \in N_G(Q) \setminus M$ such that y has order r^c for some $c \geq 1$, and let $B = \langle y, Q \rangle$. Then $B/(B \cap K)$ is non-abelian (because Y/K is centralised by M/K but not by G/K and therefore not by any element of $G \setminus M$). Also $|B| = |Q|r^c$, which is not divisible by p (since q divides $p - 1$ and r divides $q - 1$). By Maschke's theorem A is completely reducible under the action of B and, as $B/(B \cap K)$ is non-abelian, not all the irreducible summands can be of dimension 1 (over $\mathbb{F}_p$) so $A.B$ cannot be supersoluble. Hence $G = A.B$. Note that $A \cap B = \{1\}$ since $|B|$ is not divisible by p. If $X = B \cap K$ then $X \trianglelefteq B$ and X centralises A, so $X \trianglelefteq G$ and it follows that $X = \{1\}$ since $A \not\leq X$. Thus B acts faithfully on A, so in fact $A = K$ and $|Q| = q$. By construction $Q \trianglelefteq B$ and y does not centralise Q, so if x is a generator of Q then $y^{-1}xy = x^{\rho}$ for some $\rho \not\equiv 1 \pmod{q}$. On the other hand, y^r lies in M and M/A is abelian, so y^r does centralise Q: hence $\rho^r \equiv 1 \pmod{q}$, that is, ρ is a primitive r^{th} root of 1 modulo q. Moreover, the order of y^r is r^{c-1} and so, since M/A has exponent dividing $p - 1$, we have $p \equiv 1 \pmod{r^{c-1}}$. This case therefore leads to the groups of type (III) and completes the proof of the lemma.

Note that more can be said. In case (I), if $b = 1$ then a is the order of p modulo q and so a divides $q - 1$, while if $b > 1$ then $a = q$; also, we may identify A with the additive group of $\mathbb{F}_{p^a}$ and then a generator x of B acts as multiplication by a primitive $(q^b)^{\text{th}}$ root of 1 in that field. In case (II), since B is non-abelian we must have $b \geqslant 3$ and since B acts faithfully and irreducibly its centre is cyclic; also, every proper subgroup of B acts completely reducibly on A and therefore is abelian of exponent dividing $p - 1$; it follows that the centre $Z(B)$ is cyclic of order q^{b-2} so $p \equiv 1 \pmod{q^{b-2}}$; moreover, from the fact that the centre is cyclic of order q^{b-2} one can easily classify the groups B that can arise.

The lemma (combined in case (II) with the above notes) tells us that if there is a non-supersoluble group of order n then

(I) $n = p^a q^b$, where p, q are prime and a is minimal subject to the conditions that $p^a \equiv 1 \pmod{q^b}$ and $p \equiv 1 \pmod{q^{b-1}}$,

(II) $n = p^q q^b$, where p, q are prime, $b \geqslant 3$ and $p \equiv 1 \pmod{q^{b-2}}$, or

(III) $n = p^r q r^c$, where p, q, r are prime, $q \equiv 1 \pmod{r}$ and $p \equiv 1 \pmod{(q r^{c-1})}$.

In fact, for each of these orders there does exist a minimal non-supersoluble group. We leave the proof, which is not hard, to the reader.

To prove Theorem 21.29 we let R_5 be the set of n such that there is a non-supersoluble group of order n but every group whose order properly divides n is supersoluble. This is the set of numbers listed above but with multiples deleted. For example, it contains $p^2 \cdot 2^3$ when $p \equiv 9 \pmod{16}$ but not $p^2 \cdot 2^4$, even though there is a minimal non-supersoluble group of the latter order (and type (I)). Thus R_5 consists of numbers of the following forms (where p, q, r are prime numbers):

(i) $p^a \cdot q$ where a is the order of p modulo q and $a \geqslant 2$;

(ii) $p^q \cdot q^2$ if $p \equiv 1 \pmod{q}$, $p \not\equiv 1 \pmod{q^2}$ and $p^q \cdot q^3$ if $p \equiv 1 \pmod{q^2}$;

(iii) $p^r \cdot q \cdot r$ if $q \equiv 1 \pmod{r}$ and $p \equiv 1 \pmod{qr}$.

Here the numbers (i) come from some of the groups of type (I), the numbers (ii) come from some of the groups of types (I) and (II), and the numbers (iii) come from some of the groups of type (III). Note that even the union of these lists contains numbers that can be discarded to yield R_5. For example, it contains both 12 in list (i) and 36 in list (ii). We find that

$$R_5 = \{12, 56, 75, 80, 196, 200, 294, 351, 363, 405, 484, 867, 992, \dots\}.$$

Then

$$T_5 = \{n \in \mathbb{N} \mid m \in R_5 \Rightarrow m \nmid n\},$$

and so the desired result for supersolubility will follow from Corollary 21.31 if we can prove that $\sum_{r \in R_5} r^{-1}$ converges. Notice that if $r \in R_5$ then $r = p^a \cdot k$ where $k \geqslant a + 1$. Also, for given p (prime) and $a \geqslant 2$ the number of entries for p^a in list (i) is at most the number of distinct prime divisors of $p^a - 1$ that are greater than a, and this is less than $a \log p / \log a$. Similarly, the number of entries for p^a in list (ii) is at most 1, the number in list (iii) is at most 1 and the number in list (iv) is at most the number of prime divisors of $p - 1$ which are greater than a, and this is less than $\log p / \log a$. In total, therefore, the number of terms in R_5 that come from given p^a is at most $2 + (a + 1) \log p / \log a$, which is less than $2(a + 1) \log p / \log a$. Thus if $u(p, a)$ is the contribution to $\sum_{r \in R_5} r^{-1}$ from given p and a then

$$u(p, a) < \frac{2(a + 1) \log p}{\log a} \times \frac{1}{p^a (a + 1)} = \frac{2 \log p}{p^a \log a}.$$

Now for fixed p the sum $\sum 1/(p^a \log a)$ converges and

$$\sum_{a \geqslant 2} \frac{1}{p^a \log a} < \frac{1}{p(p - 1) \log 2},$$

and the sum $\sum \log p / (p(p - 1))$ converges by comparison with $\sum p^{-3/2}$. Therefore $\sum_{r \in R_5} r^{-1}$ converges, as required.

Solubility can be treated similarly. A group is soluble if and only if it has no section isomorphic to a minimal simple group, that is a simple group all of whose proper subgroups are soluble. The minimal simple groups were classified by John Thompson in the 1960s (see [90]). They are the groups $SL(2, 2^q)$, where q is prime, $PSL(2, p)$ for primes $p \equiv \pm 2 \pmod 5$, $PSL(3, 3)$ and the groups $PSL(2, 3^q)$, ${}^2B_2(2^q)$ (the Suzuki groups $Sz(2^q)$) for prime numbers $q \geqslant 3$. Define R_6 to be the set of orders of these groups, so that (with q, p as just specified),

$$R_6 = \{2^q(2^{2q} - 1), \tfrac{1}{2}p(p^2 - 1), 5616, \tfrac{1}{3}3^q(3^{2q} - 1), 2^{2q}(2^{2q} + 1)(2^q - 1)\}$$

$$= \{60, 168, 504, 1092, 2448, 5616, 6072, 9828, 25\,308, 29\,120, \dots\}.$$

Then

$$T_6 = \{n \in \mathbb{N} \mid m \in R_6 \Rightarrow m \nmid n\}.$$

In this case it is very easy to see that $\sum_{r \in R_6} r^{-1}$ converges and so the theorem follows from Corollary 21.31.

21.8 Enumerating perfect groups

A group G is said to be *perfect* if it has no non-trivial abelian quotient group, that is, if $G = G'$. Let $f_{\text{perf}}(n)$ denote the number of perfect groups of order n up to isomorphism. By modifying Higman's construction (see Section 4.2) one can prove that there exists $c > 0$ such that $f_{\text{perf}}(n) > n^{c\mu^2 - O(\mu)}$ for infinitely many values of n, where $\mu = \mu(n)$ as defined on p. 2. The idea is this.

Let $S = \text{Alt}(5)$ and let p be a prime number, $p > 5$. Then there is an irreducible 4-dimensional $\mathbb{F}_p S$-module V which is a composition factor of the natural permutation module. Let P denote the free group of exponent p, nilpotency class 2 and rank $4r$ (see Section 4.1), and let $W = P/P'$. There is an action of S on W so that as module $W \cong rV$ (a direct sum of r copies of V). This action can be lifted to give an action of S on P. The action of S on P' turns it into an $\mathbb{F}_p S$-module, and this is isomorphic to the exterior square $W^{\wedge 2}$ (see Proposition 3.5). Now

$$W^{\wedge 2} \cong (rV)^{\wedge 2} \cong rV^{\wedge 2} \oplus \tfrac{1}{2} r(r-1)(V \otimes V),$$

and since V is self-dual (being the unique summand of dimension 4 in a necessarily self-dual permutation module of dimension 5) $V \otimes V$ has a one-dimensional trivial submodule. Therefore if T is the fixed-point set of S in P' then T is an elementary abelian p-group of rank $\geqslant \tfrac{1}{2} r(r-1)$. By Maschke's theorem there is an S-invariant complement U for T in P'. Let $R = P/U$ and define Q to be the semidirect product of R by S. Then $Q = Q'$ since R/R' (which is the same as P/P') has no trivial composition factors as $\mathbb{F}_p S$-module. Consider subgroups K of index p^r (say) in R'. The construction has ensured that R' is central in Q, of order $\geqslant p^{\frac{1}{2} r(r-1)}$ and of index $60 p^{4r}$. Thus our groups K are normal in Q, of index $60 p^{5r}$, and the number of them is greater than

$$\frac{(p^N - 1)(p^N - p) \cdots (p^N - p^{r-1})}{(p^r - 1)(p^r - p) \cdots (p^r - p^{r-1})} > p^{r(N-r)},$$

where $N = \tfrac{1}{2} r(r-1)$. The group induced by $\text{Aut} \, Q$ on R' is certainly a subgroup of $\text{GL}(4r, p)$. Since $|\text{GL}(4r, p)| < p^{16r^2}$ the number of orbits of groups K of index p^r in R' under the action of this automorphism group is greater than $p^{rN - 17r^2}$, and therefore this is a lower bound on the number of non-isomorphic quotient groups of Q that have order $60 p^{5r}$. Thus if $m = 5r$, so that $N = \tfrac{1}{50} m(m-5)$, then

$$f_{\text{perf}}(60 p^m) > p^{\frac{1}{250} m^3 - O(m^2)}.$$

Obviously the constant $1/250$ could be increased significantly—we have limited ourselves to sketching an idea and have made no attempt to optimise constants.

In his delightful paper [49] Derek Holt goes into the matter with much more care, seeking both lower and upper bounds. In the language we have used earlier in this book his results may be stated in the following form:

Theorem 21.33 (Holt 1989) *Let $\lambda = \lambda(n)$ and $\mu = \mu(n)$ as on p. 2.*

(1) *There are infinitely many values of n for which $f_{\text{perf}}(n) > n^{\mu^2/54 - 2\mu/9}$.*

(2) *Also $f_{\text{perf}}(n) < n^{(\mu^2/48) + \lambda}$ for all n.*

In fact, in relation to (1), what Holt actually proves is that if $p > 3$ and p is prime then $f_{\text{perf}}(p^x(p^3 - p)) > p^{x^3/54 - x^2/6}$ for sufficiently large integers x, and our version comes from the observation that if $n = p^x(p^3 - p)$ then $\mu = x + 1$. His theorem raises two questions. First, can the gap between the coefficient $\frac{1}{54}$ in clause (1) and the coefficient $\frac{1}{48}$ in clause (2) be closed? Second, can λ in clause (2) be replaced by $O(\mu)$: as it stands this clause gives a much weaker bound than Pyber's theorem when λ is large compared with μ.

Amongst the perfect groups are those all of whose composition factors are non-abelian. These have been studied by Benjamin Klopsch in his paper [56]. He proves the following two theorems.

Theorem 21.34 *Let $F_\Sigma(n)$ be the number of groups of order $\leqslant n$ (up to isomorphism, of course) all of whose composition factors are non-abelian. There exist $B, C > 0$ such that*

$$n^{B \log \log n} \leqslant F_\Sigma(n) \leqslant n^{C \log \log n}$$

for all $n \geqslant 60$.

Theorem 21.35 *Let S be a non-abelian simple group and let $\hat{f}(k)$ be the number of isomorphism classes of groups G that have composition length k and in which all the composition factors are isomorphic to S. There exist $B_S, C_S > 0$ such that*

$$k^{B_S k} \leqslant \hat{f}(k) \leqslant k^{C_S k}$$

for all $k \geqslant 1$.

The upper bound in the first of these results, Theorem 21.34, has been refined by Jaikin-Zapirain and Pyber in [52]. They prove

Theorem 21.36 *There exists a positive real number c such that the number of d-generator finite groups of order n without abelian composition factors is at most n^{cd}.*

The upper bound in Theorem 21.34 then follows from the fact that if the non-trivial group G has order n and no abelian composition factors, then, as proved by Klopsch in [56] (Proposition 1.1), if G has composition length k then G can be generated by d elements where $d \leqslant 3 \log k + 2$. Since $n \geqslant 60$ and $k \leqslant \log n / \log 60$ we see that $d \leqslant 3(\log\log n - \log\log 60) + 2 < 3\log\log n - 15$ (remember that logarithms are to base 2), so the number of groups of order $\leqslant n$ without abelian composition factors is smaller than $n \times n^{3c(\log\log n - 5)}$, where c is as in Theorem 21.36. Therefore this number is at most $n^{c_1 \log\log n}$, where $c_1 := \max\{3c, \frac{1}{5}\}$.

22

Some open problems

In this chapter we collect open problems. Some of these have been mentioned already in the preceding text, some are mentioned here for the first time. We begin with two questions of contextual interest that were discussed in Chapter 2.

Question 22.1 *What exactly is the asymptotic behaviour of $f_{\text{semigroups}}(n)$?*

Question 22.2 *What exactly is the asymptotic behaviour of $f_{\text{latin squares}}(n)$?*

By far the most important challenges in the field of group enumeration are the next two questions.

Question 22.3 (C. C. Sims) *Is it true that, if $E_p(m)$ is defined by the equation $f(p^m) = p^{\frac{2}{27}m^3 + E_p(m)}$, then $E_p(m) = O(m^2)$?*

Charles Sims originally proved that $E_p(m) = O(m^{8/3})$ and this was improved by Newman and Seeley to $O(m^{5/2})$ (see Chapter 5). The corresponding problems for graded Lie rings and for p-groups whose nilpotency class is at most 3 have been solved positively by S. R. Blackburn (see [7] and Sections 20.3, 20.4 above) but the general question remains open—and one of the most interesting in the area.

Question 22.4 (Graham Higman: see Section 21.1) *Is the PORC conjecture true? That is to say, is it true that for fixed m the function $f(p^m)$ is a PORC function of p?*

And if not, is it approximately true:

Question 22.5 (Graham Higman: see Section 21.1) *Does there exist, for each natural number m, a PORC function F_m such that $|F_m(p) - f(p^m)|$ is significantly smaller than $p^{\frac{2}{27}m^3}$?*

As an important variant of the PORC conjecture it would be valuable to know whether the results of Higman and Evseev (see p. 223) can be extended:

Question 22.6 *Is the PORC conjecture true for nilpotent groups of class at most 2? That is to say, is it true that for fixed m the function $f_{\mathfrak{N}_2}(p^m)$ is a PORC function of p?*

Recall that for a class $\mathfrak{X}$ of groups, $f_{\mathfrak{X}}(n)$ is the number of groups of order n in $\mathfrak{X}$ up to isomorphism. The next question, restricted to varieties of p-groups, was conceived as a refinement of Sims's conjecture.

Question 22.7 (Peter M. Neumann) *Let $\mathfrak{B}$ be a variety of groups and $\mathfrak{U}$ a subvariety. Define $g(n) = f_{\mathfrak{B}}(n)/f_{\mathfrak{U}}(n)$. Is it true that $g(n) \leqslant n^{O(\mu(n))}$?*

The special case where p is an odd prime number, $\mathfrak{U}$ is the minimal non-abelian variety $\mathfrak{B}_{p,2}$ consisting of p-groups of exponent dividing p and nilpotency class at most 2, and $\mathfrak{B}$ is the variety $\mathfrak{H}_p$ of groups of Φ-class 2 (that is, groups of exponent dividing p^2 in which p^{th} powers and commutators are central) is treated in Chapter 19. That chapter contains background to the following questions.

Question 22.8 *Let p be a prime number, let $\mathfrak{B}$ be a non-abelian locally finite variety of p-groups, and define $E_{\mathfrak{B}}(m)$ by the equation $f_{\mathfrak{B}}(p^m) = p^{\frac{2}{27}m^3 + E_{\mathfrak{B}}(m)}$. Does*

$$\lim_{m \to \infty} m^{-2} E_{\mathfrak{B}}(m)$$

exist? If so, what is this limit for $\mathfrak{B}_{p,2}$, $\mathfrak{H}_p$ and other 'small' varieties of p-groups?

Related to Question 22.7 we ask

Question 22.9 *Is it true that $f(p^m)/f_{\mathfrak{B}}(p^m) = p^{O(m^2)}$ for some reasonably small variety $\mathfrak{B}$ of p-groups? In particular, is this true for $\mathfrak{B} = \mathfrak{H}_p$ or for $\mathfrak{B} = \mathfrak{B}_{p^2,2}$?*

Avinoam Mann has formulated a question of a similar kind to this in his paper [64]:

Question 22.10 (Avinoam Mann) *Is it true that most p-groups are of class* 2 ? *Or at least, that a positive proportion of all p-groups are of class* 2 ?

Returning from Mann's question to varieties, we ask:

Question 22.11 (see pp. 189, 192) *Is it true that*

$$f_{\mathfrak{H}_p}(n)/f_{\mathfrak{B}_{p,2}}(n) \geqslant p^{\frac{2}{9}m^2 - O(m)} \, ?$$

Could it perhaps even be true that

$$f_{\mathfrak{B}_{p,2}}(n) = p^{\frac{2}{27}m^3 - \frac{2}{9}m^2 + O(m)} \quad and \quad f_{\mathfrak{H}_p}(n) = p^{\frac{2}{27}m^3 + O(m)} ?$$

(See Theorems 19.3 and 19.2.)

Note that if this question has a positive answer then so does Question 22.8 in the case of $\mathfrak{B}_{p,2}$ and $\mathfrak{H}_p$.

Question 22.12 *Can the gap between the bounds in Theorem* 21.3 *be significantly reduced? In particular, is it true that if $\mathfrak{X}$ is an isoclinism class of p-groups with invariants a, c defined as in that theorem, then*

$$\lim_{m \to \infty} \frac{f_{\mathfrak{X}}(p^m)}{p(m) \, m^{(a+c)/2}}$$

exists?

Recall Jaikin-Zapirain's theorem enunciated on p. 196:

$$p^{\frac{1}{4}(d-1)m^2 + O(m^2)} \leqslant f_d(p^m) \leqslant p^{\frac{1}{2}(d-1)m^2 + O(m^2)}$$

where $f_d(p^m)$ is the number of d-generator groups of order p^m. It would be interesting to know where in the range $[\frac{1}{4}, \frac{1}{2}]$ the correct coefficient of the leading term in these estimates lies.

Question 22.13 *Does*

$$\lim_{m \to \infty} \frac{\log_p f_d(p^m)}{(d-1)m^2}$$

exist? If so, what is it?

We turn now to insoluble groups. As was mentioned in the introduction to Section 20.2, the theorem of Camina, Everest and Gagen discussed in that section is an attempt to capture the feeling that non-soluble groups are rare. The theorem of Klopsch mentioned there gives a more precise understanding of what the word 'rare' means in that context. Quoting John Thompson (see

[91, p. 2]): 'As for rarity, this depends on our choice of measure'. For a finite group G having a composition series $G = G_0 > G_1 > G_2 > \cdots > G_k = \{1\}$, with composition factors $X_i = G_{i-1}/G_i$, define I to be the set of indices i such that X_i is non-abelian. Then define J to be $\{1, \ldots, k\} \setminus I$ and

$$|G|_n = \prod_{i \in I} |X_i|, \qquad |G|_s = \prod_{j \in J} |X_j|.$$

Thus $|G|_n$ is a measure of the non-soluble part of G while $|G|_s$ measures the soluble part. The measure $\tau(G)$ defined by Camina, Everest and Gagen and used in Section 20.2 (for which Thompson uses $s(G)$) is $1/|G|_s$. The groups G for which $\tau(G) \geqslant \varepsilon$ are the groups for which $|G|_s \leqslant K$, where $K = \varepsilon^{-1}$. If they are very large these are groups which, in quite a strong sense, are *very* non-soluble. We propose a less stringent measure of solubility: define

$$\mathrm{ms}(G) = \frac{\log |G|_s}{\log |G|_n}.$$

Thus $\mathrm{ms}(G) = \infty$ if and only if G is soluble, while $\mathrm{ms}(G) = 0$ if and only if all composition factors of G are non-abelian simple groups. Also,

$$|G|_s = |G|_n^{\mathrm{ms}(G)} \quad \text{and} \quad |G| = |G|_n^{1+\mathrm{ms}(G)} = |G|_s^{1+(1/\mathrm{ms}(G))}.$$

Now define $F(n) = \sum_{m \leqslant n} f(n)$ and for $\sigma > 0$ define

$$F_\sigma(n) = \text{the number of groups } G \text{ with } |G| \leqslant n \text{ and } \mathrm{ms}(G) \leqslant \sigma.$$

With this notation we ask

Question 22.14 *Is it true that* $\dfrac{F_\sigma(n)}{F(n)} \to 0$ *as* $n \to \infty$ *for any* σ *in the range* $0 < \sigma < \infty$?

In this connection it is not hard to see that if $\mathrm{ms}(G) \geqslant 1/2$ then G must have a non-trivial soluble normal subgroup. We sketch a proof beginning with the following preparatory fact.

Fact *Let H be a finite group with no non-trivial soluble normal subgroup. Then* $\mathrm{ms}(H) < 1/2$.

Proof: Let T be the socle of H (the product of all the minimal normal subgroups of H). Then $T = X_1 \times \cdots \times X_t$ where the X_i are non-abelian simple groups (some of the non-abelian composition factors of H). Acting by conjugation, H permutes the set $\{X_1, \ldots, X_t\}$ since these are the only minimal normal subgroups of T. Let K be the kernel of this action, so that $T \leqslant K \leqslant H$ and $H/K \leqslant \mathrm{Sym}(t)$. Now $K/T \leqslant A_1 \times \cdots \times A_t$, where $A_i = \mathrm{Aut}\,(X_i)/X_i$, the outer automorphism class group of X_i. Examining the known simple groups (see, for example, [19]) one quite easily sees that $|A_i| < |X_i|^{0.251}$ (in fact $|A_i| \leqslant |X_i|^{\log 12/\log 20\,160}$, with equality if and only if $X_i \cong \mathrm{PSL}(3,4)$). Also, following the method of proof of Theorem 10.1 (Dixon's theorem), or using that result together with the fact that every finite group B contains a soluble subgroup C that covers every soluble normal factor of B (in the sense that if B_1, B_2 are normal subgroups of B with $B_2 \leqslant B_1$ and B_1/B_2 soluble then $B_1 \leqslant CB_2$), we see that if $B \leqslant \mathrm{Sym}(t)$ then $|B|_s \leqslant 24^{(t-1)/3}$. Thus

$$|H|_s \leqslant |T|^{0.251} \times 24^{(t-1)/3} < |T|^{0.251} \times 24^{t/3}.$$

Since $|X_i| \geqslant 60$ for $1 \leqslant i \leqslant t$ we must have $t \leqslant \log|T|/\log 60$, and so

$$|H|_s \leqslant |T|^{0.251+(\log 24/3\log 60)} < |T|^{0.251+0.259} = |T|^{0.51}.$$

Now in fact the exponent 0.251 can be reduced to $\log 4/\log 360$ (achieved by $\mathrm{Alt}(6)$) for all simple groups other than $\mathrm{PSL}(3,4)$, and since for this latter group the exponent 0.259 can be reduced to $\log 24/3\log 20\,160$, which is less than 0.11, we see that in fact $|H|_s < |T|^{1/2}$. Clearly, $|H|_n \geqslant |T|$ and it follows immediately that $\mathrm{ms}(H) < 1/2$.

Corollary *Let G be a finite group and let S be its soluble radical (the largest soluble normal subgroup). Define $\sigma(G) = (\mathrm{ms}(G) - \frac{1}{2})/(\mathrm{ms}(G)+1)$. Then $|S| > |G|^{\sigma(G)}$.*

Proof: Let $H = G/S$. The previous lemma applies to H and so $|H|_s < |H|_n^{1/2}$. Now $|S| = |G|_s/|H|_s$ and so $|S| > |G|_s/|H|_n^{1/2}$. But $|H|_n = |G|_n$ and $|G|_s = |G|_n^{\mathrm{ms}(G)}$. Therefore $|S| > |G|_n^{\mathrm{ms}(G)-\frac{1}{2}}$ and since $|G| = |G|_s|G|_n = |G|_n^{\mathrm{ms}(G)+1}$ the required result follows.

A different measure of solubility is $\mathrm{ms}'(G)$ defined by $\mathrm{ms}'(\{1\}) = 1$ and for non-trivial groups G,

$$\mathrm{ms}'(G) = \frac{\mathrm{complength}(G)}{\lambda(|G|)},$$

where $\mathrm{complength}(G)$ is the length of a composition series of G and $\lambda(n)$ is as defined on p. 2. We have $0 < \mathrm{ms}'(G) \leqslant 1$ for all G and $\mathrm{ms}'(G) = 1$ if and only if G is soluble. Now for $0 < \sigma \leqslant 1$ define

$F'_\sigma(n)$ = the number of groups G with $|G| \leqslant n$ and $\mathrm{ms}'(G) < \sigma$.
We ask

Question 22.15 *Is it true that* $\dfrac{F'_\sigma(n)}{F(n)} \to 0$ *as* $n \to \infty$ *for any* σ *in the range* $0 < \sigma \leqslant 1$?

The answer should be yes, and should not be difficult to prove, at least when $\sigma < 1$. The case $\sigma = 1$ is simply the question whether

$$\frac{F_{\mathrm{insol}}(n)}{F(n)} \to 0 \quad \text{as } n \to \infty,$$

where $F_{\mathrm{insol}}(n)$ is the number of insoluble groups of order $\leqslant n$.

Questions 22.14, 22.15 and the surrounding discussion are centred on the intuition that most finite groups are soluble. In fact it seems likely that by far the majority of groups are not merely soluble but in fact nilpotent. Of course it is not true that $f_{\mathfrak{N}}(n)/f(n) \to 1$ as $n \to \infty$ (where $f_{\mathfrak{N}}(n)$ denotes the number of nilpotent groups of order n) for, if n is square-free then $f_{\mathfrak{N}}(n) = 1$, whereas Hölder's work cited in Section 21.3 shows that $f(n)$ is unbounded for square-free n. The ubiquity of nilpotent groups can be formulated in a number of other ways, however. A very weak form of it is the following quite natural conjecture, whose origin we have been unable to trace satisfactorily. We have heard it attributed at various times to various people, such as Paul Erdős and Graham Higman.

Question 22.16 *Is it true that if* $1 < n < 2^m$ *then* $f(n) < f(2^m)$?

It seems very likely that the answer is yes. Some progress has been made by Ioannis Pantelidakis in his Oxford DPhil thesis (see [81]), who proves the assertion in case n is odd and $m \geqslant 3619$, and also makes a study of the following conjecture.

Question 22.17 *Is it true that* $f(2^m p) < f(2^{m+2})$ *for any prime number* p *and all natural numbers* m?

The case $p = 3$ is of course a special case of Question 22.16. And in fact the tables [6] of Besche, Eick and O'Brien suggest a much stronger conjecture, namely that for any odd prime number p

$$f(2^m p) < f(2^{m+1}) \text{ as long as } m \geqslant 5.$$

What Pantelidakis succeeds in showing is that

$$f(2^m p) < 2^{3m/2} f(2^m) + 2^{(m^2+4m-3)/2} f(2^{m-1}).$$

He comments 'The problem of showing that the right-hand side of the inequality is less than $f(2^{m+2})$, for all m, remains unsolved due to our inability to compare $f(2^{m+2})$ with $f(2^{m-1})$ and $f(2^m)$'. He defines $g(m)$ by the equation $f(2^m) = 2^{\frac{2}{27}m^3 + g(m)}$ and shows that if

$$g(m+1) - g(m) > -(m^2 - 15m)/18 \qquad (\star)$$

then $f(2^m p) < f(2^{m+2})$. The inequality $(\star)$ is very likely to be true for all m, but it may be little easier to prove than the Sims conjecture formulated above as Question 22.3.

A much stronger conjecture than that formulated in (or, technically, suggested by) Question 22.16 is the following.

Question 22.18 *Is it true that $F(2^m - 1) < f(2^m)$ for sufficiently large m, where, as before, $F(n) = \sum_{r \leqslant n} f(r)$? Is it perhaps true that*

$$F(2^m - 1) \sim f(2^{m-1}) \text{ as } m \to \infty$$

(in the sense that $F(2^m - 1)/f(2^{m-1}) \to 1$)?

Again, the tables [6] of Besche, Eick and O'Brien suggest that the first of these statements should be true as soon as $m \geqslant 7$. The second also looks plausible. If the answer is positive, an immediate consequence would be a positive answer to the following well-known old question (see Pyber [83], Mann [64]).

Question 22.19 *Is it true that the majority of finite groups are nilpotent in the sense that*

$$\frac{F_{\mathfrak{N}}(n)}{F(n)} \to 1 \quad \text{as } n \to \infty,$$

where $F_{\mathfrak{N}}(n)$ is the number of nilpotent groups of order at most n?

Indeed, if the second part of Question 22.18 has a positive answer then 2-groups would dominate: we would have (in a self-explanatory notation) $F_{2\text{-gps}}(n)/F(n) \geqslant 1 - \eta(n)$ where $\eta(n) \to 0$ as $n \to \infty$. Note that it is possible, although in our opinion unlikely, that Question 22.19 has a positive answer but Question 22.18 a negative one.

We turn now to some matters of rather narrower focus that have arisen earlier in the book. The first of these was formulated on p. 53.

Question 22.20 *Is there an elementary proof of Theorem 6.11? Is there an efficient algorithm to find a set of d generators for a permutation group of degree $n \geqslant 4$, where $d \leqslant \lfloor n/2 \rfloor$?*

For background to our next question see Section 7.4 and [67, Section 4]. We ask whether it is the case that there are many more soluble A-groups of order n than insoluble ones. Consideration of numbers $n = 60p$, where p is prime, shows that the ratio $f_{A,\text{sol}}(n)/f_A(n)$ can take the same value < 1 infinitely often – in fact, if $p \equiv -1 \pmod{60}$ then $\text{Alt}(5) \times C_p$ is the only insoluble group of order n, while there are 24 soluble groups of order n (there is $\text{Alt}(4) \times C_{5p}$, there are 15 extensions of C_{15p} by $C_2 \times C_2$, and there are 8 extensions of C_{15p} by C_4). And of course all of them are A-groups, so that

$$\frac{f_{A,\text{sol}}(n)}{f_A(n)} = \frac{f_{\text{sol}}(n)}{f(n)} = \frac{24}{25}.$$

for these values of n. It seems very unlikely, however, that this sort of behaviour can occur when n has many prime factors.

Question 22.21 *Is it the case that $f_A(n)/f_{A,\text{sol}}(n) \to 1$ as $\lambda(n) \to \infty$? How big is $f_A(n) - f_{A,\text{sol}}(n)$ compared with $f_A(n)$?*

For background to the next question see the discussion in Chapter 12, p. 112, Chapter 15, p. 139 and Chapter 18 (especially p. 184).

Question 22.22 *Define*

$$\alpha = \frac{\limsup_{n\to\infty} \log f_A(n)}{\mu(n)\log(n)}.$$

What is α? Could it perhaps be $3 - 2\sqrt{2}$?

That number $3 - 2\sqrt{2}$ comes from the study of varieties of A-groups in Chapter 18. The following stronger conjecture was formulated at the end of that chapter. Note that the minimal non-abelian varieties of A-groups are the product varieties $\mathfrak{A}_p\mathfrak{A}_q$ where p, q are distinct primes.

Question 22.23 *For which varieties $\mathfrak{B}$ of A-groups is it true that the leading term of the enumeration function $f_{\mathfrak{B}}(n)$ is equal to the leading term of $f_{\mathfrak{U}}(n)$ for some minimal non-abelian subvariety $\mathfrak{U}$ of $\mathfrak{B}$?*

Chapter 11 is devoted to bounds for the number of conjugacy classes of maximal soluble subgroups of symmetric groups $\text{Sym}(n)$. In the second comment on p. 101 we formulated the following conjectures. Recall that $m_{\text{ss}}(n)$ denotes the number of conjugacy classes of maximal soluble subgroups in $\text{Sym}(n)$.

Question 22.24 *Can the bound* $m_{ss}(n) < 2^{16n}$ *be improved to* $m_{ss}(n) < 2^n$ *?*
Is it perhaps true that $m_{ss}(n) = 2^{O(\sqrt{n})}$ *? Or even that there exists* $k > 0$ *such that* $m_{ss}(n)/p(n) \to k$ *as* $n \to \infty$ *? (Recall that* $p(n)$ *is the partition function.)*

Similarly, Chapter 13 is devoted to bounds for the number of conjugacy classes of maximal soluble subgroups of linear groups $GL(d, q)$ and in the fourth comment on p. 131 we formulated the following conjecture about $m_{lss}(d, q)$, which is the number of conjugacy classes of maximal soluble subgroups in $GL(d, q)$.

Question 22.25 *Can the estimate* $m_{lss}(d, q) < 2^{278\,833\,d}$ *be improved to*

$$m_{lss}(d, q) < 2^d d?$$

Is it perhaps true that $m_{lss}(d, q) < 2^d d^{1/2}$*?*

The next question was posed on p. 135. See also [10].

Question 22.26 (L. Pyber) *Is it the case that* $|\mathrm{Mss}(G)| \leqslant |G|$ *? That is, is it true that for any finite group* G *the number of maximal soluble subgroups is at most* $|G|$ *?*

For the next question see Chapter 15, and especially p. 139.

Question 22.27 *What is the best exponent in Theorem* 15.5*? Is it perhaps true that the number of soluble groups of order* n *with specified Sylow subgroups* $P_1, \dots, P_k$ *is at most* $n^{2\mu + o(\mu)}$ *?*

Question 22.28 *And what is the best exponent in Theorem* 16.20*? Is it perhaps true that the number of groups of order* n *with specified Sylow subgroups* $P_1, \dots, P_k$ *is also at most* $n^{2\mu + o(\mu)}$ *?*

Next we restate a question from Section 21.2.

Question 22.29 *Is there anything interesting to be said about the enumeration function* $f_{\mathcal{X}}(n)$ *for general isoclinism classes* $\mathcal{X}$ *of finite groups?*

The following is a reworking of Conjecture 21.16 in Section 21.4.

Question 22.30 *Is it true that* $f(n) < n^2$ *for cube-free integers* n *? Is it true that* $f(n)/n^2 \to 0$ *as* $n \to \infty$ *through cube-free integers* n *? What is the best upper bound one can give for* $f(n)$ *for cube-free* n *?*

Turning to Holt's enumeration of perfect groups:

Question 22.31 *Can the gap between the coefficient* $1/54$ *in clause* (1) *and the coefficient* $1/48$ *in clause* (2) *of Theorem* 21.33 *be closed? In particular, can the coefficient* $1/48$ *be reduced?*

Question 22.32 *And can the term* λ *in clause* (2) *of Theorem* 21.33 *be replaced by* $O(\mu^2)$ *or even by* $O(\mu)$ *?*

Connected with this we have

Question 22.33 (D. F. Holt, 1989, page 67) *Define*

$$\text{Imperf} = \{n \in \mathbb{N} \mid \exists\, G : |G| = n \text{ and } G = G'\}.$$

Is it true that this set has density 0? *If* $k(x) = |\text{Imperf} \cap [1, x]|$ *for* $x \in \mathbb{R}$, *approximately how large is* $k(x)$?

The theorems of Klopsch about groups all of whose composition factors are non-abelian raise the following two questions.

Question 22.34 *With* $F_\Sigma(n)$ *defined as in Theorem* 21.34, *does*

$$\lim_{n \to \infty} \frac{\log F_\Sigma(n)}{\log n \, \log \log n}$$

exist, and if so what is its value?

Question 22.35 (Klopsch, 2003) *For a finite non-abelian simple group S, with* $\hat{f}_S(k)$ *defined as in Theorem* 21.35, *does*

$$\lim_{n \to \infty} \frac{\log \hat{f}_S(k)}{k \log k}$$

exist, and if so what is its value?

Finally, the questions treated in Section 21.6:

Question 22.36 *Is the group enumeration function* $f(n)$ *surjective?*

Question 22.37 *In particular, is Keith Dennis's exceptional set E defined on p. 242 finite?*

Appendix A: Maximising two functions

This appendix maximises two functions using standard techniques from calculus. The fact that the maximum value of the functions is $\frac{2}{27}$ and less than $\frac{2}{27}$ respectively is used in Section 5.5 to complete the proof of the upper bound for the p-group enumeration function. We will implicitly use Lagrange multipliers; see Voxman and Goetschel [96], for example.

Lemma A.1 *The function* $A(x, y, z, u)$ *defined by*

$$\tfrac{1}{2}x^2(z+y-u) + \tfrac{1}{2}xyu + (uy - \tfrac{1}{2}u^2)z + \tfrac{1}{2}uz^2$$

satisfies the inequality $A(x, y, z, u) \leqslant \frac{2}{27}$ *whenever* x, y, z *and* u *are real numbers such that* $x \geqslant \frac{6}{10}$, $y \geqslant 0$, $z \geqslant 0$, $u \geqslant 0$, $x + y + z = 1$ *and* $u \leqslant y$.

Proof: We begin by showing that $A(x, y, z, u)$ has no internal critical points. Now,

$$\frac{\partial A}{\partial y} = \tfrac{1}{2}x^2 + \tfrac{1}{2}xu + uz, \tag{A.1}$$

$$\frac{\partial A}{\partial z} = \tfrac{1}{2}x^2 + uy - \tfrac{1}{2}u^2 + uz \text{ and} \tag{A.2}$$

$$\frac{\partial A}{\partial u} = -\tfrac{1}{2}x^2 + \tfrac{1}{2}xy + yz - uz + \tfrac{1}{2}z^2. \tag{A.3}$$

At an interior critical point, $\frac{\partial A}{\partial x} = \frac{\partial A}{\partial y} = \frac{\partial A}{\partial z}$ and $\frac{\partial A}{\partial u} = 0$. In particular, we may equate (A.1) and (A.2) to obtain

$$\tfrac{1}{2}xu = uy - \tfrac{1}{2}u^2.$$

Since $u \neq 0$ in the interior of our region, this implies that

$$u = 2y - x \tag{A.4}$$

269

at an interior critical point.

Substituting (A.4) into (A.3) and using the fact that $\frac{\partial A}{\partial u} = 0$ we find that

$$\tfrac{1}{2}x(y-x) + yz - 2yz + xz + \tfrac{1}{2}z^2 = 0.$$

Rearranging this formula, we obtain

$$z^2 = (x-y)(x-2z) \tag{A.5}$$

at an interior critical point.

Now, $x \geqslant \frac{6}{10}$ and $x+y+z = 1$. Since z is non-negative, $y \leqslant \frac{4}{10}$ and so $x - y \geqslant \frac{2}{10}$ on our region.

Since u is non-negative, (A.4) implies that $2y \geqslant x$ at an interior critical point; hence $y \geqslant \frac{3}{10}$ and so $z \leqslant \frac{1}{10}$. Moreover, we have that $x - 2z \geqslant \frac{4}{10}$. But then the left-hand side of (A.5) is at most $(\frac{1}{10})^2 = \frac{1}{100}$ and the right-hand side of (A.5) is at least $\frac{2}{10}\frac{4}{10} = \frac{8}{100}$. This contradiction shows that $A(x, y, z, u)$ does not have an interior critical point.

We now check the maximum value of $A(x, y, z, u)$ on the boundary of our region. We first consider the case when $u = 0$. We find

$$A(x, y, z, 0) = \tfrac{1}{2}x^2(y+z).$$

It is not difficult to show that $A(x, y, z, 0)$ maximises when $x = \frac{2}{3}$ and $y+z = \frac{1}{3}$ at the value $\frac{2}{27}$.

When $y = 0$, we have that $u = 0$ and so $A(x, 0, z, u) \leqslant \frac{2}{27}$.

In the case when $z = 0$, we find that

$$A(x, y, 0, u) = \tfrac{1}{2}x^2(y-u) + \tfrac{1}{2}xyu$$

$$= \tfrac{1}{2}x^2y + \tfrac{1}{2}xu(y-x)$$

$$< \tfrac{1}{2}x^2y$$

(since $y < x$ on our region)

$$\leqslant \tfrac{2}{27}.$$

We now consider the case when $x = \frac{6}{10}$. Let $A_1(y, z, u) = A(\frac{6}{10}, y, z, u)$. We must show that $A_1(y, z, u) \leqslant \frac{2}{27}$ when $y \geqslant 0$, $z \geqslant 0$, $y+z = \frac{4}{10}$ and $u \leqslant y$. The argument given above to show that $A(x, y, z, u)$ had no interior critical points on our region also shows that $A_1(y, z, u)$ has no interior critical point (as no use was made of $\frac{\partial A}{\partial x}$ in the argument). The previous three paragraphs show that $A_1(y, z, u) \leqslant \frac{2}{27}$ when $u = 0$, $y = 0$ or $z = 0$. So it remains to consider the case when $u = y$. Using the substitution $z = \frac{4}{10} - y$, we find that

$$A_1(y, z, y) = \frac{18}{100}z + \frac{3}{10}y^2 + \frac{1}{2}y^2z + \frac{1}{2}yz^2$$
$$= \frac{72}{1000} + \frac{1}{10}(y^2 - y),$$

which is a decreasing function of y on the region $0 \leqslant y \leqslant \frac{4}{10}$. Hence the maximum value of $\frac{72}{1000}$ is attained when $y = 0$. Since $\frac{72}{1000} < \frac{2}{27}$, we find that $A_1(y, z, u) \leqslant \frac{2}{27}$, as required.

It remains to verify that $A(x, y, z, u) \leqslant \frac{2}{27}$ when $u = y$. Define the function $A_2(x, y, z)$ by $A_2(x, y, z) = A(x, y, z, y)$, so

$$A_2(x, y, z) = \frac{1}{2}x^2z + \frac{1}{2}xy^2 + \frac{1}{2}y^2z + \frac{1}{2}yz^2.$$

We need to show that $A_2(x, y, z) \leqslant \frac{2}{27}$ on the region $x \geqslant \frac{6}{10}$, $y \geqslant 0$, $z \geqslant 0$ and $x + y + z = 1$. Since $A_2(\frac{6}{10}, y, z) = A(\frac{6}{10}, y, z, y)$, we have already shown that $A_2(x, y, z) \leqslant \frac{2}{27}$ when $x = \frac{6}{10}$. Similarly, we have already established that $A_2(x, y, z) \leqslant \frac{2}{27}$ when $y = 0$ or when $z = 0$. The argument above shows that $A_2(x, y, z) \leqslant \frac{2}{27}$ on the boundary $x = \frac{6}{10}$. Also, $A_2(x, y, z) \leqslant \frac{2}{27}$ when $y = 0$ or $z = 0$, as the corresponding inequality holds for the function $A(x, y, z, u)$. So it is sufficient to show that $A_2(x, y, z)$ has no interior critical points. At an internal critical point, $\frac{\partial A_2}{\partial y} = \frac{\partial A_2}{\partial z}$, so

$$xy + yz + \frac{1}{2}z^2 = \frac{1}{2}x^2 + \frac{1}{2}y^2 + yz.$$

Rearranging we find that $z^2 = (x - y)^2$ and so (since both z and $x - y$ are positive on our region) $z = x - y$. Substituting $y = 1 - x - z$ into this expression and rearranging shows that $x = \frac{1}{2}$. However, $x \geqslant \frac{6}{10}$, and so there are no internal maxima. Thus $A(x, y, z, y) \leqslant \frac{2}{27}$ on our region, as required.

We have shown that $A(x, y, z, u)$ has no interior critical points, and that $A(x, y, z, u) \leqslant \frac{2}{27}$ on the boundaries of our region. Hence the lemma follows.

Lemma A.2 *The function $B(x, y, z, u)$ defined by*

$$B(x, y, z, u) = \frac{1}{2}x^2(z + y - u) + (uy - \frac{1}{2}u^2)z + \frac{1}{2}uz^2$$

satisfies the inequality

$$B(x, y, z, u) \leqslant \frac{72}{1000} < \frac{2}{27}$$

whenever x, y, z and u are real numbers satisfying $x \geqslant 0$, $y \geqslant 0$, $z \geqslant 0$, $u \geqslant 0$, $x + y + z = 1$, $u \leqslant \min\{x, y\}$ and $x \leqslant \frac{6}{10}$.

Proof: We begin by showing that B has no interior critical points. At such a point, $\frac{\partial B}{\partial x} = \frac{\partial B}{\partial y} = \frac{\partial B}{\partial z}$ and $\frac{\partial B}{\partial u} = 0$. Now,

$$\frac{\partial B}{\partial y} = \tfrac{1}{2}x^2 + uz \text{ and}$$

$$\frac{\partial B}{\partial z} = \tfrac{1}{2}x^2 + uy - \tfrac{1}{2}u^2 + uz,$$

and so $uy - \tfrac{1}{2}u^2 = 0$ at a critical point. Dividing by u, we find that $2y = u$. But this contradicts the fact that $u \leqslant y$ and $u > 0$ on the interior of our region. Hence B has no interior critical points.

We now maximise B on the boundaries of our region. The same argument as above shows that B has no interior critical points when we restrict our region to the plane $x = 0$ or the plane $x = \frac{6}{10}$. So it suffices to consider the cases when $y = 0$, $z = 0$, $u = 0$, $u = x$ or $u = y$.

When $z = 0$, we find that

$$B(x, y, 0, u) = \tfrac{1}{2}x^2(y - u) \leqslant \tfrac{1}{2}x^2 y,$$

where $x + y = 1$ and $0 \leqslant x \leqslant \frac{6}{10}$. Basic calculus shows that this function is maximised at the value $\frac{72}{1000}$ when $x = \frac{6}{10}$ and $y = \frac{4}{10}$.

When $u = 0$, we find that

$$B(x, y, z, 0) = \tfrac{1}{2}x^2(z + y),$$

where $x + (z + y) = 1$ and $0 \leqslant x \leqslant \frac{6}{10}$. As in the case when $z = 0$, we find that this function attains the maximimum value of $\frac{72}{1000}$ when $x = \frac{6}{10}$ and $z + y = \frac{4}{10}$. Since $u = 0$ whenever $y = 0$, this argument also covers the case when $y = 0$.

Suppose that $u = x$. Note that when $u = x$, the inequality $u \leqslant \min\{x, y\}$ is equivalent to the inequality $x \leqslant y$. Define

$$B_1(x, y, z) = B(x, y, z, x) = \tfrac{1}{2}x^2 y - \tfrac{1}{2}x^3 + xyz + \tfrac{1}{2}xz^2.$$

We must show that $B_1(x, y, z)$ is at most $\frac{72}{1000}$ on the region defined by $x \geqslant 0$, $y \geqslant 0$, $z \geqslant 0$, $x \leqslant \frac{6}{10}$, $x + y + z = 1$ and $x \leqslant y$. The condition $x \leqslant \frac{6}{10}$ is implied by the remaining conditions (indeed, we must have $x \leqslant \frac{1}{2}$) and so we may ignore it.

At an internal maxima, we have that $\frac{\partial B_1}{\partial y} = \frac{\partial B_1}{\partial y}$, and so

$$x\left(\tfrac{1}{2}x + z\right) = x(y + z).$$

Since $x > 0$ at an interior point, this implies that $x = 2y$. But the condition that $x \leqslant y$ now shows that $x = y = 0$, and so B_1 has no internal maxima.

Clearly $B_1(x, y, z) = 0$ when $x = 0$. Since B_1 is a restriction of B, it follows that $B_1(x, y, z) \leqslant \frac{72}{1000}$ when $y = 0$ or $z = 0$. So to establish our bound on B_1 it remains to consider the situation when $x = y$. Define

$$B_2(x, z) = B_1(x, x, z) = x^2 z + \tfrac{1}{2} x z^2.$$

We need to show that $B_2(x, z) \leqslant \frac{72}{1000}$ on the region defined by $x \geqslant 0$, $z \geqslant 0$ and $2x + z = 1$.

At an interior critical point, $\frac{\partial B_2}{\partial x} = 2 \frac{\partial B_2}{\partial z}$, so

$$2xz + \tfrac{1}{2} z^2 = 2 \left(x^2 + xz \right).$$

Hence $z^2 = 4x^2$, and therefore $z = 2x$. Since $2x + z = 1$, we find that there is a unique interior critical point when $x = \frac{1}{4}$ and $z = \frac{1}{2}$, where B_2 takes on the value $\frac{1}{16} < \frac{72}{1000}$. Clearly $B_2(x, z) = 0$ on the boundary of the region, and so B_2 is bounded above by $\frac{72}{1000}$. This implies that B_1 is bounded above by $\frac{72}{1000}$, as required.

It remains to consider the function B in the case when $u = y$. Note that when $u = y$ the inequality $u \leqslant \min\{x, y\}$ is equivalent to the inequality $y \leqslant x$. Define

$$B_3(x, y, z) = B(x, y, z, y) = \tfrac{1}{2} x^2 z + \tfrac{1}{2} y^2 z + \tfrac{1}{2} y z^2.$$

We must show that $B_3(x, y, z)$ is at most $\frac{72}{1000}$ on the region defined by $0 \leqslant x \leqslant \frac{6}{10}$, $y \geqslant 0$, $z \geqslant 0$, $x + y + z = 1$ and $y \leqslant x$.

Since $B_3(x, y, z)$ is a restriction of $B(x, y, z)$, we find that $B_3(x, y, z) \geqslant \frac{72}{1000}$ whenever $y = 0$ or $z = 0$. When $x = 0$ we have that $y = 0$, and so this case follows trivially. When $y = x$ then B_3 is the restriction of B to the case when $u = x = y$, and we have dealt with the case when $u = x$ above. So it only remains to show that $B_3(x, y, z) \leqslant \frac{72}{1000}$ on the interior of the region, and that $B_3(\frac{6}{10}, y, z) \leqslant \frac{72}{1000}$.

At an interior critical point, we have that $\frac{\partial B_3}{\partial x} = \frac{\partial B_3}{\partial y} = \frac{\partial B_3}{\partial z}$. Now,

$$\frac{\partial B_3}{\partial x} = xz,$$

$$\frac{\partial B_3}{\partial y} = yz + \tfrac{1}{2} z^2 \text{ and}$$

$$\frac{\partial B_3}{\partial z} = \tfrac{1}{2} x^2 + \tfrac{1}{2} y^2 + yz.$$

The equality $\frac{\partial B_3}{\partial x} = \frac{\partial B_3}{\partial y}$ shows that $x = y + \tfrac{1}{2} z$; together with $x + y + z = 1$ this implies that $y = 3x - 1$ at a critical point. The equality $\frac{\partial B_3}{\partial y} = \frac{\partial B_3}{\partial z}$ implies that $z^2 = x^2 + y^2$. Substituting $z = 1 - x - y$ and then $y = 3x - 1$ into this expression and solving the resulting equation in x shows that the unique critical point occurs when

$$x = \frac{5-\sqrt{7}}{6}, y = \frac{3-\sqrt{7}}{2} \text{ and } z = \frac{2\sqrt{7}-4}{3}.$$

At this point, B_3 takes the value $\frac{7\sqrt{7}-17}{27} < \frac{72}{1000}$.

Define

$$B_4(y, z) = B_3\left(\frac{6}{10}, y, z\right) = \frac{18}{100}z + \frac{1}{2}y^2z + \frac{1}{2}yz^2.$$

To prove the lemma, it suffices to show that the maximum value of B_4 on the region defined by $y \geqslant 0$, $z \geqslant 0$ and $y + z = \frac{4}{10}$ is at most $\frac{72}{1000}$.

At an interior critical point, $\frac{\partial B_3}{\partial y} = \frac{\partial B_3}{\partial z}$ and so

$$yz + \frac{1}{2}z^2 = \frac{18}{100} + \frac{1}{2}y^2 + yz.$$

But $z \leqslant \frac{4}{10}$ on our region, and so at a critical point we have

$$y^2 = z^2 - \frac{36}{100} < 0.$$

This contradiction implies that there are no critical points. Finally, when $z = 0$ we have that B_3 takes the value 0; when $y = 0$ we have that B_3 takes the value $\frac{72}{1000}$. Thus B_3 has the properties we require, and so the lemma follows.

References

[1] M. Aschbacher, *Finite Group Theory*, Cambridge University Press, Cambridge, 1986.

[2] M. Aschbacher and R. Guralnick, 'Solvable generation of groups and Sylow subgroups of the lower central series', *J. Algebra* **77** (1982) 189–201.

[3] George E. Andrews, *The Theory of Partitions*, Addison-Wesley, Reading, MA, 1976.

[4] Tom M. Apostol, *Introduction to Analytic Number Theory*, Springer-Verlag, New York, 1976.

[5] Giuseppe Bagnera, 'La composizione dei gruppi finiti il cui grado è la quinta potenza di un numero primo', *Ann. di Mat. pura e applicata* (3) **1** (1898) 137–228.

[6] Hans Ulrich Besche, Bettina Eick and E. A. O'Brien, 'A millenium project: constructing small groups', *Int. J. Algebra and Computation* **12** (2002) 623–44.

[7] Simon R. Blackburn, *Group Enumeration*, DPhil Thesis, Oxford, 1992.

[8] S. R. Blackburn, 'Enumeration within isoclinism classes of groups of prime power order', *J. London Math. Soc.* (2) **50** (1994) 293–304.

[9] Simon R. Blackburn, 'Groups of prime power order with derived subgroup of prime order', *J. Algebra* **219** (1999) 625–57.

[10] A. V. Borovik, L. Pyber and A. Shalev, 'Maximal subgroups in finite and profinite groups', *Trans. American Math. Soc.* **348** (1996) 3745–61.

[11] R. M. Bryant and L. G. Kovács, 'Lie representations and groups of prime power order', *J. London Math. Soc.* (2) **17** (1978) 415–21.

[12] D. W. Brydon, *Enumeration Functions in Varieties of Groups*, MSc Thesis, Oxford, October 1998.

[13] Peter J. Cameron, *Combinatorics: Topics, Techniques, Algorithms*, Cambridge University Press, Cambridge, 1994.

[14] Peter J. Cameron, Ron Solomon and Alexander Turull, 'Chains of subgroups in symmetric groups', *J. Algebra* **127** (1989) 340–52.

[15] A. R. Camina, G. R. Everest and T. M. Gagen, 'Enumerating non-soluble groups – a conjecture of John G. Thompson', *Bull. London Math. Soc.* **18** (1986) 265–8.

[16] P. M. Cohn, *Algebra, Vol. 1* (2nd Edition), John Wiley & Sons, Chichester, 1982.

[17] P. M. Cohn, *Algebra, Vol. 3* (2nd Edition), John Wiley & Sons, Chichester, 1991.

[18] M. J. Collins, *Representations and Characters of Finite Groups*, Cambridge University Press, Cambridge, 1990.

[19] J. H. Conway, R. T. Curtis, S. P. Norton, R. A. Parker and R. A. Wilson, *Atlas of Finite Groups: Maximal Subgroups and Ordinary Characters for Simple Groups*, Clarendon Press, Oxford, 1985.

[20] Charles W. Curtis and Irving Reiner, *Representation Theory of Finite Groups and Associative Algebras*, John Wiley & Sons, New York, 1962.

[21] Gabrielle A. Dickenson, 'On the enumeration of certain classes of soluble groups', *Quart. J. Math. Oxford* (2) **20** (1969) 383–94.

[22] Heiko Dietrich and Bettina Eick, 'On the groups of cube-free order', *J. Algebra* **292** (2005) 122–37.

[23] John D. Dixon, 'The Fitting subgroup of a linear solvable group', *J. Austral. Math. Soc.* **7** (1967) 417–24.

[24] Marcus du Sautoy, 'Counting *p*-groups and nilpotent groups', *Inst. Hautes Études Sci. Publ. Math.* **92** (2000) 63–112.

[25] Marcus du Sautoy, 'Counting subgroups in nilpotent groups and points on elliptic curves', *J. reine angew. Math.* (Crelle's Journal) **549** (2002) 1–21.

[26] G. P. Egoryčev, 'A solution of van der Waerden's permanent problem', *Dokl. Akad. Nauk SSSR* **258** (1981) 1041–4 (in Russian). Translated in *Soviet Math. Dokl.* **23** (1981) 619–22.

[27] P. Erdős, 'On an elementary proof of some asymptotic formulas in the theory of partitions', *Ann. Math.* **43** (1942) 437–50.

[28] P. Erdős, 'Some asymptotic formulas in number theory', *J. Indian Math. Soc. (New Series)* **12** (1948) 75–8.

[29] P. Erdős, M. Ram Murty and V. Kumar Murty, 'On the enumeration of finite groups', *J. Number Theory* **25** (1987) 360–78.

[30] P. Erdős and G. Szekeres, 'Über die Anzahl der Abelscher Gruppen gegebener Ordnung und über ein verwandtes zahlentheoretisches Problem', *Acta Acad. Sci. Math. (Szeged)* **7** (1935) 97–102.

[31] Anton Evseev, *On Higman's PORC Conjecture*. Preprint, Mathematical Institute, Oxford, 2005.

[32] D. I. Falikman, 'A proof of van der Waerden's conjecture on the permanent of a doubly stochastic matrix', *Mat. Zametki* **29** (1981) 931–8 (in Russian). Translated in *Math. Notes* **29** (1981) 475–9.

[33] Walter Feit and John G. Thompson, 'Solvability of groups of odd order', *Pacific J. Math.* **13** (1963) 775–1029.

[34] G. Frobenius, 'Über auflösbare Gruppen', *Sitzungsber. Kön. Preuss. Akad. Wiss. Berlin*, 1893, pp. 337–45 = *Ges. Abh.* (J.-P. Serre, ed.), Springer-Verlag, Berlin, 1968, Vol. II, pp. 565–73.

[35] Wolfgang Gaschütz, 'Zu einem von B. H. und H. Neumann gestellten Problem', *Math. Nachrichten* **14** (1956) 249–52.

[36] Daniel Gorenstein, *Finite Groups*, Harper & Row, New York, 1968.

[37] Daniel Gorenstein and John H. Walter, 'The characterization of finite groups with dihedral Sylow 2-subgroups, I, II, III', *J. Algebra* **2** (1965) 85–151, 218–70, 354–93.

[38] R. Guralnick, 'Generation of simple groups', *J. Algebra* **103** (1986) 381–401.

[39] R. Guralnick, 'On the number of generators of a finite group', *Arch. Math. (Basel)* **53** (1989) 521–3.

[40] Marshall Hall, Jr, 'Distinct representatives of subsets', *Bull. Amer. Math. Soc.* **54** (1948) 922–6.

[41] Marshall Hall, Jr, *The Theory of Groups*, Macmillan, New York, 1959.

[42] P. Hall, 'The classification of prime-power groups', *J. reine angew. Math.* (Crelle's Journal) **182** (1940) 130–41 = *Collected Works of Philip Hall* (K. W. Gruenberg and J. E. Roseblade, eds), Clarendon Press, Oxford, 1988, pp. 265–76.

[43] Richard R. Hall, *Sets of Multiples*, Cambridge University Press, Cambridge, 1996.

[44] G. H. Hardy and S. A. Ramanujan, 'Asymptotic formulae in combinatory analysis', *Proc. London Math. Soc.* **17** (1918) 75–115.

[45] Graham Higman, 'Enumerating p-groups. I: Inequalities', *Proc. London Math. Soc.* (3) **10** (1960) 24–30.

[46] Graham Higman, 'Enumerating p-groups, II: Problems whose solution is PORC', *Proc. London Math. Soc.* (3) **10** (1960) 566–82.

[47] Otto Hölder, 'Die Gruppen der Ordnungen p^3, pq^2, pqr, p^4', *Math. Annalen* **43** (1893) 301–412.

[48] Otto Hölder, 'Die Gruppen mit quadratfreier Ordnungszahl', *Nachr. Gesellsch. Wiss. zu Göttingen. Math.-phys. Klasse*, 1895, pp. 211–29.

[49] D. F. Holt, 'Enumerating perfect groups', *J. London Math. Soc.* (2) **39** (1989) 67–78.

[50] B. Huppert, *Endliche Gruppen I*, Springer-Verlag, Berlin, 1967.

[51] A. Jaikin-Zapirain, 'The number of finite p-groups with bounded number of generators', *Finite Groups 2003* (C. Y. Ho, P. Sin, P. H. Tiep and A. Turull, eds), Walter de Gruyter, Berlin, 2004, pp. 209–17.

[52] A. Jaikin-Zapirain and L. Pyber, 'Random generation of finite and profinite groups and group enumeration'. Preprint (submitted August 2006 for publication).

[53] Gordon James and Martin Liebeck, *Representations and Characters of Groups* (2nd Edition), Cambridge University Press, Cambridge, 2001.

[54] D. G. Kendall and R. A. Rankin, 'On the number of abelian groups of a given order', *Quart. J. Math. Oxford* **18** (1947) 197–208.

[55] Daniel J. Kleitman, Bruce L. Rothschild and Joel H. Spencer, 'The number of semigroups of order n', *Proc. Amer. Math. Soc.* **55** (1976) 227–32.

[56] Benjamin Klopsch, 'Enumerating finite groups without abelian composition factors', *Israel J. Math.* **137** (2003) 265–84.

[57] Benjamin Klopsch, 'Enumerating highly non-soluble groups', *Finite Groups 2003* (C. Y. Ho, P. Sin, P. H. Tiep and A. Turull, eds), Walter de Gruyter, Berlin, 2004, pp. 219–27.

[58] John Knopfmacher, *Abstract Analytic Number Theory*, North-Holland, Amsterdam, 1975.

[59] L. G. Kovács, 'On finite soluble groups', *Math. Z.* **103** (1968) 37–9.

[60] Hong-Quan Liu, 'On the number of abelian groups of a given order (supplement)', *Acta Arith.* **64** (1993) 285–96.

[61] Alexander Lubotzky, 'Enumerating boundedly generated finite groups', *J. Algebra* **238** (2001) 194–9.

[62] Andrea Lucchini, 'A bound on the number of generators of a finite group', *Arch. Math.* (*Basel*) **53** (1989) 313–17.

[63] Avinoam Mann, 'Enumerating finite groups and their defining relations', *J. Group Theory* **1** (1998) 59–64.

[64] Avinoam Mann, 'Some questions about p-groups', *J. Austral. Math. Soc. Series A* **67** (1999) 356–79. (*Note*: An updated version of this paper is available as a preprint, Hebrew University, Jerusalem, August 2006.)

[65] Avinoam Mann, 'Enumerating finite groups and their defining relations II', *J. Algebra* **302** (2006) 586–92.

[66] Michael E. Mays, 'Counting abelian, nilpotent, solvable, and supersolvable group orders', *Arch. Math. (Basel)* **31** (1979) 536–8.

[67] Annabelle McIver and Peter M. Neumann, 'Enumerating finite groups', *Quart. J. Math. Oxford* (2) **38** (1987) 473–88.

[68] Brendan D. McKay and Ian M. Wanless, 'On the number of Latin squares', *Ann. Comb.* **9** (2005) 335–44.

[69] Henryk Minc, *Permanents*, Addison-Wesley, Reading, MA, 1978.

[70] Henryk Minc, *Nonnegative Matrices*, John Wiley & Sons, New York, 1988.

[71] M. Ram Murty and V. Kumar Murty, 'On the number of groups of a given order', *J. Number Theory* **18** (1984) 178–91.

[72] M. Ram Murty and V. Kumar Murty, 'On groups of squarefree order', *Math. Annalen* **267** (1984) 299–309.

[73] M. Ram Murty and S. Srinivasan, 'On the number of groups of squarefree order', *Canad. Math. Bull.* **30** (1987) 412–20.

[74] B. H. Neumann, 'Identical relations in groups I', *Math. Annalen* **114** (1937) 506–23.

[75] Hanna Neumann, *Varieties of Groups*, Springer-Verlag, Berlin, 1967.

[76] Peter M. Neumann, 'An enumeration theorem for finite groups', *Quart. J. Math. Oxford* (2) **20** (1969) 395–401.

[77] M. F. Newman, personal communication.

[78] M. F. Newman, E. A. O'Brien and M. R. Vaughan-Lee, 'Groups and nilpotent Lie rings whose order is the sixth power of a prime', *J. Algebra* **278** (2004) 383–401.

[79] E. A. O'Brien and M. R. Vaughan-Lee, 'The groups with order p^7 for odd prime p', *J. Algebra* **292** (2005) 243–58.

[80] P. P. Pálfy, 'A polynomial bound for the orders of primitive solvable groups', *J. Algebra* **77** (1982) 127–37.

[81] Ioannis Pantelidakis, *On the Number of Non-isomorphic Groups of the Same Order*, DPhil Thesis, Oxford, August 2003.

[82] L. Pyber, 'Enumerating finite groups of a given order', *Ann. Math.* **137** (1993) 203–20.

[83] L. Pyber, 'Group enumeration and where it leads us', *Proc. European Math. Congress, Budapest, 1996, Vol. II* (A. Balog, G. O. H. Katona, D. Szász and A. Recski, eds), Progress in Mathematics Vol. 169, Birkhäuser, Basel, 1998, pp. 187–99.

[84] I. Schur, 'Über die Darstellung der endlichen Gruppen durch gebrochene lineare Substitutionen', *J. reine angew. Math.* (Crelle's Journal) **127** (1904) 20–50.

[85] E. J. Scourfield, 'An asymptotic formula for the property $(n, f(n)) = 1$ for a class of multiplicative functions', *Acta Arith.* **29** (1976) 401–23.

[86] Charles C. Sims, 'Enumerating p-groups', *Proc. London Math. Soc.* (3) **15** (1965) 151–66.

[87] Charles C. Sims, *Computation with Finitely Presented Groups*, Cambridge University Press, Cambridge, 1994.

[88] D. A. Suprunenko, *Matrix Groups*, American Mathematical Society, Providence, MI, 1976.

[89] Michio Suzuki, *Group Theory I*, Springer-Verlag, New York, 1982.

[90] John G. Thompson, 'Nonsolvable finite groups all of whose local subgroups are solvable, I–VI', *Bull. Amer. Math. Soc.* **74** (1968) 383–437; *Pacific J. Math.* **33** (1970) 451–536; **39** (1971) 483–534; **48** (1973) 511–92; **50** (1974) 215–97; **51** (1974) 573–630.

[91] John G. Thompson, 'Finite non-solvable groups', *Group Theory. Essays for Philip Hall* (K. W. Gruenberg and J. E. Roseblade, eds), Academic Press, London, 1984, pp. 1–12.

[92] Michael Vaughan-Lee, *The Restricted Burnside Problem*, Clarendon Press, Oxford, 1990.

[93] Geetha Venkataraman, *Enumeration of Types of Finite Groups*, DPhil Thesis, Oxford, 1993.

[94] Geetha Venkataraman, 'Enumeration of finite soluble groups with abelian Sylow subgroups', *Quart. J. Math. Oxford* (2) **48** (1997) 107–25.

[95] Geetha Venkataraman, 'Enumeration of finite soluble groups in small varieties of A groups and associated topics', Tech. Report, Centre for Mathematical Sciences, St. Stephen's College, University of Delhi, 1999.

[96] William L. Voxman and Roy H. Goetschel, Jr, *Advanced Calculus. An Introduction to Modern Analysis*, Marcel Dekker, New York, 1981.

[97] T. R. Wolf, 'Solvable and nilpotent subgroups of GL(n, q^m)', *Can. J. Math.* **34** (1982) 1097–111.

Index